C/MATH Toolchest
for Engineering
and Scientific Applications

Charles P. Bernardin

93-1671

PTR Prentice Hall
Englewood Cliffs, New Jersey 07632

Library of Congress Cataloging-in-Publication Data

Bernardin, Charles P.
    C/math toolchest for engineering and scientific applications /
Charles P. Bernardin.
        p.   cm.
    Includes bibliographical references and index.
    ISBN 0-13-605866-3
    1. Engineering mathematics--Data processing.   2. Signal
processing--Digital techniques--Data processing.   3. C/MATH
Toolchest   I. Title
TA345.B48   1993
510'.285'5133--dc20

92-40205
CIP

**Editorial production/management
  and interior design: bookworks**
Cover design: Jerry Votta
Buyer: Mary Elizabeth McCartney
Acquisitions editor: Michael Hays
Photo Credit: Image Bank ©Steven Hunt

©1993 by PTR Prentice Hall, Inc.
A Simon & Schuster Company
Englewood Cliffs, New Jersey 07632

The publisher offers discounts on this book when ordered
in bulk quantities. For more information, contact:

Corporate Sales Department
PTR Prentice Hall
113 Sylvan Avenue   Phone: 201-592-2863
Englewood Cliffs, NJ 07632   FAX: 201-592-2249

Printed in the United States of America

10 9 8 7 6 5 4 3 2 1

ISBN 0-13-605866-3

Prentice-Hall International (UK) Limited, *London*
Prentice-Hall of Australia Pty. Limited, *Sydney*
Prentice-Hall Canada, Inc., *Toronto*
Prentice-Hall Hispanoamericana, S.A., *Mexico*
Prentice-Hall of India Private Limited, *New Delhi*
Prentice-Hall of Japan, Inc., *Tokyo*
Simon & Schuster Asia Pte. Ltd., *Singapore*
Editora Prentice-Hall do Brasil, Ltda., *Rio de Janeiro*

Contents

PREFACE   **vii**

CHAPTER 1   INTRODUCTION   **1**

    1.1 Installation   1
    1.2 Source Code Included   2
          1.2.1   C/Math Library Source   2
          1.2.2   C/Math GRAFIX Source   3
    1.3 Compiling with MATHLIB.H   3
    1.4 Linking with MATHLIB.LIB   4
    1.5 Function Categories   4
    1.6 Data Types   9
          1.6.1   Real   9
          1.6.2   Complex   10
          1.6.3   Real__Vector and Complex__Vector   10
          1.6.4   Real__Matrix and Complex__Matrix   12
          1.6.5   Vector and Matrix Arrays   14
    1.7 Dynamic Memory Management   15
          1.7.1   Recovering Memory   16
    1.8 Error Handing   17
          1.8.1   Overriding Default Error Handling   18
          1.8.2   The matherr__ Function   19
          1.8.3   Floating Point Exceptions   20

1.9   Global Variables   21
    1.9.1   Numerical Constants   21
    1.9.2   Error Variables   22
    1.9.3   Miscellaneous Variables   23

CHAPTER 2   GRAFIX PROGRAM   **24**

2.1   Creating Data   24
    2.1.1   Data Files without Headers   25
    2.1.2   Data Files with Headers   26
2.2   Loading Curves   31
2.3   Plotting Curves   33
    2.3.1   Zooming   33
    2.3.2   Changing Axes Range   35
    2.3.3   Printing Screen   37
    2.3.4   GPRINT Program   39
2.4   Editing Curves   39
2.5   Deleting Curves   41
2.6   Outputting Curves   42
2.7   Labeling Plot   44
2.8   Interpolation   45
2.9   Regression   46
2.10 Options   49

CHAPTER 3   COMPLEX NUMBERS   **53**

CHAPTER 4   PROBABILITY   **55**

4.1   Combinatorial Analysis   55
    4.1.1   Permutations and Factorial   56
    4.1.2   Combinations   56
    4.1.3   Binomial Expansion and Pascal's Triangle   56
4.2   Discrete Random Variables   57
    4.2.1   Binomial Distribution   57
    4.2.2   Hypergeometric Distribution   58
    4.2.3   Poisson Distribution   60
4.3   Continuous Random Variables   60
    4.3.1   Normal Distribution   61
    4.3.2   Uniform Distribution   64

CHAPTER 5   STATISTICS   **68**

5.1   Expectation and Random Variables   68
    5.1.1   Expectation of a Function   69
    5.1.2   Mean and Variance of a Random Variable   69
    5.1.3   Chebyshev's Theorem   70
5.2   Regression and Least-Squares   71
    5.2.1   Generalized Least-Squares   71
    5.2.2   Polynomial Regression   72
    5.2.3   Linear Regression   72

5.2.4   Logarithmic Regression   73
5.2.5   Fourier Regression   73
5.2.6   Nonlinear Regression   74

CHAPTER 6   LINEAR ALGEBRA   77

6.1   Matrices   77
    6.1.1   Matrix Addition and Subtraction   78
    6.1.2   Scaling a Matrix   78
    6.1.3   Matrix Multiplication   78
    6.1.4   Matrix Transpose   79
    6.1.5   Some Special Square Matrices   79
    6.1.6   Matrix Trace   81
    6.1.7   Matrix Determinant   81
    6.1.8   Matrix Singularity and Rank   81
    6.1.9   Matrix Inverse   82
6.2   Simultaneous Linear Equations   82
    6.2.1   Pseudo-Inverse   83
6.3   Vectors   84
    6.3.1   Dot Product   84
    6.3.2   Cross Product   84
    6.3.3   Vector Magnitude   85
6.4   Eigenanalysis   85
    6.4.1   Real Symmetric Matrices   85
    6.4.2   Complex Hermitian Matrices   86

CHAPTER 7   NUMERICAL ANALYSIS   88

7.1   Interpolation   89
    7.1.1   Piecewise Interpolation   89
    7.1.2   Linear Interpolation   89
    7.1.3   Lagrange n-Point Interpolation   89
    7.1.4   Cubic Spline Interpolation   89
    7.1.5   Coordinates of a Centroid   90
7.2   Integration   91
    7.2.1   Rectangular Rule   91
    7.2.2   Trapezoidal Rule   92
    7.2.3   Romberg Integration   92
    7.2.4   Simpson's Rule   92
7.3   Differentiation   93
    7.3.1   First Order Differences   93
    7.3.2   Higher Order Differences   93
    7.3.3   Examples of Difference Calculations   94
    7.3.4   Improving the Difference Estimates   95
7.4   Function Minimization   96
    7.4.1   Finding Roots of One-Dimensional Functions   96
    7.4.2   Finding Roots of Polynomials   98
    7.4.3   Minimization of Multi-Dimensional Functions   98

CHAPTER 8   DIGITAL SIGNAL PROCESSING   **100**

8.1   Background   100
    8.1.1   The Sampling Theorem   101
    8.1.2   The Discrete Fourier Transform   101
    8.1.3   The Discrete Cosine Transform   102
    8.1.4   Convolution   103
    8.1.5   Windowing   105
8.2   Digital Filtering   106
    8.2.1   Finite Impulse Response (FIR) Filters   107
    8.2.2   Infinite Impulse Response Filters   114
    8.2.3   Data Smoothing   116
    8.2.4   Median Filtering   116
8.3   Sample Rate Conversion   117
    8.3.1   Decimation   117
    8.3.2   Interpolation   117
8.4   Spectral Analysis   118
    8.4.1   The Fast Fourier Transform   118
    8.4.2   The Power Spectrum   119
    8.4.3   Correlation and Covariance   120
8.5   Adaptive Signal Processing   121
    8.5.1   Wiener Filtering   122
    8.5.2   The LMS Algorithm   124
    8.5.3   Adaptive Interference Canceling   125
    8.5.4   Adaptive Line Enhancement   125

CHAPTER 9   FUNCTIONS   **129**

APPENDIX A   FUNCTION CATEGORIES   **467**

APPENDIX B   REFERENCES   **473**

INDEX   **477**

Preface

"The mass of mathematical truth is obvious and imposing;
its practical applications, the bridges and steam-engines,
and dynamos obtrude themselves on the dullest imagination.
The public does not need to be convinced that there is
something in mathematics."

G.H. Hardy

Through "hands-on" experience, computers have given us an unrivaled insight into the work of history's great mathematicians. Indeed, the post-WWII period spawned its own group of great numerical mathematicians such as Clenshaw, Hastings, Hart, Wilkinson, and Knuth to name only a few. The advent of the computer has given us an unprecedented opportunity to expand our mathematical horizons through applied analysis. In the last forty years, applied analysis has evolved mostly on mainframe computers in FORTRAN.

About 1975, the C programming language (designed by Kernighan and Ritchie) emerged from Bell Labs. Although originally used to develop the Unix operating system, C is now used for developing a wide range of applications. The popularity of C is primarily due to its elegant combination of high level structure with low level power. In addition, the ANSI standard for C has made it possible to develop programs that can be easily ported from one computer architecture to another.

Although C is quite capable of scientific computation, unlike FORTRAN, it was not designed specifically for that purpose. Several deficiencies are obvious to anyone who has attempted to do any extensive scientific programming in C. The lack of a complex data type is just one example. Another is the lack of support for matrix operations. It is not that C cannot support complex data types or matrix operations, it is just that these features were not defined as part of the standard language. Fortunately, C was designed so that you can easily add extensions to the language.

The purpose of this product is to add the extensions to C that are necessary to make it a powerful language for the development of scientific and engineering applications. As such, this product is aimed directly at the engineer in industry, the undergraduate and graduate student in "hard-science," and anyone in applied research and development who wants to use the C programming language.

Some discretion was used in selecting the functions that went into this package. **This is a results oriented package,** so an attempt has been made to choose the best approaches, not only based on the numerous textbooks in applied analysis, but also based on many years of experience and many hours of computer verification. Despite all precautions, it is impossible to please all users, and any suggestions for improving this package are welcomed.

ABOUT THE AUTHOR

Charles P. Bernardin, president of Pete Bernardin Software, received a B.S. in mathematics from Villanova University in 1973 and a Ph.D. in biomedical engineering from The Johns Hopkins University in 1979. His graduate work included research in the auditory system, applying mathematics and digital signal processing in investigation of the neural processes of hearing.

He has had over 15 years of extensive experience in solving engineering and mathematical problems with digital methods. He has worked in the Radar and Digital Systems division of Texas Instruments, and the Advanced Technology Laboratory of RCA. He has also worked as an image processing consultant for NASA, as a digital signal processing staff specialist for the Tandy Corporation and as a telecommunications engineer at Bell Northern Research.

Dr. Bernardin is the author of several research papers, covering a broad range of technical areas which include image processing, radar, seismic signal processing and spectral estimation.

1

Introduction

The **C/Math Toolchest Library** should work with any ANSI compatible C or C++ compiler. Prebuilt libraries for several popular compilers are included. Although the library takes advantage of many nonstandard extensions that are supported by some compilers, the use of nonstandard features can be disabled by changing a single line in a header file. Therefore, it is very simple to make the library work with any ANSI compatible C compiler.

1.1 INSTALLATION

The C/Math Toolchest Library is supplied on two 3 1/2" (720K) disks. Before proceeding, you should make a backup copy of each supplied disk. In order to save space, most of the files on the supplied disk(s) are stored in a compressed format. The compressed files (.ARC) will be automatically decompressed by the INSTALL program. Place the supplied disk 1 into drive A or drive B, then type the following to install the C/Math Toolchest Library files.

A:INSTALL

or

B:INSTALL

The INSTALL program will ask you a series of questions. For example, you will be asked to select one of the supported compilers. All the questions have default responses that may be selected by simply pressing the **Enter** key. In most cases, the defaults will provide you with reasonable responses. If you don't like the default response, simply type in an appropriate response, and then press the **Enter** key.

After installation is complete, please store the supplied disks in a safe place. The files created by the installation process are all that you need. You should notice that a file named **READ.ME** was created during the installation process. Before proceeding any further, you should examine the READ.ME file, as this file may contain important information that is not present in this printed documentation. The READ.ME file may be viewed by using any ASCII text editor, or you may copy its contents to your printer by typing COPY READ.ME PRN, and then pressing the **Enter** key.

1.2 SOURCE CODE INCLUDED

The **C/Math Library Source** and **C/Math GRAFIX Source** are included with the C/Math Toolchest Library. The **C/Math Library Source** is the C source code for all of the functions in the C/Math Toolchest Library. With the library source code, you have the ability to:

1. Create libraries for other memory models, such as the small and large memory models. The prebuilt library supplied is a medium memory model library.
2. Create libraries that use single rather than double precision arithmetic. The prebuilt library supplied uses double precision.
3. Create libraries that support individual vectors and matrices larger than 64K in size. The prebuilt library supplied limits the size of individual vectors and matrices to 64K.
4. Create libraries for any ANSI standard C or C++ compiler not currently supported, or for any operating system not currently supported.

The **C/Math GRAFIX Source** is the C source code for the graphical data analysis program included with the C/Math Toolchest Library. This program is discussed in the next chapter. With the source to the GRAFIX program, you have the ability to:

1. Add additional features to the program to satisfy your personal needs.
2. Use the text-based video library to add menus to your own programs.
3. Use the graphics-based video library to add graphics to your own programs (including the ability to print graphics screens).

1.2.1 C/Math Library Source

The INSTALL program automatically creates a subdirectory named **SOURCE**, and copies all the library source files into this subdirectory. These C source files are ready to be compiled using the compiler you selected during installation. If you look in the subdirectory named **SOURCE**, you should notice that a file named **READ.ME** was created during the installation process. This file describes how to compile the library with your particular compiler.

If you use a compiler other than the one that you selected during installation, then you will have to modify the control files for your particular compiler. The control files are the files named MATHLIB.BAT, MATHLIB.MAK or MATHLIB.PRJ, and MATHLIB.CTL. These files simply control the building of the libraries. You may also need to edit the header file named MATHLIB.H and change the line **#define STDC 0** to **#define STDC 1**. This disables the use of all nonstandard C language features. After modifying the control files and the MATHLIB.H file, you will need to rebuild the libraries for use with your particular compiler.

1.2.2 C/Math GRAFIX Source

The INSTALL program automatically creates a subdirectory named **GRAFIX**, and copies all of the GRAFIX program source files into this subdirectory. These C source files are ready to be compiled using the compiler you selected during installation. If you look in the subdirectory named **GRAFIX**, you should notice that a file named **READ.ME** was created during the installation process. This file describes how to compile the GRAFIX program with your particular compiler.

If you use a compiler other than the one that you selected during installation, then you will have to modify the control files for your particular compiler. The control files are the files named MAKEGRAF.BAT, GRAFIX.MAK or GRAFIX.PRJ, and GRAFIX.CTL. These files simply control the building of the GRAFIX data analysis program.

Unlike the C/Math Library Source, the C/Math GRAFIX Source contains functions that are hardware or operating system dependent. This is unavoidable since the program is graphics oriented. Although the system-dependent functions are isolated to make it easy to add support for additional compilers, it may require substantial effort to make the GRAFIX program work with some compilers or operating systems that are not currently supported.

1.3 COMPILING WITH MATHLIB.H

After installation, you should notice that a file named **MATHLIB.H** was created. This is the header file that must be included by any program that uses one or more of the C/Math Toolchest Library functions. At the beginning of any file that contains a call to a C/Math Toolchest Library function, the following line should be present:

#include "mathlib.h"

Typically the above line should immediately follow any lines that include standard C header files, such as *#include <stdio.h>*. In order for your compiler to locate MATHLIB.H, this header file must either be in the current directory, or in a directory in which your compiler already searches for header files. Otherwise your compiler will be unable to locate the file and will issue an error message when it encounters *#include "mathlib.h"*.

The READ.ME file contains suggestions for configuring your compiler so that it can locate the MATHLIB.H header file.

1.4 LINKING WITH MATHLIB.LIB

After installation, you should also notice that a file named **MATHLIB.LIB** was created. This is the library file that contains all the C/Math Toolchest Library functions. Any program that uses one or more of the C/Math Toolchest Library functions must be linked with MATHLIB.LIB.

The supplied MATHLIB.LIB file was created using the medium memory model. Therefore, any object files that you link with MATHLIB.LIB must be created using the medium memory model. Otherwise, your linker will report a link error such as "wrong memory model"; or worse, some linkers may not report an error at all, but your program won't work. The medium memory model supports unlimited code size, but limits the data size to 64K. However, the C/Math Toolchest Library overcomes the data size limitation by storing data in **far** memory. This allows all the available memory to be utilized, no matter which memory model is used. Therefore, the medium memory model is typically the most appropriate memory model to use. However, in some circumstances you may wish to use a different memory model, such as the small or large memory model. If so, then you will need to compile the C/Math Library Source to create additional libraries for the other memory models.

A compiler/linker automatically searches only the standard C libraries. Therefore, you must explicitly instruct your compiler/linker to search the MATHLIB.LIB file. Otherwise, you will receive link errors such as "unresolved external references," meaning that one or more functions and/or global variables could not be found. Most compilers require that you use a *make* or *project* file to link external libraries. In some cases, you may be able to utilize an environment variable.

The READ.ME file contains suggestions for configuring your compiler/linker so that it will locate and search the MATHLIB.LIB library file.

1.5 FUNCTION CATEGORIES

The C/Math Toolchest Library provides a broad spectrum of mathematical functions that are frequently used in scientific and engineering applications. This section lists all of the functions in their respective categories, and provides a brief description of each.

	Complex Arithmetic Functions
Cadd	Add two complex numbers
Cdiv	Divide two complex numbers
Cmag	Compute the magnitude of a complex number
Cmul	Multiply two complex numbers
Complx	Form a complex number from its imaginary and real parts
Conjg	Conjugate a complex number
CscalR	Multiply a complex number by a real number
Csqrt	Compute the principal square root of a complex number
Csub	Subtract two complex numbers

Real Matrix and Vector Functions	
levinson	Levinson's recursion to solve a Toeplitz system of equations
lineqn	Solve system of simultaneous linear equations
mxadd	Add two matrices
mxcentro	Centroid a unimodal 2-D array of data
mxcopy	Copy one matrix to another
mxdeterm	Compute the determinant of a matrix
mxdup	Duplicate and allocate a matrix
mxeigen	Compute the eigenvalues and eigenvectors of a real symmetric matrix
mxhisto	Compute the histogram of 2-D data to any resolution
mxident	Create the identity matrix of specified size
mxinit	Initialize a matrix to a floating point value
mxinv	Invert a matrix of any size with LU decomposition
mxinv22	Invert a two by two matrix
mxinv33	Invert a three by three matrix
mxmaxval	Find the max value of a matrix and the indices where it occurs
mxminval	Find the min value of a matrix and the indices where it occurs
mxmul	Multiply two matrices
mxmul1	Multiply the transpose of a matrix by another matrix
mxmul2	Multiply a matrix by the transpose of another matrix
mxscale	Multiply a matrix by a floating point scalar
mxsub	Subtract two matrices
mxtrace	Compute the trace of a matrix
mxtransp	Transpose a general matrix (in place, if square, and if desired)
pseudinv	Solve an over-constrained system of simultaneous linear equations
vadd	Add two vectors
v_centro	Centroid a unimodal vector of data
vcopy	Copy one vector to another
vcross	Form the cross product between two vectors
vdot	Compute the dot product between two vectors
vdup	Duplicate and allocate a vector
vectomat	Multiply two vectors into a matrix
vinit	Initialize a vector to a floating point value
vmag	Compute the magnitude of a real vector
vmaxval	Find the max value of a vector and the index where it occurs
vminval	Find the min value of a vector and the index where it occurs
vmxmul	Multiply a matrix by a vector to form another vector
vmxmul1	Multiply the transpose of a matrix by a vector into another vector
vscale	Multiply a vector by a floating point scalar
vsub	Subtract two vectors

Complex Matrix and Vector Functions

Clineqn	Solve system of simultaneous linear equations
Cmxadd	Add two matrices
Cmxconjg	Conjugate a complex matrix
Cmxcopy	Copy one matrix to another
Cmxeigen	Compute the eigenvalues and eigenvectors of a Hermitian matrix
Cmxdeter	Compute the determinant of a matrix
Cmxdup	Duplicate and allocate a matrix
Cmxinit	Initialize a matrix to a complex value
Cmxinv	Invert a matrix of any size with LU decomposition
Cmxinv22	Invert a two by two matrix
Cmxinv33	Invert a three by three matrix
Cmxmag	Compute the magnitude of a complex matrix
Cmxmaxvl	Find the max value of a matrix and the indices where it occurs
Cmxminvl	Find the min value of a matrix and the indices where it occurs
Cmxmul	Multiply two matrices
Cmxmul1	Multiply the transpose of a matrix by another matrix
Cmxmul2	Multiply a matrix by the transpose of another matrix
CmxscalC	Multiply a matrix by a complex scalar
CmxscalR	Multiply a matrix by a real scalar
Cmxsub	Subtract two matrices
Cmxtrace	Compute the trace of a matrix
Cmxtrans	Transpose a general matrix (in place, if square, and if desired)
Cvadd	Add two vectors
Cvconjg	Conjugate a complex vector
Cvcopy	Copy one vector to another
Cvdot	Compute the dot product between two vectors
Cvdup	Duplicate and allocate a vector
Cvectomx	Multiply two vectors into a matrix
Cvinit	Initialize a vector to a complex value
Cvmag	Compute the magnitude of a complex vector
Cvmaxval	Find the max value of a vector and the index where it occurs
Cvminval	Find the min value of a vector and the index where it occurs
Cvmxmul	Multiply a matrix by a vector to form another vector
Cvmxmul1	Multiply the transpose of a matrix by a vector into another vector
CvscalC	Multiply a vector by a complex scalar
CvscalR	Multiply a vector by a real scalar
Cvsub	Subtract two vectors

Probability and Statistics Functions

binomdst	Cumulative Binomial distribution function; given k0, find P(k < k0)
curvreg	Polynomial regression routine, y(x) = a + b*x + c*x^2 + d*x^3 + ...
hyperdst	Cumulative Hypergeometric distrib. function; given k0, find P(k < k0)
invprob	Inverse cumulative normal distribution funct; given P(x), find x
least_sq	Generalized least-squares regression, y(x) = a*f0(x) + b*f1(x) + ...
linreg	Linear regression routine, y(x) = m*x + b
normal	Normal (Gaussian) random number of specified mean and variance
nprob	Cumulative normal distribution function; given x0, find P(x < x0)
poissdst	Cumulative Poisson distribution function; given k0, find P(k < k0)
poisson	Poisson random number of specified mean and variance
stats	Compute the mean and variance of an array of data
urand	Uniform random number generator (0...1)

Numerical Analysis Functions

conjgrad	Minimize n-dimensional differentiable function f(x1, x2,..., xn)
deriv	Derivative of equidistant data array with differentiating filter
deriv1	Differentiate a user-defined (analytic) function f(x) at x = x0.
integrat	Integrate equidistant array of data with Simpson's rule
interp	Interpolate equidistant array with 5th degree Lagrange polynomial
p_roots	Compute real and imaginary roots of polynomials with real coefficients
interp1	Interpolate data with nth degree Lagrange polynomial, 0 < n < 10
newton	Find the zeros of the 1-dimensional function; i.e. x, where f(x) = 0
romberg	Romberg integration of a user-defined function f(x) over [x0, x1]
spline	Interpolate data with a natural cubic spline

Signal Processing Functions

auto2dft	2D-autocorrelation of an array via 2D-FFT
autocor	Autocorrelation of a real-time series
autofft	Fast autocorrelation of a real-time series via FFT
bandpass	Design a (FIR) bandpass filter with a Kaiser-Bessel window
bilinear	Bi-linear transformation of general s-plane function to z-plane

(continues)

Signal Processing Functions (Continued)

---

conv2dft	2-D convolution of two real arrays via 2-D FFT
convofft	Fast convolution of two time real series via FFT
convolve	Implement a finite impulse response (FIR) filter with convolution
Cpowspec	Frequency analysis via Welch modified periodogram power spectrum
cros2dft	Fast 2D cross correlation of two real arrays via 2D-FFT
crosscor	Cross correlation of two real-time series
crossfft	Fast cross correlation of two time real series via FFT
dct	1-dimensional discrete cosine transform
dct2d	2D-discrete cosine transform via fast block matrix approach
downsamp	Lower the sampling rate of (i.e., decimate) an equidistant array
fft2d	2-D fft generalized for complex data
fft2d_r	Forward 2-D fft optimized for real data
fft42	Cooley–Tukey radix-"4 + 2" Fast Fourier Transform
fftrad2	Cooley–Tukey radix-2 Fast Fourier Transform
fftreal	Forward fft optimized for real data
fftr_inv	Inverse fft optimized for real data
highpass	Design a (FIR) highpass filter with a Kaiser–Bessel window
iirfiltr	Filter real data with an IIR type filter
lmsadapt	Least-mean-square algorithm (for line-enhancement and noise cancelling)
lowpass	Design a (FIR) lowpass filter with a Kaiser–Bessel window
median	Remove shot-noise or noise from digital drop-out with a median filter
powspec	Frequency analysis via Welch modified periodogram power spectrum
resample	Resample a digital signal to any new integral rate
smooth	Smooth equidistant array of data with a lowpass filter
tdwindow	9 spectral windows: Hanning, Hamming, Blackman, etc.
whitnois	White noise generator

Input/Output Functions

Cvread	Read a complex vector from a binary disk file
Cvwrite	Write a complex vector to a binary disk file
hwdclose	Close a Hypersignal Waveform Data file
hwdcreate	Create a new Hypersignal Waveform Data file
hwdopen	Open an existing Hypersignal Waveform Data file
hwdread	Read data from a Hypersignal Waveform Data file
hwdwrite	Write data to a Hypersignal Waveform Data file
iir_read	Read coefficients from a Hypersignal IIR or FIR data file
vread	Read a real vector from a binary disk file
vwrite	Write a real vector to a binary disk file
xyinfo	Read the header information of a binary or ASCII (LOTUS) disk file
xyread	Read data from a binary or ASCII (LOTUS) disk file
xywrite	Write data to a binary or ASCII (LOTUS) disk file

Miscellaneous Functions	
acosh	Inverse hyperbolic cosine function
asinh	Inverse hyperbolic sine function
atanh	Inverse hyperbolic tangent function
besi0	Modified Bessel function of zeroth order, I0(x)
besi1	Modified Bessel function of first order, I1(x)
besin	Modified Bessel function of nth order, In(x)
besk0	Modified Bessel function of zeroth order, K0(x)
besk1	Modified Bessel function of first order, K1(x)
beskn	Modified Bessel function of nth order, Kn(x)
besj0	Bessel function of first kind and zeroth order, J0(x)
besj1	Bessel function of first kind and first order, J1(x)
besjn	Bessel function of first kind and nth order, Jn(x)
besy0	Bessel function of second kind and zeroth order, Y0(x)
besy1	Bessel function of second kind and first order, Y1(x)
besyn	Bessel function of second kind and nth order, Yn(x)
chebser	Evaluate a Chebyshev series with finite coefficients
combin	Combinations function, N!/(r!*(N - r)!)
fact	Factorial function, N!
logn	Logarithm function of general base, n
permut	Permutations function, N!/(N - r)!
pseries	Evaluate a power series with finite coefficients

1.6 DATA TYPES

The MATHLIB.H header file contains the definitions of several data types that are used extensively by the functions in this library.

1.6.1 Real

The most fundamental data type defined is the type **Real**. By default, the type **Real** is defined to be equivalent to the standard type **double**. However, the type **Real** can easily be changed to be equivalent to the standard type **float**. That is why all the library functions use the type **Real** rather than the type **double**. By simply redefining the type **Real** in MATHLIB.H to be equivalent to the type **float**, you can create a new MATHLIB.LIB file that performs single precision rather than double precision floating point arithmetic. To change the library from double precision to single precision, you would simply edit MATHLIB.H and change the line #define REAL double to #define REAL float, and then compile the C/Math Library Source to create a new MATHLIB.LIB file.

Following is an example program that uses the data type **Real**.

```
#include <stdio.h>
#include "mathlib.h"

main()
{
    Real x = 0.0;
    if (sizeof(x) == sizeof(double))
        printf("Real is equivalent to double\n");
    else if (sizeof(x) == sizeof(float))
        puts("Real is equivalent to float\n");
    printf("x = %f\n", x);
    return 0;
}
```

1.6.2 Complex

Higher level data types are defined from the fundamental data type **Real**. The type **Complex** is defined as a structure containing both a real and an imaginary part. Both members of this data structure are defined as type **Real**:

typedef struct {Real r; Real i;} Complex;

Assume that a variable named x is declared as type **Complex**. Then the real part of the complex variable would be accessed as $x.r$, and the imaginary part as $x.i$. Following is an example program that uses the data type **Complex**.

```
#include <stdio.h>
#include "mathlib.h"
Complex x = {1.0, -1.0};
main()
{
    printf("The real part of x = %f\n", x.r);
    printf("The imaginary part of x = %f\n", x.i);
    return 0;
}
```

1.6.3 Real_Vector and Complex_Vector

From the **Real** and **Complex** data types, two higher level data types are defined: **Real_Vector** and **Complex_Vector**. The **Real_Vector** type is a single dimensioned array of **Real** values. The **Complex_Vector** type is a single dimensioned array of **Complex** values. A variable declared as either of these data types is dynamic. This means that memory for the variable must be allocated at run-time, rather than being statically allocated at compile-time.

The advantage of run-time memory allocation is that the available memory can be utilized much more efficiently. For example, if a program reads data from a file into an array, typically the size of the array needed to store the data is not known until the file has been read. In other words, the size of the array is not known at compile-time. The size is determined only after the program is executed (for example, at run-time). With static or compile-time allocation, an array must be over-dimensioned to accommodate the largest possible data size (for example, wasted space). But with dynamic or run-time allocation, an array can be dimensioned the exact size needed to store the data (for example, no wasted space).

A variable of type **Real_Vector** or **Complex_Vector** is really just a pointer. The actual memory for the single dimensioned array of **Real** or **Complex** values must be allocated by the *valloc* or *Cvalloc* functions, respectively. The *valloc* function allocates memory for a specified number of **Real** values, while the *Cvalloc* function allocates memory for a specified number of **Complex** values. The prototypes for these two functions are:

> **Real_Vector valloc(Real_Ptr** *address*, **unsigned** *npts*);
>
> **Complex_Vector Cvalloc(Complex_Ptr** *address*, **unsigned** *npts*);

Notice that each function requires two arguments, *address* and *npts*. The *npts* argument specifies the number of points in the vector (for example, the number of **Real** or **Complex** values). The *address* argument is typically the value **NULL**, but may optionally be the name of a single dimensioned array of type **Real** or **Complex**. The data types **Real_Ptr** and **Complex_Ptr** are simply defined as pointers to the types **Real** and **Complex**, respectively.

If the value of *address* is **NULL**, then space for the array of **Real** or **Complex** values is allocated from the available memory, and a pointer to the allocated space is returned. However, if the name of an array is specified for *address*, then no space is allocated. Instead, the *valloc* and *Cvalloc* functions simply return a pointer to the beginning of the specified array. Since dynamically allocated arrays cannot be pre-initialized, this provides a mechanism for using initialized arrays.

The individual values of a vector are accessed like a single dimension array. The variable name is immediately followed by an index value enclosed by square brackets. For example, $v[i]$ would access the ith value in vector v. If vector v contains n values, then $v[0]$ accesses the first value, and $v[n-1]$ accesses the last value in the vector.

Following is an example program that uses both the **Real_Vector** and **Complex_Vector** data types.

```
#include <stdio.h>
#include "mathlib.h"
main()
{
    int i;
    Real_Vector v;
    Complex_Vector cv;
    v = valloc(NULL, 5);      /* allocate space for 5 Real values */
    cv = Cvalloc(NULL, 5);    /* allocate space for 5 Complex values */
    for (i = 0; i < 5; i++) {
        v[i] = (Real) i;      /* initialize Real vector */
        cv[i].r = (Real) i;   /* initialize real part of Complex vector */
        cv[i].i = (Real) -i;  /* initialize imaginary part of Complex vector */
        printf("v[%d] = %f, ", i, v[i]);
        printf("cv[%d].r = %f, ", i, cv[i].r);
        printf("cv[%d].i = %f\n", i, cv[i].i);
    }
    return 0;
}
```

The following example is identical to the previous example, except that preinitialized arrays are used.

```
#include <stdio.h>
#include "mathlib.h"

Real V[5] = {0.0, 1.0, 2.0, 3.0, 4.0};
Complex CV[5] = {0.0, 0.0, 1.0, -1.0, 2.0, -2.0, 3.0, -3.0, 4.0, -4.0};

main()
{
    int i;
    Real_Vector v;
    Complex_Vector cv;
    v = valloc(V, 5);        /* allocate space for 5 Real values */
    cv = Cvalloc(CV, 5);     /* allocate space for 5 Complex values */
    for (i = 0; i < 5; i++) {
        printf("v[%d] = %f, ", i, v[i]);
        printf("cv[%d].r = %f, ", i, cv[i].r);
        printf("cv[%d].i = %f\n", i, cv[i].i);
    }
    return 0;
}
```

1.6.4 Real_Matrix and Complex_Matrix

The **Real_Matrix** and **Complex_Matrix** data types are similar to the **Real_Vector** and **Complex_Vector** data types, except that a matrix represents an array of two dimensions rather than a single dimension. The **Real_Matrix** type is a two-dimensional array of **Real** values. The **Complex_Matrix** type is a two-dimensional array of **Complex** values. A variable of type **Real_Matrix** or **Complex_Matrix** is dynamic. This means that memory for the variable must be allocated at run-time, rather than being statically allocated at compile-time.

A variable of type **Real_Matrix** or **Complex_Matrix** is really just a pointer. The actual memory for the two-dimensional array of **Real** or **Complex** values must be allocated by the *mxalloc* or *Cmxalloc* functions, respectively. The *mxalloc* function allocates memory for a specified number of **Real** values, while the *Cmxalloc* function allocates memory for a specified number of **Complex** values. The prototypes for these two functions are as follows:

> **Real_Matrix mxalloc(Real_Ptr *address*,**
> **unsigned *rows*, unsigned *cols*);**
> **Complex_Matrix Cmxalloc(Complex_Ptr *address*,**
> **unsigned *rows*, unsigned *cols*);**

Notice that each function requires three arguments, *address*, *rows*, and *cols*. The *rows* and *cols* arguments specify the number of rows and columns in the matrix. Space is allocated for *rows*∗*cols* **Real** or **Complex** values. The *address* argument is typically the value **NULL**, but may optionally be the name of a two-dimensional array of type **Real** or **Complex**. The data types **Real_Ptr** and **Complex_Ptr** are simply defined as pointers to the types **Real** and **Complex**, respectively.

If the value of *address* is **NULL**, then space for the two-dimensional array of **Real** or **Complex** values is allocated from the available memory, and a pointer to the allocated space is returned. However, if the name of an array is specified for *address*, then only an array of pointers is allocated and initialized to point to the beginning of each row in the specified two-dimensional array. Since dynamically allocated arrays cannot be preinitialized, this provides a mechanism for using initialized arrays.

The individual values of a matrix are accessed like a two-dimensional array. The variable name is immediately followed by two index values, each enclosed by square brackets. For example, $mx[i][j]$ would access the ith row and jth column of matrix mx. If matrix mx contains m rows and n columns, then $mx[0][0]$ accesses the first value, and $mx[m-1][n-1]$ accesses the last value in the matrix.

Following is an example program that uses both the **Real_Matrix** and **Complex_Matrix** data types.

```
#include <stdio.h>
#include "mathlib.h"
main()
{
    int i, j;
    Real_Matrix m;
    Complex_Matrix cm;
    /* allocate space for 10 Real values, 2 rows by 5 columns */
    m = mxalloc(NULL, 2, 5);
    /* allocate space for 10 Complex values, 2 rows by 5 columns */
    cm = Cmxalloc(NULL, 2, 5);
    for (i = 0; i < 2; i++) {        /* for each row */
        for (j = 0; j < 5; j++) {    /* for each column */
            /* initialize Real matrix */
            m[i][j] = (Real) j;
            /* initialize real part of Complex matrix */
            cm[i][j].r = (Real) j;
            /* initialize imaginary part of Complex matrix */
            cm[i][j].i = (Real) -j;
            printf("m[%d][%d] = %f, ", i, j, m[i][j]);
            printf("cm[%d][%d].r = %f, ", i, j, cm[i][j].r);
            printf("cm[%d][%d].i = %f\n", i, j, cm[i][j].i);
        }
    }
    return 0;
}
```

The following example is identical to the one above, except that preinitialized arrays are used.

```
#include <stdio.h>
#include "mathlib.h"

Real M[2][5] = {
    0.0, 1.0, 2.0, 3.0, 4.0,                                 /* first row  */
    0.0, 1.0, 2.0, 3.0, 4.0                                  /* second row */
};
Complex CM[2][5] = {
    0.0, 0.0, 1.0, -1.0, 2.0, -2.0, 3.0, -3.0, 4.0, -4.0,    /* first row  */
    0.0, 0.0, 1.0, -1.0, 2.0, -2.0, 3.0, -3.0, 4.0, -4.0     /* second row */
};

main()
{
    int i, j;
    Real_Matrix m;
    Complex_Matrix cm;
    /* allocate space for 10 Real values, 2 rows by 5 columns */
    m = mxalloc(M, 2, 5);
    /* allocate space for 10 Complex values, 2 rows by 5 columns */
    cm = Cmxalloc(CM, 2, 5);
    for (i = 0; i < 2; i++) {                /* for each row */
        for (j = 0; j < 5; j++) {            /* for each column */
            printf("m[%d][%d] = %f, ", i, j, m[i][j]);
            printf("cm[%d][%d].r = %f, ", i, j, cm[i][j].r);
            printf("cm[%d][%d].i = %f\n", i, j, cm[i][j].i);
        }
    }
    return 0;
}
```

1.6.5 Vector and Matrix Arrays

In addition to declaring simple variables of type **Real_Vector**, **Complex_Vector**, **Real_Matrix**, and **Complex_Matrix**, arrays of these data types may also be declared. For example, the following declares an array of 10 Real vectors:

Real_Vector vector[10];

With the above declaration, *vector*[0] accesses the first vector, and *vector*[9] accesses the last vector in the array. The first value of *vector*[0] is accessed as *vector*[0][0], and the first value of *vector*[9] is accessed as *vector*[9][0].

Likewise, we could declare an array of 10 Real matrices.

Real_Matrix matrix[10];

With the above declaration, *matrix*[0] accesses the first matrix, and *matrix*[9] accesses the last matrix in the array. The first value of *matrix*[0] is accessed as *matrix*[0][0][0], and the first value of *matrix*[9] is accessed as *matrix*[9][0][0].

Notice that when arrays of vectors or matrices are declared, accessing the individual values of a particular vector or matrix within the array requires additional indices, one for each dimension of the array. First, the array must be indexed appropriately to access the particular vector or matrix of interest; then the vector or matrix must be indexed to access the particular value of interest.

Following is an example program that declares and initializes single dimensioned arrays of these data types.

```c
#include <stdio.h>
#include "mathlib.h"

#define COUNT    10   /* number of vectors or matrices in the arrays */
#define POINTS   20   /* number of values in a vector                */
#define ROWS      4   /* number of rows in a matrix                  */
#define COLS      5   /* number of cols in a matrix                  */

main()
{
    int i, j, k;
    Real_Vector vect[COUNT];
    Complex_Vector cvect[COUNT];
    Real_Matrix mat[COUNT];
    Complex_Matrix cmat[COUNT];
    for (i = 0; i < COUNT; i++) {           /* allocate memory */
        vect[i] = valloc(NULL, POINTS);
        cvect[i] = Cvalloc(NULL, POINTS);
        mat[i] = mxalloc(NULL, ROWS, COLS);
        cmat[i] = Cmxalloc(NULL, ROWS, COLS);
        for (j = 0; j < POINTS; j++) {      /* initialize vectors */
            vect[i][j] = 0.0;
            cvect[i][j].r = 0.0;
            cvect[i][j].i = 0.0;
        }
        for (j = 0; j < ROWS; j++) {        /* initialize matrices */
            for (k = 0; k < COLS; k++) {
                mat[i][j][k] = 0.0;
                cmat[i][j][k].r = 0.0;
                cmat[i][j][k].i = 0.0;
            }
        }
    }
    puts("vectors and matrices allocated and initialized");
}
```

1.7 DYNAMIC MEMORY MANAGEMENT

As previously mentioned, the **Real_Vector**, **Complex_Vector**, **Real_Matrix**, and **Complex_Matrix** data types are dynamic. This means that the memory for variables of these data types must be allocated dynamically at run-time, rather than statically at compile-time. The *valloc*, *Cvalloc*, *mxalloc*, and *Cmxalloc* functions are used to allocate memory for these data types.

One advantage of run-time memory allocation is that the available memory can be used much more efficiently. For example, rather than overdimensioning an array to accommodate the largest possible data size, an array can be dimensioned to exactly match the size of the data. Another important advantage of run-time memory allocation is that the extended keywords **far** and **huge** can be utilized to access all available memory, no matter which memory model is being used. Most compilers that are designed to work with the Intel 80×86 microprocessors support the **far** and **huge** keywords, as well as various memory models such as *small*, *medium*, and *large*. Since the small and medium memory models are much more efficient than the large memory model, it is usually preferable to avoid using the large memory model. By utilizing the **far** and **huge** keywords, the C/Math Toolchest Library provides support for very large vectors and matrices, even with the more efficient small and medium memory models.

The **Real_Vector**, **Complex_Vector**, **Real_Matrix**, and **Complex_Matrix** data types may be defined as standard pointers, **far** pointers, or **huge** pointers. On computers with an 80×86 CPU, the amount of memory available for variables of these data types depends on the pointer definition. Assuming that the medium memory model is being used, the amount of memory available for vectors and matrices is as follows. With standard pointers, a maximum of 64K of memory is available for all vectors and matrices. With **far** or **huge** pointers, all the memory between the end of the data segment and the 640K boundary is available for storing vectors and matrices. With **far** pointers, each individual vector or matrix is limited to 64K in size. With **huge** pointers, any individual vector or matrix may be as large as the total available memory.

By default, the **Real_Vector**, **Complex_Vector**, **Real_Matrix**, and **Complex_Matrix** data types are defined as **far** pointers. Therefore, all the available memory below the 640K boundary may be used for storing vectors and matrices, but each individual vector or matrix is limited to 64K in size. To change the definition of these data types to standard pointers, you would simply edit the MATHLIB.H file and change the line #define STDC 0 to #define STDC 1. To change the definition to **huge** pointers, you would change the following lines in MATHLIB.H as indicated:

```
#define OBJECT far        ->   #define OBJECT huge
#define FAR_OBJECT 1      ->   #define FAR_OBJECT 0
#define HUGE_OBJECT 0     ->   #define HUGE_OBJECT 1
```

After changing the MATHLIB.H file, you would need to compile the C/Math Library Source to create a new MATHLIB.LIB file.

1.7.1 Recovering Memory

Many of the functions in the C/Math Toolchest Library require a vector or matrix as an input argument. Typically, these input vectors or matrices are explicitly allocated using the *valloc*, *Cvalloc*, *mxalloc*, or *Cmxalloc* functions. In addition, many of the functions return a vector or matrix as the result. These output vectors or matrices are implicitly allocated by the library function itself, and then a pointer to the allocated memory is returned as the function result. In either case, whether the memory is explicitly or implicitly allocated, the memory should be recovered after a vector or matrix is no longer needed.

The *vfree*, *Cvfree*, *mxfree*, and *Cmxfree* functions are used to recover the memory allocated to variables of type **Real_Vector**, **Complex_Vector**, **Real_Matrix**, and **Complex_Matrix**, respectively. The prototypes for these functions are

> **void vfree(Real_Vector** *v***);**
> **void Cvfree(Complex_Vector** *v***);**
> **void mxfree(Real_Matrix** *m***);**
> **void Cmxfree(Complex_Matrix** *m***);**

When one of the above functions is called with the name of a vector or matrix variable as the argument, the memory allocated to the specified vector or matrix is recovered (i. e., freed for other use). If you fail to free an allocated vector or matrix after it is no longer needed, and subsequently allocate memory for additional vectors or matrices, then you may run out of memory.

All these memory recovery functions compare the argument passed against an internal list of addresses corresponding to the currently allocated vectors and matrices. The memory is freed only if the argument matches the address of a previously allocated vector or matrix. Once freed, the address of the vector or matrix is removed from the list. Therefore, it does no harm to free a vector or matrix more than once.

Following is an example program that uses the memory recovery functions.

```
#include <stdio.h>
#include "mathlib.h"
main()
{
    Real_Vector v;
    Complex_Vector cv;
    Real_Matrix m;
    Complex_Matrix cm;

    /* These functions are used to allocate space */
    v = valloc(NULL, 100);        /* allocate vector of 100 Real values */
    cv = Cvalloc(NULL, 100);      /* allocate vector of 100 Complex values */
    m = mxalloc(NULL, 10, 10)     /* allocate matrix of 100 Real values */
    cm = Cmxalloc(NULL, 10, 10)   /* allocate matrix of 100 Complex values */

    /* These functions are used to recover the allocated space */
    vfree(v);                     /* free vector of 100 Real values */
    Cvfree(cv);                   /* free vector of 100 Complex values */
    mxfree(m);                    /* free matrix of 100 Real values */
    Cmxfree(cm);                  /* free matrix of 100 Complex values */
}
```

1.8 ERROR HANDLING

There are many types of errors that can occur when a C/Math Toolchest Library function is called. For example, an argument passed to a function may be outside the domain of the function, or there may not be enough memory to allocate storage space for a vector or matrix. When an error is detected by one of the library functions, an error message is printed to **stderr** (i. e., the screen), and then the program is automatically aborted i. e., terminated via the statement *exit*(1). The error message has the following form:

> $* * * >$ **Mathlib error in function:** *funcname*
>   **Appropriate error message displayed here**

where *funcname* is the name of the library function that detected the error.

Now let's create a program that intentionally causes an error to occur. The default maximum size of an individual vector or matrix is limited to 64K (i. e., 65,535 bytes). Therefore, the following program should demonstrate the default error processing by attempting to allocate a vector of 80,000 bytes.

```
#include <stdio.h>
#include "mathlib.h"
#define SIZE 10000
main()
{
    Real_Vector v;
    printf("allocating space for %u Real values (i.e. %ld bytes)\n",
            SIZE, (long) SIZE * sizeof(Real));
    v = valloc(NULL, SIZE);
    puts("This should not be printed");
}
```

Since by default a program is automatically aborted when an error occurs, it is not necessary to test for errors after each call to a C/Math Toolchest Library function. Because the error checking code may be eliminated, this default error handling makes it a little easier to write programs. For most programs, the default error handling should be quite acceptable. However, for some programs, you may not want an error message to be printed, and you may not want the program to be aborted when an error occurs. Therefore, the C/Math Toolchest Library provides a method to override the default error handling.

1.8.1 Overriding Default Error Handling

The error-handling process is controlled by two global integer variables named **math_errmsg** and **math_abort**. The **math_errmsg** variable determines whether or not an error message is printed, and the **math_abort** variable determines whether or not the program is aborted. By default, both of these global variables have nonzero values, which causes an error message to be printed and the program to be aborted when an error occurs. However, if both variables are set to zero, then no error message will be printed and the program will not abort when an error occurs.

If the default error handling is overridden, then your program must test the return values of the C/Math Toolchest Library functions to determine whether or not an error occurred. For functions that return a vector or matrix as the result, a return value of **NULL** indicates that an error occurred. For most other functions, a return value of −**1** indicates that an error occurred. For some functions, no errors can occur, such as a function of type **void** (i. e., no return value).

If an error occurs, a corresponding error number is stored in the global integer variable named **math_errno**. The **math_errno** variable always contains the number of the last error that occurred. A value greater than zero indicates that an error occurred. A value of zero indicates that no error occurred. The *math_err* function may be used to print the error message associated with the error number stored in the **math_errno** variable. See the description of the *math_err* function for a list of the error numbers and meanings.

Following is a program that uses the **math_errmsg** and **math_abort** variables to override the default error processing.

```
#include <stdio.h>
#include "mathlib.h"
#define VECTORS 20
#define VECTOR_SIZE 5000
main()
{
    int i, j;
    Real_Vector v[VECTORS]; /* declare an array of vectors */
    math_errmsg = 0;        /* turn off error messages */
    math_abort = 0;         /* turn off program abort */
    for (i = 0; i < VECTORS; i++) {
        v[i] = valloc(NULL, VECTOR_SIZE);  /* allocate memory for vector */
        if (v[i] != NULL) {                /* test for error */
            /* no error */
            for (j = 0; j < VECTOR_SIZE; j++)
                v[i][j] = 0.0;             /* initialize vector */+
            printf("memory for vector %2d successfully allocated\n", i);
        }
        else {
            /* error */
            printf("unable to allocate memory for vector %2d\n", i);
        }
    }
    if (math_errno != 0) math_err();  /* print error message */
}
```

1.8.2 The matherr_ Function

All the C/Math Toolchest Library functions call an internal library function named *matherr_* when an error occurs. This is the function that actually prints an error message to **stderr** and then aborts the program when an error occurs. Therefore, another method to override the default error handling is to write your own *matherr_* function. The prototype for the *matherr_* function is

void matherr_(char **funcname***, int** *errnum***);**

where *funcname* is the name of the library function that detected the error, and *errnum* is the error number.

The global variable named **math_errs** is an array of pointers to character strings, each character string corresponding to the text of an error message. The error messages printed by the *math_err* function are stored in the **math_errs** array. As an alternative to calling the *math_err* function, the error messages may be accessed directly by indexing the **math_errs** array with the appropriate error number.

Following is an example program that uses its own *matherr_* function for error handling.

```
#include <stdio.h>
#include "mathlib.h"

void matherr_(char *funcname, int errnum)
{
    printf("Error occurred in library function: %s\n", funcname);
    printf("Error number: %d\n", errnum);
    printf("Error message: %s\n", math_errs[errnum]);
    exit(1); /* exit program */
}


main()
{
    Real_Vector v;
    v = valloc(NULL, 10000);                /* cause an error to occur */
    printf("This should not be printed\n");
}
```

1.8.3 Floating Point Exceptions

Another type of error that can occur is called a floating point exception. A floating point exception is an arithmetic error such as underflow, overflow, or divide by 0. Arithmetic errors can occur during floating point operations such as addition, multiplication, or division. Since floating point exceptions are typically fatal errors, it is usually not possible to continue program execution after an error of this type has occurred. Therefore, it is typical for a floating point exception to cause a program to abort, perhaps after displaying an appropriate error message.

In some applications, you may want to trap floating point exceptions in order to display an appropriate error message and perform clean-up operations before allowing the program to terminate. For example, if your program has changed the video mode to graphics, then it is probably desirable to have the program change the video mode back to text before terminating.

The ANSI standard C function named *signal* may be used to trap events such as floating point exceptions. The *signal* function requires two arguments. For trapping signals caused by floating point exceptions, the first argument is **SIGFPE**, which is defined in the standard header file SIGNAL.H. The second argument is the name of the function that will receive control when a floating point exception occurs. This is a signal handling function that you must write to report the error and perform any desired clean-up operations. The prototype for the signal handling function is

void *sighandler*(int);

The signal handling function requires a single integer argument. For floating point exceptions, this argument will be the value **SIGFPE**. After handling the error condition, the signal handling function can continue program execution at the point of the error by simply returning, or continue at some other point in the program by calling the standard *longjmp* function, or terminate the program by calling the standard *exit* or *abort* functions. Since floating point exceptions are often fatal, the signal handling function should typically not allow program execution to continue after this type of error has occurred. Rather, after displaying an appropriate error message and performing any desired clean-up operations, the signal handling function should simply terminate the program.

Following is an example program that uses the standard C *signal* function to trap floating point exceptions.

```
#include <stdio.h>
#include <signal.h>
#include "mathlib.h"

void fpe_handler(int signal_type)
{
    if (signal_type == SIGFPE)
        puts("A floating point exception occurred (e.g. underflow, "
            "overflow, divide by 0)");
    else puts("Signal type was not a floating point exception");
    exit(1);    /* exit program */
}

main()
{
    Real x = 1.0, y;
    signal(SIGFPE, fpe_handler);  /* setup signal handler */
    puts("fpe_handler assigned to handle floating point exceptions");
    y = x / (1.0 - x);            /* cause divide by 0 error */
    puts("This should not be printed");
}
```

1.9 GLOBAL VARIABLES

The MATHLIB.H header file contains external declarations for several global variables. Many of these global variables are used internally by the library. However, you may wish to reference these variables in your own programs.

1.9.1 Numerical Constants

Several global variables are used to store frequently used numerical constants. Using variables rather than actual constants is more efficient, since it prevents the compiler from having to decode the value each time it is used in an expression. The following numerical constants are declared as external double precision variables in the MATHLIB.H header file:

extern double pi_	= **3.14159265358979323846;**
extern double twopi_	= **6.28318530717958647 69;**
extern double e_	= **2.71828182845904523536;**
extern double sqrt2_	= **1.41421356237309504880;**
extern double sqrt3_	= **1.73205080756887729352;**
extern double euler_	= **0.57721566490153286061;**

The following example program simply prints the value of each of these global variables.

```
#include <stdio.h>
#include "mathlib.h"
main()
{
    printf("PI      = %.14f\n", pi_);
    printf("PI*2    = %.14f\n", twopi_);
    printf("e       = %.14f\n", e_);
    printf("sqrt(2) = %.14f\n", sqrt2_);
    printf("sqrt(3) = %.14f\n", sqrt3_);
    printf("Euler   = %.14f\n", euler_);
}
```

1.9.2 Error Variables

Several global variables are used for processing errors. These variables were described in the section on **Error Handling**. They are declared in the MATHLIB.H header file as follows:

> **extern int math_errmsg;**
> **extern int math_abort;**
> **extern int math_errno;**
> **extern char *math_errs[ ];**

The **math_errmsg** variable controls whether or not an error message is printed to **stderr** when an error occurs. A value of zero turns printing off, while a nonzero value turns it on. The default value is nonzero (on).

The **math_abort** variable controls whether or not a program is automatically terminated when an error occurs. A value of zero turns program termination off, while a nonzero value turns it on. The default value is nonzero (on).

The **math_errno** variable contains the error number of the last error that occurred. A value of zero indicates that no errors have occurred, while a value greater than zero indicates that one of the C/Math Toolchest Library functions detected an error.

The **math_errs[]** variable is an array of pointers to error message strings. This array may be indexed by the **math_errno** variable to access the error message associated with the last error that occurred. The following program demonstrates the use of each of these error variables.

```
#include <stdio.h>
#include "mathlib.h"
main()
{
    math_errmsg = 0;   /* turn off automatic error message printing */
    math_abort = 0;    /* turn off automatic program termination */
    atanh(2.0);        /* cause error to occur */
    if (math_errno != 0) {
        printf("*** Error occurred in atanh function ***\n");
        printf("%s\n", math_errs[math_errno]);
    }
    else printf("No errors detected\n");
}
```

1.9.3 Miscellaneous Variables

Several global variables are used for various purposes. These miscellaneous variables are declared in MATHLIB.H as:

> **extern Real determ_;**
> **extern Complex Cdeterm_;**
> **extern int useinput_;**
> **extern int math_digits;**
> **extern int math_linesize;**

The **determ_** and **Cdeterm_** variables are used to store the Real and Complex values of a matrix determinant. The **determ_** variable is used by the *lineqn, mxdeter, mxinv, mxinv22,* and *mxinv33* functions. The **Cdeterm_** variable is used by the *Clineqn, Cmxdeter, Cmxinv, Cmxinv22,* and *Cmxinv33* functions. All these functions perform matrix operations, and in the process store the determinant of the resulting matrix in the **determ_** or **Cdeterm_** **variables**.

The **useinput_** variable controls whether or not an input vector or matrix is overwritten by a resulting output vector or matrix. Many of the functions in the library perform vector or matrix operations. These functions typically require an input argument that is either a vector or matrix. The library function then performs an operation on the input vector or matrix, and returns a resulting vector or matrix. By default, the **useinput_** variable is zero, which causes the library function to allocate memory for the resulting vector or matrix. Therefore, the input vector or matrix is unchanged after calling the library function. However, if the **useinput_** variable is set to a nonzero value before calling the library function, then the original contents of the input vector or matrix is lost (i. e., the library function stores the resulting vector or matrix in the memory allocated to the input vector or matrix). The **useinput_** variable should be used only if a program would otherwise run out of memory.

The **math_digits** and **math_linesize** variables are used only by *xywrite*, a library function that outputs one or more vectors to a disk file. These variables affect ASCII formatted output only (i. e., not binary output). The **math_digits** variable specifies the number of digits to print for floating point values (default = 6). The **math_linesize** variable specifies the maximum length of each line in the file (default = 80). This may be important if you intend to edit the output file.

2

GRAFIX Program

The GRAFIX program is a graphical data analysis tool. Its primary purpose is to plot the data that you create with the C/Math Toolchest Library functions. In addition, the GRAFIX program allows you to edit data, draw smooth curves through data points using cubic spline or Lagrange interpolation, or fit curves to a set of data points using regression. It also allows you to zoom, scale, and print graphs.

2.1 CREATING DATA

Before attempting to use the GRAFIX program, you must first create some data files. By default, the GRAFIX program assumes that any file with a **DAT** extension is a data file. Before proceeding, make sure that the directory where you installed the C/Math Toolchest Library files is the current directory.

The GRAFIX program can read data files in several different formats. Both ASCII and binary formats are supported. The advantage of ASCII formatted data is that it is readable (by humans) and portable from one program to another, whereas binary data is nonreadable (by humans) and often nonportable. Binary data is typically nonportable because there are many different ways to represent floating point values in binary format. However, a big advantage of binary data is that it requires much less disk space, and can be read or written much faster than ASCII formatted data. The speed difference is most apparent when one is dealing with large data files.

In ASCII formatted data files, floating point values are represented as follows, where [] means "optional," | means "or," and *d* is one or more digit characters between **0** and **9**:

[+|−] [*ddd*] [.[*ddd*]] [e|E[+|−]*ddd*]

In binary data files, floating point values are bit patterns, with each value represented as a sequence of **sizeof(Real)** bytes in the standard IEEE format. However, if you create a new GRAFIX program by recompiling the GRAFIX Source, then the format of binary floating point values may change. For example, the **Real** data type may be defined as either **float** or **double** (default = double). Also, some compilers may use a floating point format other than the standard IEEE format. The format of the floating point values in a binary data file must match the format used by the compiler that created the GRAFIX program. Otherwise, the GRAFIX program will be unable to successfully read the data file. For more information about creating binary data files for the GRAFIX program, please see the description of the *xywrite* function.

A data file may contain a header at the beginning of the file. A header provides information about the data stored in the file. For example, a header can specify the format (ASCII or binary), the number of curves, the number of points in each curve, and, optionally, labels to identify the curves. A binary data file is required to have a header. However, an ASCII formatted data file may or may not have a header.

2.1.1 Data Files without Headers

The simplest data file is an ASCII formatted file containing data for a single curve. We'll call this the XY format. For the XY format, each line in the file must contain a single pair of X and Y values (X value followed by Y value). The X and Y values on each line must be separated by one or more blank or tab characters (i. e., ' ' or '\t'). For example, a file containing the following five lines represents a single curve with five data points, the X values ranging from 1.0 to 5.0 and the Y values ranging from 10.0 to 15.0.

File Containing Single Curve with Five Data Points

```
1.0     10.0
2.0     11.0
3.0     12.0
4.0     13.0
5.0     15.0
```

A simple extension of the XY format may be used to represent multiple curves, with each curve represented by the same X values. We'll call this the XYY format. For the XYY format, each line in the file must contain a single X value followed by one or more Y values. For example, a file containing the following five lines represents three curves with five data points each. The X values for each curve range from 1.0 to 5.0, while the Y values for the first curve range from 10.0 to 15.0, the Y values for the second curve range from 16.0 to 20.0, and the Y values for the third curve range from 21.0 to 25.0.

File Containing Three Curves with Five Data Points Each

1.0	10.0	16.0	21.0
2.0	11.0	17.0	22.0
3.0	12.0	18.0	23.0
4.0	13.0	19.0	24.0
5.0	15.0	20.0	25.0

You can create ASCII data files in either the XY or XYY format using a program such as an editor or spreadsheet. However, most data files will probably be created by one of your own C programs.

If you compile and execute the sample program named BESSEL.C, it creates an ASCII data file named BESSEL.DAT. The BESSEL.DAT file contains two curves with sixteen data points each, representing the bessel functions $j0$ and $y0$ over the range 1 to 16.

BESSEL.C

```c
#include <stdio.h>
#include "mathlib.h"
main()
{
    int i;
    double x;
    FILE *fp;
    fp = fopen("bessel.dat", "w");     /* create BESSEL.DAT */
    if (fp != NULL) {
        for (x = 1.0; x <= 16.0; x += 1.0) {
            fprintf(fp, "%5.2f %10.6f %10.6f\n",
                    x, besj0(x), besy0(x));
        }
        puts("bessel.dat created");
        fclose(fp);
    }
    else puts("*** unable to create bessel.dat ***");
}
```

2.1.2 Data Files with Headers

A header can be added at the beginning of a data file to provide additional information about the data. The additional information supplied by a header allows the GRAFIX program to process a data file more efficiently. For example, without a header, the GRAFIX program must count the number of data points in the file before it can allocate the appropriate amount of memory needed to store the data. However, since a header specifies the number of points in the data file, there is no need to count the points. Also, a header provides more flexibility for storing multiple curves in a single file. For example, without a header, only one curve may be stored in a file using the XY format. Remember that the XYY format allows multiple curves, but each curve must have identical X values. However, since a header can specify both the number of curves and the number of points in each curve, multiple curves with independent X values may be stored using the XY format.

A header is a string containing at least three essential pieces of information about the data: the format, the number of curves, and the number of points in each curve. The format can be ASCII or binary, and it can be XY or XYY. Therefore the possible formats are **XY_ASCII**, **XY_BINARY**, **XYY_ASCII**, or **XYY_BINARY**. By default, the number of curves is limited to five and the number of points in each curve is limited to 8192. However, these limitations may change if the GRAFIX Source is recompiled.

In the previous section, a listing of the contents of two ASCII data files without headers was presented. One contained a single curve in the XY format and the other contained three curves in the XYY format. With headers, those same two data files would appear as follows.

File Containing Single Curve with Five Data Points

```
"Format=XY_ASCII Curves=1 Points=5"
   "X0"      "Y0"
   1.0       10.0
   2.0       11.0
   3.0       12.0
   4.0       13.0
   5.0       15.0
```

File Containing Three Curves with Five Data Points Each

```
"Format=XYY_ASCII Curves=3 Points=5"
   "X0"      "Y0"      "Y1"      "Y2"
   1.0       10.0      16.0      21.0
   2.0       11.0      17.0      22.0
   3.0       12.0      18.0      23.0
   4.0       13.0      19.0      24.0
   5.0       15.0      20.0      25.0
```

The first line in each of the previous two data files is the header. Notice that the header assigns values to three keywords: **Format**, **Curves**, and **Points**. The second line is a comment line that *must* be present. In this case, the comment line is used to identify each column of values. The column of X values is identified as "X0." The columns of Y values are identified as "Y0" through "Y2." The actual data values immediately follow the comment line.

The data file containing three curves could also be represented using the XY format rather than the XYY format. Since all three curves have the same X values, this would be wasteful of both memory and disk space. However, it does illustrate that the XY format can be used to represent multiple curves with independent X values.

File Containing Three Curves with Five Data Points Each

```
"Format=XY_ASCII Curves=3 Points=5,5,5"
    "X0"    "Y0"
    1.0     10.0
    2.0     11.0
    3.0     12.0
    4.0     13.0
    5.0     15.0
    "X1"    "Y1"
    1.0     16.0
    2.0     17.0
    3.0     18.0
    4.0     19.0
    5.0     20.0
    "X2"    "Y2"
    1.0     21.0
    2.0     22.0
    3.0     23.0
    4.0     24.0
    5.0     25.0
```

Notice that the header line in the previous file defines **Points** with three values rather than one. The XY format allows each curve to have a different number of data points. Therefore, the number of points must be defined for each curve. The first value assigned to **Points** defines the number of points in the first curve, and so on. Notice that the X and Y values for the first curve are identified on the comment line as "X0" and "Y0." The data points for each subsequent curve *must* also be preceded by a comment line. Notice that the comment line preceding the second curve is "X1" "Y1," and the comment line preceding the third curve is "X2" "Y2."

If you compile and execute the sample program named HYPER.C, it creates an ASCII data file named HYPER.DAT. The HYPER.DAT file contains three curves with 15 data points each, representing the inverse hyperbolic sine, cosine, and tangent functions. The XY format is used so that the range of X values can be different for each function.

The HYPER.C program uses the library function named *xywrite* to create the data file. The *xywrite* function creates data files with headers. You should notice that this program also adds label information to the header. The GRAFIX program will display the label information when the data is plotted. You should examine the HYPER.DAT file if you are interested in how the label information is stored in the header.

HYPER.C

```
#include <stdio.h>
#include "mathlib.h"
#define CURVES 3
Labels labels = {"INVERSE HYPERBOLIC FUNCTIONS",    /* graph title        */
                "X",                                /* x axis title       */
                "Y",                                /* y axis title       */
                "asinh function",                   /* label for 1st curve */
                "acosh function",                   /* label for 2nd curve */
                "atanh function"};                  /* label for 3rd curve */
double (*func[CURVES])(double) = {asinh, acosh, atanh}; /* function to plot */
unsigned        points[CURVES] = {   15,    15,     15}; /* number of points */
double         start_x[CURVES] = { -3.0,   1.0, -0.95}; /* starting x value */
double           end_x[CURVES] = {  3.0,   3.0,  0.95}; /* ending x value   */
main()
{
    int i, j;
    Real x, dx;
    Real_Vector vx[CURVES], vy[CURVES];
    for (i = 0; i < CURVES; i++) {                  /* allocate vectors */
        vx[i] = valloc(NULL, points[i]);
        vy[i] = valloc(NULL, points[i]);
    }
    for (i = 0; i < CURVES; i++) {                  /* initialize vectors */
        x = start_x[i];
        dx = (end_x[i] - x) / (points[i] - 1);
        for (j = 0; j < points[i]; j++, x += dx) {
            vx[i][j] = x;
            vy[i][j] = (Real) (*func[i])(x);
        }
    }                                               /* output vectors */
    xywrite("hyper.dat", XY_ASCII, CURVES, points, &labels, vx, vy);
    for (i = 0; i < CURVES; i++) {                  /* free vectors */
        vfree(vx[i]);
        vfree(vy[i]);
    }
    puts("hyper.dat created");
}
```

If you compile and execute the sample program named FILTER.C, it creates three ASCII data files named LOWPASS.DAT, HIGHPASS.DAT, and BANDPASS.DAT. Each of these files contains a single curve with 257 data points, representing a lowpass, highpass, and bandpass filter response.

Again, the *xywrite* function is used to create the three data files. Since only a single X and Y vector is output to each data file, there is no need to declare an array of vectors as was done in the previous example. Notice that the **points**, **mag**, and **freq** variables in the following example are declared as simple variables rather than arrays. Because these variables are not arrays, notice that the **&** symbol must precede each of them in the call to the *xywrite* function.

FILTER.C

```c
#include <stdio.h>
#include <string.h>  /* prototype for strcpy function */
#include <math.h>    /* prototypes for log10 and sqrt functions */
#include "mathlib.h"
#define FFTSIZE 512
#define NWEIGHTS 31
Labels labels = {"Frequency Response",     /* graph title */
                "Normalized Frequency",    /* x axis text */
                "Magnitude (dB)",          /* y axis text */
                ""};                       /* curve text  */
main()
{
    unsigned points = FFTSIZE/2 + 1;
    Real fl = .20,              /* normalized low frequency cutoff    */
         fh = .30,              /* normalized high frequency cutoff   */
         db = 30.0;             /* stopband attenuation (in decibels) */
    Real_Vector freq, mag;
    Real weights[NWEIGHTS];
    Complex_Vector fft;
    char filename[13];
    int i, filter, curves = 1;
    mag = valloc(NULL, points);     /* allocate a vector for magnitude */
    freq = valloc(NULL, points);    /* allocate a vector for frequency */
    fft = Cvalloc(NULL, FFTSIZE);   /* allocate complex vector for FFT */
    for (filter = 0; filter < 3; filter++) {        /* for each filter */
        switch (filter) {
            /* define file name, curve text, and filter weights */
            case 0: /* Lowpass Filter */
                    strcpy(filename, "lowpass.dat");
                    strcpy(labels.curve, "Lowpass Filter");
                    lowpass(weights, NWEIGHTS, fh, db, 0);
                    break;
            case 1: /* Highpass Filter */
                    strcpy(labels.curve, "Highpass Filter");
                    strcpy(filename, "highpass.dat");
                    highpass(weights, NWEIGHTS, fl, db, 0);
                    break;
            case 2: /* Bandpass Filter */
                    strcpy(labels.curve, "Bandpass Filter");
                    strcpy(filename, "bandpass.dat");
                    bandpass(weights, NWEIGHTS, fh, fl, db, 0);
                    break;
        }
        /* initialize complex FFT vector with filter weights */
        for (i = 0; i < NWEIGHTS; i++) {fft[i].r = weights[i]; fft[i].i = 0.0;}
        /* initialize excess FFT points to 0 */
        for (i = NWEIGHTS; i < FFTSIZE; i++) {fft[i].r = 0.0; fft[i].i = 0.0;}
        /* Transform filter weights using FFTSIZE (512 point) FFT */
        fft42(fft, FFTSIZE, -1);
```

(continued)

```
/* Calculate the magnitude (in dB) of the transformed data
   over the normalized frequency range of 0 to .5          */
for (i = 0; i < points; i++) {
    freq[i] = (Real) i / FFTSIZE;
    mag[i] = 20 * log10(sqrt(fft[i].r * fft[i].r + fft[i].i * fft[i].i));
}
/* output magnitude/frequency response for each filter */
xywrite(filename, XY_ASCII, curves, &points, &labels, &freq, &mag);
fputs(filename, stdout); fputs(" created\n", stdout);
}
vfree(mag);   /* free magnitude vector */
vfree(freq);  /* free frequency vector */
Cvfree(fft);  /* free complex FFT vector */
}
```

2.2 LOADING CURVES

After creating some data files, we are now ready to execute the GRAFIX program. To execute the program, simply type the following and press the **Enter** key.

GRAFIX

If this is the first time that the GRAFIX program has been executed, then a menu titled *Choose Monitor Type* should appear. The menu choices are *Color* or *Black and White*. Press **C** if you have a color monitor, or press **B** for black and white. The screen should now look like Figure 2.1.

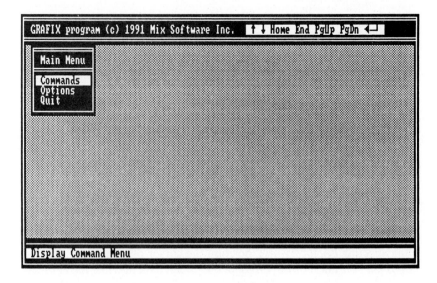

Figure 2.1   Main Menu

The GRAFIX program is menu driven. Notice that the outermost menu is titled *Main Menu*. This menu contains three items: *Commands*, *Options*, and *Quit*.

To the right of the copyright notice is a list of keys that may be used in menus. The **up** and **down** arrow keys move the cursor up or down. The **Home** and **End** keys position the cursor to the first or last item in a menu. The **PgUp** and **PgDn** keys scroll the menu items up or down, but only if the menu contains more items than are currently displayed. The **Enter** key selects the item under the cursor. As an alternative, items may also be selected by pressing a hot key. Notice that **C**, **O**, and **Q** are the hot keys for the *Main Menu*.

There is a help line at the bottom of the screen. This line provides a brief description of the item under the cursor. As the cursor is moved within a menu, the help message changes.

Now position the cursor over *Commands* and press the **Enter** key. The *Command Menu* will appear. Notice that **Esc** appears to the right of the list of keys on the top line of the screen. The **Esc** key is used to return to the previous menu. Press **Esc** to return to the *Main Menu*, then press **Enter** to select *Commands* once again.

Notice that the cursor is positioned over the item *Load curve(s)*. Press the **Enter** key and a list of the data (.dat) files should appear. The screen should look like Figure 2.2. These are the data files that were created in the previous section. Notice that the help line now contains a blinking **.**. This indicates that the files being displayed are in the current directory. By default, the GRAFIX program loads data files from only the current directory. However, as you will see later, one of the options allows you to specify the name of the directory from which data files are loaded.

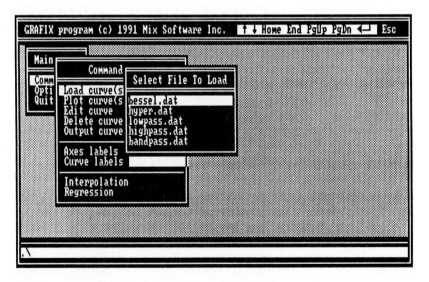

Figure 2.2   Loading Curves

After selecting *Load curves(s)*, position the cursor over *hyper.dat* and press the **Enter** key to load the curves from this data file. Small data files like the HYPER.DAT file are loaded almost instantly. However, large ASCII data files can take a while to load. For large files, the message "Loading File—Please Wait..." will appear on the help line until the file has finished loading.

If an error occurs while a file is being loaded, an error message window will appear in the center of the screen. An appropriate message describing the error will appear inside the window. The error message window will disappear when any key is pressed. If no error message appears after the HYPER.DAT file is loaded, then you can assume that the file was successfully loaded.

You should now have three curves loaded into memory from the HYPER.DAT file. If you now attempt to load another data file, a menu titled *Replace or Combine with existing curves* will appear after you select the data file. This gives you the option of combining curves from multiple data files. However, there can be no more than five curves loaded into memory at any one time.

With the *Command Menu* displayed, press **L** to load another data file, position the cursor over *hyper.dat*, then press the **Enter** key. A menu with the two items *Replace* and *Combine* should appear. Press **R** to replace the curves. This simply reloads the curves from the HYPER.DAT file.

If you had pressed **C** to combine the curves, the GRAFIX program would have attempted to combine the three curves from the HYPER.DAT file with the three curves already stored in memory. Since there can be no more than five curves loaded at any one time, this would have produced an error.

2.3 PLOTTING CURVES

We can now plot the three curves that were loaded from the HYPER.DAT file. With the *Command Menu* displayed, position the cursor over *Plot curve(s)* and press the **Enter** key, or simply press **P**. A menu titled *Scale Axes* containing the items *Auto scale* and *Specify scale* will appear. Selecting *Auto scale* automatically scales the plot to display all of the data points in all of the curves. Selecting *Specify scale* allows you to specify the minimum and maximum values for the X and Y axes.

With the *Scale Axes* menu displayed, press **A** to select *Auto scale*. The screen will switch to graphics mode and the three curves from the HYPER.DAT file will be plotted. The screen should now look like Figure 2.3. This is a plot of the three inverse hyperbolic functions: *asinh*, *acosh*, and *atanh*. Notice that rectangles identify the data points on the *asinh* curve, triangles on the *acosh* curve, and diamonds on the *atanh* curve.

Notice the horizontal menu at the top of the screen. This is the main plot menu. It contains six items: *Return*, *Zoom*, *Auto scale*, *Specify scale*, *Change scale*, and *Print*. Notice the cursor over the *Return* item. The **left** and **right** arrow keys move the cursor left or right. The **Home** and **End** keys position the cursor over the first or last item in the menu. Any item may be selected by positioning the cursor over the item and pressing the **Enter** key, or by simply pressing the hot key identified by the highlighted character in each menu item.

Selecting *Return* simply returns control back to the *Command Menu*. Pressing **Esc** has the same effect except that the next time *Plot curve(s))* is selected, the cursor will return to its current position, not necessarily over the *Return* item.

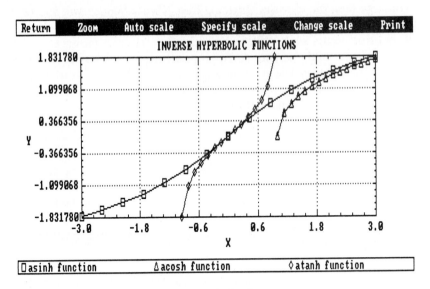

Figure 2.3 Plotting Curves

2.3.1 Zooming

Zoom allows you to plot a rectangular area within the current plot. This provides an easy way to view a particular area of the plot in greater detail.

Press **Z** to select *Zoom*. A rectangular box should appear in the center of the plot. Notice the line that appears just below the X axis. This line shows the minimum and maximum values along the X and Y borders of the zoom box. The *Xmin* and *Ymin* values are the coordinates of the lower left corner of the zoom box. The *Xmax* and *Ymax* values are the coordinates of the upper right corner of the zoom box. Also notice the new zoom menu at the top of the screen. It contains three items: *Plot zoom area*, *Size zoom area*, and *Move zoom area*.

Press **P** to select *Plot zoom area*. The area inside the zoom box will be plotted. The coordinates of the X and Y axes should now be equal to the previous coordinates of the zoom box.

Press **Z** to select *Zoom* once again. The screen should now look like Figure 2.4. Before plotting the area inside the zoom box again, let's first change the size and position of the box.

Press **S** to select *Size zoom area*. Notice that the line at the top of the screen is now displaying a help message. The arrow keys change the size of the box while keeping the box centered at its current position. To change the horizontal size of the box, use the **left** arrow key to make it smaller or the **right** arrow key to make it larger. To change the vertical size of the box, use the **down** arrow key to make it smaller or the **up** arrow key to make it larger. The **Home, End, PgUp,** and **PgDn** keys change the size of the box by simply changing the position of the upper right corner of the box. The **Home** key moves the upper right corner to the left, **End** moves it to the right, **PgUp** moves it up, and **PgDn** moves it down. After changing the box to the desired size, press any key to return to the zoom menu.

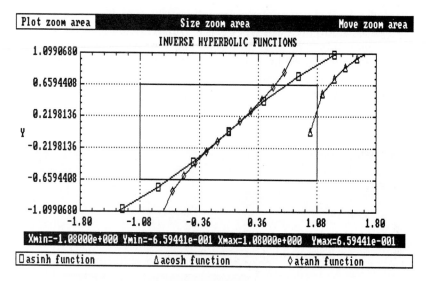

Figure 2.4   Zooming

Press **M** to select *Move zoom area*. Notice that the same keys that were used to size the zoom box are also used to move it. The **left** and **right** arrow keys move the box to the left or right. The **up** and **down** arrow keys move the box up or down. The **Home** and **End** keys move the box to the upper left or lower left corner of the plot. The **PgUp** and **PgDn** keys move the box to the upper right or lower right corner of the plot. The number **5** key moves the box to the center of the plot. After moving the box to the desired position, press any key to return to the zoom menu.

You can alternately select *Size zoom area* and *Move zoom area* until the zoom box has the desired size and position. Then simply press **P** to plot the area inside the zoom box. If desired, you may press the **Esc** key to return to the main plot menu without changing the plot.

2.3.2 Changing Axes Range

The *Auto scale*, *Specify scale*, and *Change scale* items on the main plot menu allow you to change the range of the X and Y axes. *Auto scale* automatically scales the axes so that all of the data points in all of the curves are visible. *Specify scale* allows you to specify the minimum and maximum values for each axis. *Change scale* allows you to increment or decrement the range of each axis.

Press **A** to select *Auto scale*. All the data points from the HYPER.DAT file should now be visible again. The minimum and maximum values on the X and Y axes should be the same as when the curves were first plotted from the *Command Menu*. Remember that from the *Command Menu*, we selected **P** for *Plot curve(s)* followed by **A** for *Auto scale*.

Press **S** to select *Specify scale*. A menu that allows you to specify the minimum and maximum values for the X and Y axes will appear. The cursor should be positioned over *X minimum*. Notice that the value for *X minimum* is displayed on the help line at the bottom of the screen. In this case, the minimum value for the X axis is −3. Press the **down** arrow key to position the cursor over *X maximum*. Notice that the maximum value for the X axis is 3. Press the **down** arrow key again to position the cursor over *Y minimum*. The value for *Y minimum* is −1.83178. Since this is not a nice round number, let's change it to −2.

With the cursor over *Y minimum*, press the **Enter** key to display a window for editing the value. The screen should look like Figure 2.5. You are now being prompted to edit the value for *Y minimum*. Notice that the current value is −1.83178.

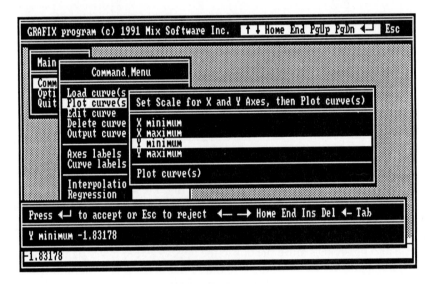

Figure 2.5   Specifying Scale

Above the current value in the edit window is a list of the keys that are available for editing. The **left** and **right** arrow keys move the cursor left or right. The **Home** and **End** keys move the cursor to the beginning or end of the value. The **Ins** key toggles insert mode on or off (default = off). When insert mode is turned on, the cursor changes from an underline to a block. The **Del** key deletes the character under the cursor. The **backspace** key deletes the character to the left of the cursor. The **Tab** key deletes all characters to the right of the cursor, including the character under the cursor. The **Enter** key accepts the changed value and the **Esc** key discards it.

Press the **Tab** key to delete the current value, then type −2 and press the **Enter** key. The edit window will disappear. Notice that the value displayed on the help line for *Y minimum* is now −2.

Press the **down** arrow key to position the cursor over *Y maximum*. Let's change the value 1.83178 to 2. Press the **Enter** key to display the edit window, type **2**, press the **Tab** key, then press the **Enter** key again. Notice that the value of *Y maximum* is now 2. Press **P** to plot the inverse hyperbolic curves again. Notice that the values on the Y axis now range from −2 to 2.

Press **C** to select *Change scale* and notice the help line that appears at the top of the screen. The **PgUp/PgDn** keys increase/decrease the range of both the X and Y axes simultaneously. The **Right/Left** arrow keys increase/decrease the range of the

X axis only. The **Up/Down** arrow keys increase/decrease the range of the Y axis only. Any other key simply returns control to the main plot menu.

Press **PgDn** to decrease the range of both the X and Y axes. The screen should look like Figure 2.6. Press **PgUp** to increase the range of both the X and Y axes. This should return the X and Y axes back to their original ranges.

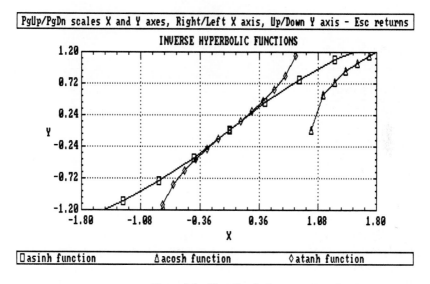

Figure 2.6   Changing Scale

Press the **right** arrow key to increase the range of the X axis only. Press the **up** arrow key to increase the range of the Y axis only. This combination is equivalent to pressing **PgUp**.

Press the **left** arrow key to decrease the range of the X axis only. Press the **down** arrow key to decrease the range of the Y axis only. This combination is equivalent to pressing **PgDn**.

Press the **Esc** key to return to the main plot menu.

2.3.3 Printing Screen

The GRAFIX program supports Epson and LaserJet compatible printers. Most printers are compatible with one of these two types of printers. By default, the program assumes that you are using an Epson compatible printer.

If you are using a LaserJet compatible printer, press **Esc** to return to the *Command Menu*. Press **Esc** again to return to the *Main Menu*. Press **O** to select *Options* and **P** to select *Printer types*. The screen should now look like Figure 2.7. You are now being prompted to enter the printer type (i. e., Epson or LaserJet). Notice that the current value is Epson. To change the printer type to LaserJet, simply type **L** and press the **Enter** key. Notice that the help line at the bottom of the screen now shows LaserJet for the printer type. Press **S** to save the new configuration; then press **Esc**, **C**, **P**, **A** to plot the curves once again.

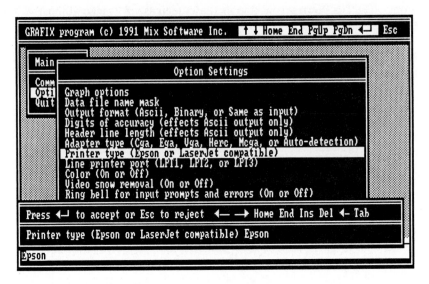

Figure 2.7   Printer Types

With the main plot menu displayed, press **P** to select *Print*. A menu containing the items *Small size print*, *Medium size print*, and *Large size print* will appear. This menu determines the size of the printed graph. The large size print approximately fills an 8 1/2" by 11" sheet of paper.

Press **M** to select *Medium size print*. A menu containing the items *copy screen to Printer* and *copy screen to File* will appear. This menu gives you the option of copying the screen directly to the printer, or to a file.

Press **P** to copy the screen directly to the printer. A message appears to inform you that the output will be sent to an Epson or LaserJet compatible printer at LPT1, the default printer port. The *Options* selection in the *Main Menu* allows you to set the printer port to LPT1, LPT2, or LPT3. This may be done in similar fashion to changing the printer type, as described earlier.

If you have a printer of the correct type connected to your computer, press **Y** to print the screen. Otherwise press **N** to return to the main plot menu. If you press **Y**, you will notice a scan line moving from right to left across the screen. When the scan line reaches the left side of the screen, the printing is finished.

Press **P** to print the screen once again. This time press **L** to select *Large size print*, and then press **F** to select *copy screen to File*. A menu containing the items *Write to file* and *Append to file* will appear. This menu gives you the option of writing or appending to the file.

By default, the screen is copied to a file named GRAFIX.LPT. If you select *Write to file*, a new GRAFIX.LPT file is created (i. e., any previously created file is lost). If you select *Append to file*, the screen image is appended to the end of the GRAFIX.LPT file. This allows you to store more than one graphics screen image in a single file. If the GRAFIX.LPT file does not exist, then *Write to file* and *Append to file* are equivalent.

Press **W** to select *Write to file*. A message appears to inform you that Epson or LaserJet compatible output will be written to the GRAFIX.LPT file, the default output file. The *Graph options* selection in the *Option Settings* menu allows you to specify the name of the file for storing graphics images.

If you have at least 45,000 bytes of storage space available in the current directory, press **Y** to copy the screen to the GRAFIX.LPT file. Otherwise press **N** to return to the main plot menu. The graphics image of a single screen requires between 40,000 and 45,000 bytes of disk space. For Epson compatible output in the medium size, the disk space required is approximately 76,000 bytes.

2.3.4 GPRINT Program

The files created by *Print* may be printed using the GPRINT program. The GPRINT program requires a single command-line argument to specify the name of the file to print. The program then simply copies the contents of the file to the **LPT1** parallel printer port. If your printer is connected to a different port, you may optionally specify **LPT2** or **LPT3** as a second command-line argument. You may also redirect **LPT1** to a **COM** serial port using the DOS **mode** command.

Press **Esc** twice to get back to the *Main Menu*, then press **Q** to quit the GRAFIX program.

If you were successful at using *Print* to create the GRAFIX.LPT file, then type the following to print the file.

GPRINT GRAFIX.LPT

2.4 EDITING CURVES

The GRAFIX program also allows you to view and/or change the values of the individual data points in a curve.

Type the following to execute the **GRAFIX** program and load the LOWPASS.DAT file. This file contains a single curve representing a lowpass filter response.

GRAFIX LOWPASS.DAT

Press **C** to display the *Command Menu*, and then press **P** followed by **A** to see what a lowpass filter response looks like. The screen should look like Figure 2.8. Notice that the curve starts out as almost a straight horizontal line, and then sharply dips downward. This is the characteristic shape of a lowpass filter. Low frequency signals pass through the filter with little or no attenuation, while high frequency signals are strongly attenuated. This particular filter was designed for a normalized cutoff frequency of 0.3. The cutoff frequency is the point at which the magnitude is approximately −6 dB.

Press **R** to return to the *Command Menu*, and then press **E** to edit the data points of the lowpass filter. The screen should look like Figure 2.9. The data points are displayed in columns of X and Y values. To the left of each X and Y value is the data point number (i. e., 1 for the first, 2 for the second, etc.). Notice that the cursor is positioned over the X value of the first data point. Above the columns of X and Y values is a list of keys that may be used to position the cursor.

The **left** and **right** arrow keys move the cursor back and forth between the X and Y columns. The **up** and **down** arrow keys move the cursor up or down to the next data

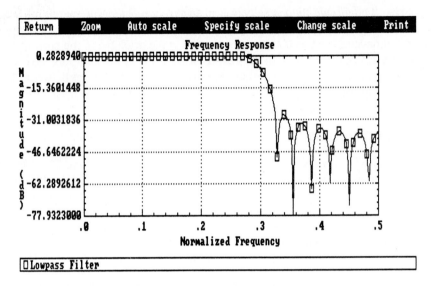

Figure 2.8   Lowpass Filter Response

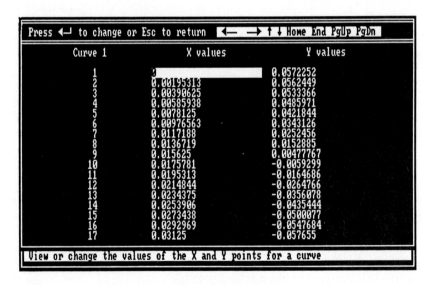

Figure 2.9   Editing Curves

point. The **Home** and **End** keys move the cursor to the first or last data point in the window, then to the first or last data point in the curve. The **PgUp** and **PgDn** keys page the window up or down through the data points, leaving the cursor at its current position.

Press the **End** key twice to position the cursor over the last data point in the curve. Notice that this curve contains 257 total data points. Press the **Home** key twice to position the cursor back to the first data point.

Press the **left** or **right** arrow key to move the cursor to the column of Y values. Press the **down** arrow key repeatedly to move the cursor toward the last data point in the curve. Notice that all of the Y values are fairly close to 0.

Press the **PgDn** key to page through the data more quickly. Notice that around the 145th data point, the Y values start decreasing rapidly. At the 155th data point, the Y value (magnitude) is approximately −6 (dB). Notice that the X value (normalized frequency) at this point is approximately 0.3, the designed cutoff frequency of the lowpass filter.

Press the **Esc** key to return to the *Command Menu*, then press **L** to load another data file. Position the cursor over *highpass.dat* and press the **Enter** key. Press **C** to combine the highpass curve with the lowpass curve. Now there should be two curves loaded into memory.

Press **P** and then **A** to plot the curves. Notice that the highpass filter response is the opposite of the lowpass filter response. For a highpass filter, high frequency signals pass through the filter with little or no attenuation, while low frequency signals are strongly attenuated. This highpass filter was designed with a cutoff frequency of 0.2.

Press the **Esc** key to return to the *Command Menu*, and then press **E** to edit a curve. Since there are now two curves, a menu appears asking you to *Select Curve to Edit*. Press **2** to select *curve 2*. Curve 2 also has a total of 257 data points.

The title line of the edit window suggests that you may press the **Enter** key to change a value, or the **Esc** key to return to the *Command Menu*. To change a value, you simply position the cursor over the value and press the **Enter** key. When the **Enter** key is pressed, the top line of the screen shows a list of keys that may be used to edit the value. These are the same keys that were used previously to edit the minimum and maximum values for the Y axis of the inverse hyperbolic curves (see Specifying Scale figure). After changing a value, you may press the **Enter** key to accept the change, or the **Esc** key to reject it.

Press the **Esc** key to return to the *Command Menu*.

2.5 DELETING CURVES

If more than one curve is loaded into memory, you may decide to delete a curve so that it won't appear when the curves are plotted. For example, let's delete the curve corresponding to the lowpass filter response.

With the *Command Menu* displayed, press **D** to select *Delete curve*. A menu titled *Select Curve to Delete* will appear. Notice that the menu contains the items *curve 1* and *curve 2*. Since the LOWPASS.DAT file was loaded before the HIGHPASS.DAT file, the lowpass filter response is *curve 1* and the highpass filter response is *curve 2*. Press **1** to delete *curve 1*.

Now there should be only one curve loaded in memory. Press **P** followed by **A** to plot the curve. The screen should look like Figure 2.10. Notice that only one curve is plotted.

Press **Esc** to return to the *Command Menu*; then press **D** again and notice that only *curve 1* is listed. The highpass filter used to be curve 2. However, since the lowpass

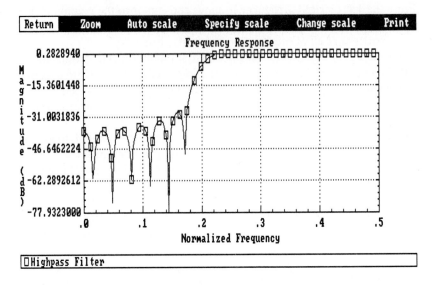

Figure 2.10 Highpass Filter Response

filter was deleted, the highpass filter has now become curve 1. Press the **Enter** key to delete the only curve. There should now be no curves loaded in memory.

2.6 OUTPUTTING CURVES

You can output the curves in memory to a specified disk file.

With the *Command Menu* displayed, press **L** to load another data file. Position the cursor over *bandpass.dat* and press the **Enter** key. If the *Replace* or *Combine with existing curves* menu appears, press **R** to replace any existing curves. Press **P** followed by **A** to see what a bandpass filter response looks like. The screen should look like Figure 2.11.

Notice that a bandpass filter response looks like the combined response of a lowpass and highpass filter. Both low and high frequency signals are strongly attenuated, while signals between the two cutoff frequencies pass through the filter with little or no attenuation. To further illustrate this point, let's combine the bandpass filter with the lowpass and highpass filters.

Press **Esc** to return to the *Command Menu*, and then press **L** to load another data file. Position the cursor over *lowpass.dat* and press the **Enter** key. Press **C** to combine the lowpass filter with the bandpass filter. Press **L** again, position the cursor over *highpass.dat*, press the **Enter** key, and then press **C** to combine the highpass filter with the bandpass and lowpass filters. We can now save all three curves in a single file.

Press **O** to select *Output curve(s)*. The screen should look like Figure 2.12. You are now being prompted to enter the name of the file where the curves will be saved. Notice that the default file name is *.\ bandpass.dat*. The *.\ specifies the current directory. The default file name is *bandpass.dat* because this was the first file loaded.

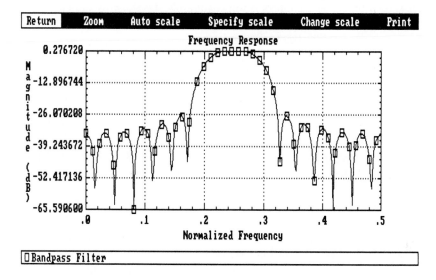

Figure 2.11   Bandpass Filter Response

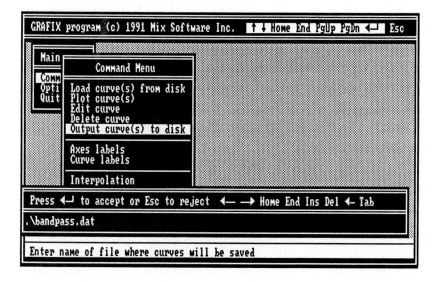

Figure 2.12   Outputting Curves

Let's save the three filter responses in a file named *filter.dat* in the current directory. Press the **Tab** key to delete the default response, and then type **filter.dat** and press the **Enter** key. The three curves should now be saved.

Press **L**, then position the cursor over *filter.dat* and press the **Enter** key. Press **R** to replace the curves in memory with the curves from the FILTER.DAT file. Press **P** followed by **A** to plot the three filter responses. Notice that the combined response of the highpass and lowpass filters is approximately the same as the bandpass filter response.

Press **Esc** to return to the *Command Menu*.

2.7 LABELING PLOT

All of the plots so far have contained labels. Each plot has had a label for the main title, a label for the X and Y axis, and labels for each curve. For example, remember the plot of the curves from the HYPER.DAT file. The main title was *INVERSE HYPERBOLIC FUNCTIONS*, the label for the X axis was *X*, the label for the Y axis was *Y*, and the labels for each of the three curves were *asinh function, acosh function,* and *atanh function*.

Now let's load the BESSEL.DAT file, which contains no label information. With the *Command Menu* displayed, press **L** to load a data file. Position the cursor over *bessel.dat* and press the **Enter** key. Press **R** to replace any existing curves in memory. There should now be two curves loaded into memory.

Press **P** followed by **A** to plot the two curves. The screen should look like Figure 2.13. Notice that the plot contains no label information.

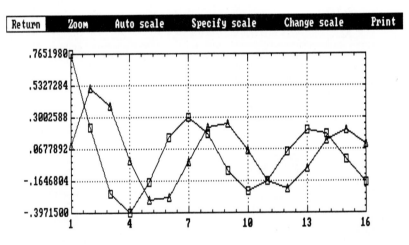

Figure 2.13   Plot without Labels

Press the **Esc** key to return to the *Command Menu*, and then press **A** to select *Axes labels*. A menu containing the items *Title, X axis,* and *Y axis* will appear.

Press **T** to enter a label for the plot title. Since the two curves loaded from the BESSEL.DAT file are bessel functions, let's use that as the title of the plot. Type **BESSEL FUNCTIONS** and press the **Enter** key.

Now let's enter labels for the X and Y axes. Press **X**, type **X axis**, and press the **Enter** key. Then press **Y**, type **Y axis**, and press the **Enter** key. The *Title, X axis,* and *Y axis* labels can be viewed from the *Axes Labels* menu. Use the **up** and **down** arrow keys to move the cursor over these three menu items. Notice that the help line at the bottom of the screen displays the labels just entered.

Press **Esc** to return to the *Command Menu*, and then press **C** to select *Curve labels*. A menu containing the items *curve 1* and *curve 2* will appear.

Press **1** to enter a label for the first curve. Since the first curve loaded from the BESSEL.DAT file was created using the *besj0* function, let's use that to label the curve. Type **besj0 function** and press the **Enter** key.

Press **2** to enter a label for the second curve. Since the second curve loaded from the BESSEL.DAT file was created using the *besy0* function, let's use that to label this curve. Type **besy0 function** and press the **Enter** key.

The labels for *curve 1* and *curve 2* can be viewed from the *Curve Labels* menu. Use the **up** and **down** arrow keys to move the cursor over these two menu items. Notice that the help line at the bottom of the screen displays the labels just entered.

Press **Esc** to return to the *Command Menu*; then press **P** followed by **A** to plot the two bessel curves again. The screen should look like Figure 2.14. Notice that the labels just entered are now displayed on the plot.

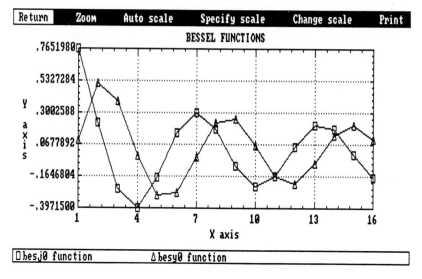

Figure 2.14   Plot with Labels

Press **Esc** to return to the *Command Menu*, and then press **O** followed by the **Enter** key to output a new BESSEL.DAT file with labels.

2.8 INTERPOLATION

You probably noticed that the bessel functions plotted in the previous section were not smooth curves. The reason is that the BESSEL.DAT file contains only 16 data points. When *Plot curve(s)* is selected, the curves are drawn by simply connecting each data point with a straight line. When there are too few data points the curves look jagged, as is the case with the two bessel curves.

When you have a sparse data set, interpolation can be used to draw smooth curves through the data points. The GRAFIX program supports two types of interpolation: *Cubic spline* and *Lagrange polynomial*. The *spline* and *spline0* library functions are used to implement cubic spline interpolation. The *interp1* library function is used to implement Lagrange polynomial interpolation. See the descriptions of these functions for further details.

With the *Command Menu* displayed, press **I** to select *Interpolation*. A menu containing the two interpolation types will appear. Press **C** to select *Cubic spline* interpolation, and then press **A** to plot the curves with autoscaling. The screen should look like Figure 2.15. Notice that the bessel functions now appear as smooth curves.

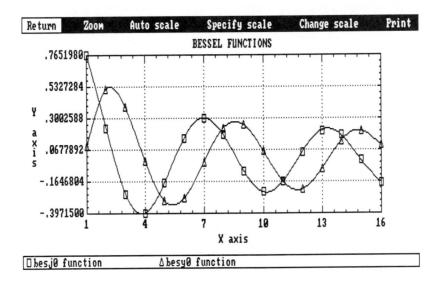

Figure 2.15   Smoothed by Interpolation

Press **R** to return to the *Command Menu*, and then press **I** to select *Interpolation* again. This time press **L** to select *Lagrange polynomial*. You will then be prompted to enter the degree of the polynomial (default = 2). The degree must be between 1 and 9. A degree of 1 means linear interpolation (i. e., straight lines connect the data points).

Type **3** and press the **Enter** key, and then press **A** to plot the curves. The curves will now be interpolated by a Lagrange polynomial of degree 3. Since the cubic spline method also uses a third degree polynomial, the plot should look very similar to Figure 2.15.

Press **R** to return to the *Command Menu*.

2.9 REGRESSION

Regression is a technique by which you attempt to find an expression that approximately fits or models a set of data points. If you can find an expression that closely fits a measured or observed set of data points, then you can use that expression to predict a result at a given observation point [(for example, given x, you can determine $f(x)$].

The GRAFIX program supports four types of regression: *Polynomial, Exponential, Logarithmic*, and *Sinusoidal*. These are expressions of the following forms:

$$\text{Polynomial}\ :\quad f(x) = c_0 + c_1 \cdot x + c_2 \cdot x^2 + c_3 \cdot x^3 \dots$$
$$\text{Exponential}\ :\quad f(x) = c_0 + c_1 \cdot \exp(x)$$
$$\text{Logarithmic}\ :\quad f(x) = c_0 + c_1 \cdot \log(x)$$
$$\text{Sinusoidal}\ :\quad f(x) = c_0 + c_1 \cdot \sin(x) + c_2 \cdot \cos(x)$$

When you select one of these regression types, the coefficients $c0$, $c1$, etc. are calculated to minimize the mean-square error between the resulting function $f(x)$ and the actual data points. A small mean-square error means that $f(x)$ is a good approximation of the actual data. A large mean-square error means that $f(x)$ is not a good model for the data.

Polynomial regression is the most general type. Given enough terms, a polynomial expression can provide a reasonably good approximation of almost any function. Exponential, logarithmic, and sinusoidal regression are not so general. The data must actually be exponential, logarithmic, or sinusoidal for these expressions to provide a good approximation. The GRAFIX program uses the library function *least_sq* to calculate the coefficients of these expressions.

With the *Command Menu* displayed, press **R** to select *Regression*. Since there are two curves loaded (i. e., the two bessel functions), you will be prompted to select a curve. Press **1** to select curve 1, the *besj0* function. You will then be prompted to select one of the four regression types. Press **P** to select *Polynomial* and you will be prompted to enter the degree of the polynomial expression. Press the **Enter** key to select the default degree of 2. The screen should now look like Figure 2.16. Notice that the mean-square error is greater than 1, and that the three coefficients $c0$, $c1$, and $c2$ have been calculated. The number of terms in the polynomial is always one greater than the degree of the polynomial.

Press **A** to auto-scale and plot this three-term polynomial expression. Since a polynomial of degree 2 is parabolic, you will notice that it doesn't do a very good job of approximating the *besj0* function. Perhaps a higher degree polynomial will do better.

Press **R** to return to the *Command Menu*, and then press **R** to select *Regression* again. Press **1** to select curve 1, and then **P** to select *Polynomial* regression. Type **7** for the degree and press the **Enter** key. With a degree of 7, notice that the mean-square error is now much less than 1. That means that this 8-term polynomial expression has done a much better job of modeling the data. To see just how well it did, press **A** to plot the expression. The screen should look like Figure 2.17. Notice that the curve now passes through most of the data points, and also closely resembles the shape of the *besj0* function. This 7th degree polynomial has done a pretty good job of modeling the data. Try a 9th degree polynomial to see if you can do even better.

Since the bessel curves have a shape similar to that of a sine wave, you might want to experiment with sinusoidal regression. With the *Command Menu* displayed, press **R** to select *Regression*, **2** to select curve 2, then **S** to select *Sinusoidal* regression. Notice that the mean-square error is less than 0.1, which is not terribly bad. Press **A** to plot the curve and you will notice that the three-term sinusoidal expression does a fairly good job of modeling the data. Assuming that our data had been measured,

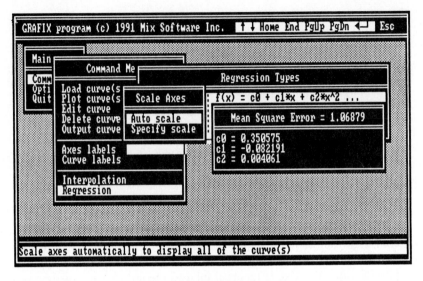

Figure 2.16 Polynomial Regression

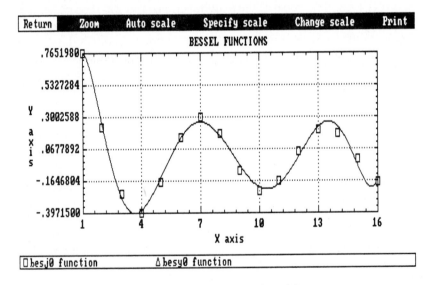

Figure 2.17 7th Degree Polynomial

and assuming that there might have been some error in our measurements, we could possibly conclude that the data represented a sine wave signal.

For regression techniques to work well, there must be a sufficient number of data points. If there are too few data points, the resulting expression may not be an accurate model for the data. For measured data, the accuracy of the expression may also be degraded by any measurement errors.

2.10 OPTIONS

The *Options* selection in the *Main Menu* allows you to set various options for the GRAFIX program. When the GRAFIX program is executed, it configures itself based on the option settings stored in the file named GRAFIX.CFG.

Press the **Esc** key until only the *Main Menu* is displayed, and then press **O** to select *Options*. The *Option Settings* menu will appear. The screen should look like Figure 2.18. As you move the cursor over the menu items, notice that the current settings are displayed on the help line at the bottom of the screen.

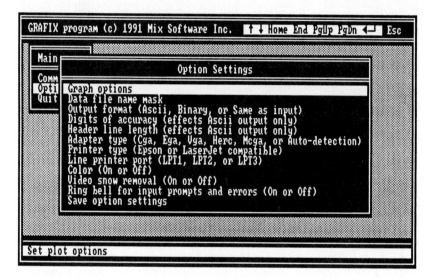

Figure 2.18   Option Settings

To change a value, position the cursor over one of the menu items and press the **Enter** key, or simply press one of the highlighted hot keys. The current value of the item selected will appear in an edit window, allowing you to change it. After changing the value in the edit window, simply press the **Enter** key to accept the changed value, or press the **Esc** key to reject it.

Press **G** to select *Graph options*. The *Graph Option Settings* menu will appear. The screen should now look like Figure 2.19. All these options are related to graphics. Move the cursor over these menu items and notice the current settings displayed on the help line at the bottom of the screen.

Following is a brief description of each option.

Option Settings

Graph Options.   Selects the *Graph Option Settings* menu.

Data File Name Mask.   This option specifies the directory and the extension for the data files that are listed when *Load curve(s)* is selected from the *Command Menu* (default = .*.dat*). The default file name mask causes the GRAFIX program to display all files in the current directory that have an extension of DAT.

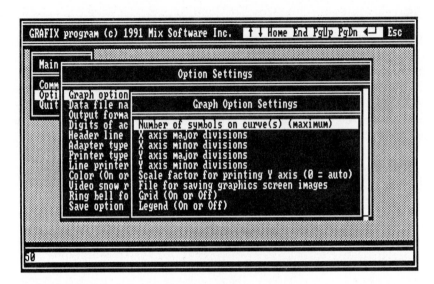

Figure 2.19 Graph Option Settings

Output Format (Ascii, Binary, or Same As Input).   This option specifies the format of the output files created when *Output curve(s)* is selected from the *Command Menu* (default = *Same*). The default is to output files in the same format as the first file that is loaded.

Digits of Accuracy (Affects Ascii Output Only).   This option specifies the number of digits in the floating point values that are output when *Output curve(s)* is selected from the *Command Menu* (default = 6). The default is to output six significant digits for each floating point value. This has no effect if the output format is binary.

Header Line Length (Affects Ascii Output Only).   This option specifies the maximum number of characters per line for the header that is output when *Output curve(s)* is selected from the *Command Menu* (default = 80). The default is to output no more than 80 characters per line. This has no effect if the output format is binary.

Adapter Type (Cga, Ega, Vga, Herc, Mcga, or Auto-detection).   This option specifies the type of video adapter that is installed in your computer (default = *Auto*). The default is to let the GRAFIX program automatically detect the type of video adapter.

Printer Type (Epson or LaserJet Compatible).   This option specifies the type of printer that is connected to your computer (default = *Epson*). The default is to assume that your printer is compatible with the Epson graphics command set.

Line Printer Port (LPT1, LPT2, LPT3).   This option specifies the port to which your printer is connected (default = *LPT1*). The default is to assume that your printer is connected to the first parallel line printer port.

Color (On or Off).   This option turns color on or off (default = *On*). The default is to use color if the video adapter supports it. If you have a black and white monitor connected to a video adapter that supports color, then you should turn color off.

Video Snow Removal (On or Off).   This option turns video snow removal on or off (default = *Off*). The default is to make no attempt at removing the snow that appears on the screen when some of the older CGA adapters are used. It takes longer to display text on the screen when this option is turned on.

Ring Bell for Input Prompts and Errors (On or Off).   This option turns sound on or off (default = *On*). The default is to sound an alert when the GRAFIX program is requesting input, or when an error occurs.

Save Option Settings.   This option saves all of the option settings (including the graph option settings) in a configuration file named GRAFIX.CFG. The GRAFIX program automatically configures itself based on the options stored in this file.

Graph Option Settings

Number of Symbols On Curve(s) (Maximum).   This option specifies the maximum number of symbols (e.g., rectangle, triangle, diamond, etc.) to display on each curve (default = 50). The default is to plot no more than 50 symbols on any single curve. The symbols identify the actual data points. If a curve has fewer than 50 data points, then all data points are identified by a symbol. Otherwise, some of the data points are not identified. If this option is set to 0, then no symbols will appear on the curves.

X Axis Major Divisions.   This option specifies the number of major divisions along the X axis (default = 5). The default is to divide the X axis into five major divisions. The major divisions are the points at which the axis is labeled.

X Axis Minor Divisions.   This option specifies the number of minor divisions between each major division along the X axis (default = 5). The default is to divide each major division into five minor divisions.

Y Axis Major Divisions.   This option specifies the number of major divisions along the Y axis (default = 5). The default is to divide the Y axis into five major divisions. The major divisions are the points at which the axis is labeled.

Y Axis Minor Divisions.   This option specifies the number of minor divisions between each major division along the Y axis (default = 5). The default is to divide each major division into five minor divisions.

Scale Factor for Printing Y Axis (0 = Auto).   This option specifies the factor that is used to scale the Y axis when a graph is printed (default = 0). The default is to automatically choose a scale factor that provides the best aspect ratio. Larger values for this option make the printed graph taller along the Y axis. This option does not affect the screen display.

File for Saving Graphics Images. This option specifies the name of the file that is used when the screen is printed to a file (default = *grafix.lpt*). The default is to store the graphics screen image(s) in a file named GRAFIX.LPT.

If you have a serial printer connected to one of the COM ports, you can specify a file name of *LPT1*, and then use the DOS **mode** command to redirect *LPT1* to the appropriate COM port. Then when you print to the file, the output will go directly to the serial printer.

Grid (On or Off). This option turns the plot grid on or off (default = *On*). The default is to display grid lines at each of the major divisions along the X and Y axes.

Legend (On or Off). This option turns the plot legend on or off (default = *On*). The default is to display a legend below the plot to identify each of the curves on the plot. If none of the curves is labeled, then the legend is not displayed.

3

Complex Numbers

Complex numbers are the most general numbers used in algebra. Any number that can be expressed in the form $a + bi$, where a and b are real numbers and $i^2 = -1$ is a complex number. This may be confusing to anyone unfamiliar with this definition, since, for example, calculators cannot compute the square root of -1. This is because calculators do only real arithmetic. The existence of i is of fundamental importance to mathematics. Engineers often use the "j" notation, $j = i$.

Complex numbers are not an abstraction of theoretical mathematics. Without complex numbers, many polynomial equations would have no solution. One very simple example is $x^2 = -1$. The solution to this equation ($x = i$) cannot be represented by a real number. Complex numbers have very many applications in applied math, physics, and engineering.

A complex number can be thought of as a two-dimensional vector (a, b), where a is the **real part** and b is the **imaginary part**. The term "imaginary" is an unfortunate misnomer left over from the seventeenth century, when mathematicians were still uncomfortable with the concept of complex numbers. The imaginary part is every bit as real as the real part of the complex number. Sometimes complex numbers are represented in the **standard form** $a + bi$. Another common representation is the **polar** or **trigonometric** form

$$(a, b) = z = r[\sin(a) + i\cos(a)]$$

The **magnitude** of a complex number is the square root of the sum of the squares of its real and imaginary part:

$$|(a, b)| = (a^2 + b^2)^{0.5}$$

The **conjugate** of the complex number (a, b) is $(a, -b)$. The complex conjugate is sometimes denoted as $(a, b)^*$, where

$$(a, b)^* = (a, -b)$$

Complex Arithmetic.   The arithmetic of complex numbers is defined as follows:

addition	$(a, b) + (c, d) = (a + c, b + d)$
subtraction	$(a, b) - (c, d) = (a - c, b - d)$
multiplication	$(a, b) * (c, d) = (ac - bd, ad - bc)$
division	$(a, b)/(c, d) = (ac + bd, ad + bc)/(c^2 + d^2)$

Powers of i.   Some important identities involve the powers of i:

$$i^1 = i^5 = i^9 \ = \ldots = i$$
$$i^2 = i^6 = i^{10} = \ldots = -1$$
$$i^3 = i^7 = i^{11} = \ldots = -i$$
$$i^4 = i^8 = i^{12} = \ldots = +1$$

Note that when the powers of i are simplified, they cycle in steps of four.

Powers and Roots of Complex Numbers.   Both the nth power and the nth root of a complex number are also complex numbers, which are best represented in **polar form**:

nth power	$\{r[\cos(\theta) + i\sin(\theta)]\}^n = r^n[\cos(n\theta) + i\sin(n\theta)]$
nth root	$z^{1/n} = r^{1/n}[\cos(\theta/n) + i\sin(\theta/n)]$

For both the nth power and the nth root, the angle θ must be evaluated modulo $360°$. This is a direct consequence of the periodicity of the sine and cosine functions. That is, suppose θ is the angle (in degrees) that resolves an nth root (or power) of z. Then for any integer, k, the larger angle $\theta + k*360$ also resolves the root (or power).

This ambiguity leads to the definition of a **principal nth root** of a complex number. A **principal nth root** of z is a root with a polar angle between 0 and 360 degrees:

$$0 \leq \theta/n \leq 360$$

There are always n unique principal nth roots of a complex number. For example, there are three principal cube roots of $2i$:

$$2^{1/3}[\cos(30) + i\sin(30)]$$
$$2^{1/3}[\cos(150) + i\sin(150)]$$
$$2^{1/3}[\cos(270) + i\sin(270)]$$

All other cube roots are redundant.

See Functions: Cadd(), Cdiv(), Cmag(), Complx(), Conjg(), Cmul(), CscalR(), Csqrt(), Csub().

4

Probability

The **probability** of an event is a ratio of the number of ways the event can occur to the total number of possible outcomes of the random experiment. For example, in a single coin toss the probability of a head is

$$P(\text{HEADS}) = (\# \text{ of ways a HEAD can occur})/(\text{total } \# \text{ of outcomes}) = 1/2$$

For this simple example, it is easy to count the total number of outcomes.

This chapter describes the application of probability theory to more complex random experiments. The concept of random variables is introduced because of its importance to probability theory. A **random variable** is any function that maps all the events of a probabilistic experiment (the sample space) to the x-axis. Each random variable has its own characteristic **probability density function** and **probability distribution function**. These functions are very useful in determining probabilities and are also defined in this chapter.

Each probability density function has **mean** and **variance**. The mean is the average or expected value of the random variable. The variance determines the spread of events about the mean. The **standard deviation** is the square root of the variance. These concepts are defined in more detail in the statistics chapter.

4.1 COMBINATORIAL ANALYSIS

Many applications of probability involve experiments that are much more complex than the coin toss. In many games of chance, the number of outcomes is purposely large to make the game interesting. For these cases, it may be difficult to count each and every outcome. For many problems, **combinatorial analysis** can be used to provide closed form expressions of the number of outcomes without having to enumerate them.

4.1.1 Permutations and Factorial

A **permutation** is any possible arrangement of a set of distinct items. The order of the arrangement of the items is important. For example, *abcd* is different from permutation *dcba*.

The number of different ways of arranging (permuting) n distinct objects is

$$n! = (n) * (n-1) * (n-2) * \ldots 1$$

where n is an integer.

The expression $n!$ is called n-*factorial*. Zero factorial is defined to be unity (i. e., $0! = 1$).

The number of different orderings of r objects from a total of n, $_nP_r$, is given by:

$$_nP_r = n!/(n-r)!$$

where $r < n$ and both n and r are integers. Note that if $r = n$ then $_nP_n = n!$.

See Functions: permut(), fact().

4.1.2 Combinations

Often only the number of occurrences of a specific outcome is important, not the order. For example, while a poker player usually hopes to be dealt four aces, the order in which the aces are received is not important. The number of *combinations* refers to the different ways of choosing r objects from a total of n, without regard to their order. The number of such combinations is usually denoted by $_nC_r$.

$$_nC_r = n!/(r! * (n-r)!)$$

where both n and r are integers and $r \leq n$.

There are several useful identities involving the combinations function. Perhaps the most important of these is

$$_nC_r = {_nC_{n-r}}$$

This formula is intuitively clear, since when we explicitly decide to choose r objects at a time, we are implicitly deciding to exclude $n - r$ objects at a time.

See Function: combin().

4.1.3 Binomial Expansion and Pascal's Triangle

Sometimes the combinations expressions, $_nC_r$, are referred to as binomial coefficients because of their relation to the binomial expansion:

$$(x + y)^n = {_nC_0}x^ny^0 + {_nC_1}x^{n-1}y^1 + \ldots + {_nC_n}x^0y^n$$

For different powers of n, the coefficients of this expansion form a pattern that is called **Pascal's Triangle**, as shown in the table.

n	Pascal's Triangle and the Binomial Coefficients					
0				1			
1			1		1		
2			1	2	1		
3		1	3		3	1	
4	1	4		6		4	1
5	1	5	10		10	5	1

4.2 DISCRETE RANDOM VARIABLES

Discrete random variables are useful in describing events that can have only a finite or countable (e. g., integer) number of outcomes. For a discrete random variable to be properly defined, the sum of the probabilities of all the discrete events must be one. Collectively, the probabilities are referred to as the **probability density function**, $f(k)$, since their envelope describes the distributional characteristics of the random variable.

The **cumulative probability distribution function**, $F(k_0)$, of a discrete random variable is defined as the sum of the density $f(k)$ over k, where $k \leq k_0$. For a given random variable, these two functions are all that is needed to calculate the probability of any outcome.

4.2.1 Binomial Distribution

The **binomial** distribution is frequently used to predict the outcome of repetitive sampling from a population. The underlying assumption behind the binomial distribution is that after each sample is selected, it is returned to the population before the next sampling occurs. Hence the phrase, "sampling with replacement" is often used in describing binomial distributed events.

Each individual sampling is called a **Bernoulli trial**. A Bernoulli trial consists of two mutually exclusive outcomes. One (the success) has a probability, p, and the other (the failure) has a probability of $1 - p$. These probabilities remain the same for all of the n repetitions (e. g., 0.5 for heads or tails in a coin flip).

The binomial density function represents the probability of i successes out of n trials:

$$b(n, p, i) = {}_nC_i p^i (1 - p)^{n-i}$$

where $i = 0, 1, 2, \ldots, n$.

It can be shown that the mean u and variance s^2 of the binomial density are

$$u = np \quad \text{and} \quad s^2 = np(1 - p)$$

The outcome of a multiple coin toss experiment is a binomial random variable. For example, consider the number of heads after 10 coin tosses. The binomial probability density function for this case is plotted in Figure 4.1. The cumulative binomial distribution function can be used to compute the probabilities of consecutive Bernoulli trials. After n trials, the probability that a binomial random variable, X, is less than or equal to $k0$ is

$$P(X \le k0) = \sum_{i=0}^{k0} \binom{n}{i} p^i (1 - p)^{n-i}$$

where $\binom{n}{i} = {}_nC_i$, and $k0 = 0, 1, \ldots n$.

DISCRETE BINOMIAL DENSITY

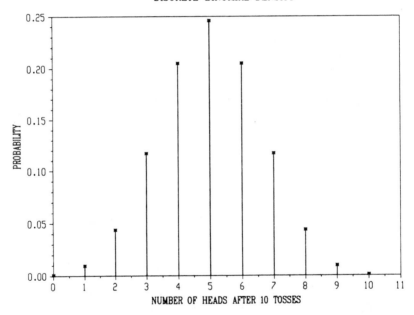

Figure 4.1   Binomial Probability Density

The cumulative distribution is computed by summing (up to $k0$) the individual probabilities shown in Figure 4.1. For example, out of ten flips of a coin the probability that four or less are heads is

$$b(10, 0.5, 0) + b(10, 0.5, 1) + \ldots + b(10, 0.5, 4) = 0.377$$

See Function: binomdst().

4.2.2 Hypergeometric Distribution

The **hypergeometric** distribution is frequently used to compute the probability of outcomes of repetitive sampling from a small population. However, unlike the binomial distribution, the samples are **not** returned to the population. Hence, the sampling is done **without** replacement.

There are two mutually exclusive classes for each random sampling (e. g., an ace or "not an ace"). The total number of elements in one class is $r1$. The total population size is r. Thus, the number of elements in the other class is $r - r1$.

The hypergeometric density function represents the probability of i successes out of n trials:

$$h(n, r, r1, i) = {}_{r1}C_i * {}_{(r-r1)}C_{(n-i)}/{}_rC_n$$

where $i = 0, 1, 2, \ldots, n$.

It can be shown that the mean u and variance s^2 of the hypergeometric density are

$$u = n * r1/r$$
$$s^2 = n * r1 * (r - r1) * (r - n)/[r^2 * (r - 1)]$$

The hypergeometric distribution can be used to compute probabilities of a poker hand dealt from a fair deck of cards. Let $n = 5$ be the number of cards in a poker hand, and $r1 = 4$ be the number of aces in a deck. The total number of cards in the deck is $r = 52$. The hypergeometric probability density function for this example is shown in Figure 4.2. For example, the odds of being dealt exactly four aces is about 0.000018 (although the odds are much better in Hollywood).

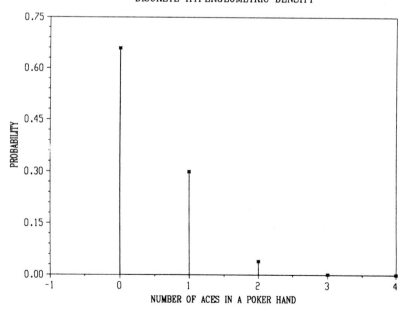

Figure 4.2   Hypergeometric Probability Density

The cumulative distribution is computed by summing (up to $k0$) the individual probabilities shown in Figure 4.2. After n trials, the probability that a hypergeometric random variable, X, is less than or equal to $k0$ is

$$P(X \leq k0) = \sum_{i=0}^{k0} \frac{\binom{r1}{i}\binom{r-r1}{n-i}}{\binom{r}{n}}$$

where $k0 = 0, 1, \ldots n$.

For example, the odds of being dealt one or less ace is

$$h(n, r, r1, 0) + h(n, r, r1, 1) = 0.9583.$$

See Function: hyperdst().

4.2.3 Poisson Distribution

The **Poisson** distribution can be used to predict the number of discrete random events that occur in a fixed time interval. This distribution has many applications in queueing theory. The events are assumed to be independent from occurrences in prior intervals. The rate at which events occur is assumed to be constant. The expected number of arrivals λ_T is the product of the average rate of arrivals of the process with the interval of interest, T.

A Poisson density function with parameter λ_T represents the probability of i arrivals in T units of time:

$$p(\lambda_T, i) = (\lambda_T)^i e^{-\lambda_T}/i!$$

where $i = 0, 1, 2, \ldots$.

For the Poisson density, it can be shown that the mean u and variance s^2 are equal:

$$u = s^2 = \lambda_T$$

The Poisson density can be used to forecast automobile traffic flow. If cars are arriving at a stop sign at a rate of one per minute, and the interval of interest is $T = 20$ minutes, then the parameter of the process is $\lambda_T = 20$. The Poisson probability density function for this example is shown in Figure 4.3. For example, the probability that exactly 20 cars arrive in 20 minutes is 0.0888.

The cumulative Poisson distribution is computed by summing (up to $k0$) the individual probabilities shown in Figure 4.3. After n trials, the probability that a Poisson random variable, X, is less than or equal to $k0$ is

$$P(X \leq k0) = \sum_{i=0}^{k0} \frac{(\lambda_T)^i e^{-\lambda_T}}{i!}$$

where $k0 = 0, 1, 2, \ldots$.

For example, the probability that 21 or less cars arrive in 20 minutes is

$$p(20, 0) + p(20, 1) + p(20, 2) + \ldots + p(20, 21) = 0.6437$$

See Functions: poissdst(), poisson().

4.3 CONTINUOUS RANDOM VARIABLES

Many statistical events are well modeled by continuous random variables. For these cases, the probability densities are continuous functions. The domain of the proba-

DISCRETE POISSON DENSITY

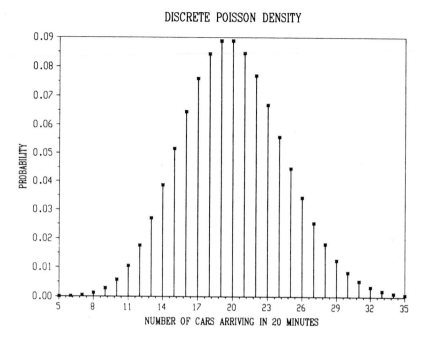

Figure 4.3   Poisson Probability Density

bility density function is the set of possible random outcomes (x values), sometimes called the **sample space**.

An outcome of a continuous random experiment can be any real value. For a continuous random variable to be properly defined, the area under its probability density function must be one. That is, the integral of the density function over the entire sample space must be unity.

The **cumulative probability distribution function** $F(x_0)$ of a continuous random variable is defined as the integral of the density $f(x)$ over x, where $x \le x_0$.

This section describes some of the properties of the two most commonly used continuous random variables: the **normal** and the **uniform** distributions.

4.3.1 Normal Distribution

The **normal** distribution (the "bell curve") is the most frequently used function in applied probability. This distribution is also called the Gaussian distribution. In its **standard** form, the normal distribution function is given by

$$p0 = P(X \le x_0) = \frac{1}{\sqrt{2\pi}} \int_{-\infty}^{x_0} e^{-x^2/2} dx$$

This form is particularly convenient because of the widespread availability of the standard normal tables. These tables list approximations to the above integral which facilitate probabilistic computations.

It can be shown that the mean of the standard normal density is zero and the variance is one. However, the real usefulness of the normal distribution is in modeling random events of arbitrary mean u and variance s^2. This is allowed by a simple transformation of variables:

$$X' = s * X + u$$

where X is the standard normal random variable and X' is the normal random variable with mean u and variance s^2. A useful shorthand notation in referring to normal variables is to say that X' is $N(u, s^2)$ and X is $N(0, 1)$.

The normal probability density function for the random variable X' is plotted in Figure 4.4. Because this is a valid probability density, the area under this curve is one. Approximately 68 percent of the area lies within one standard deviation (s) of the mean (u). This is the area in between the dashed lines of Figure 4.4.

NORMAL PROBABILITY DENSITY

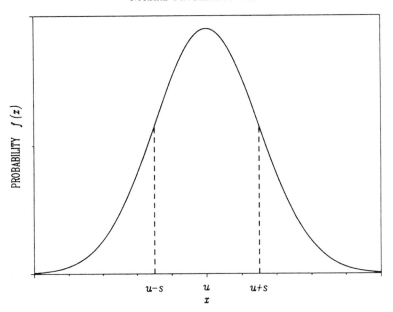

Figure 4.4   Normal Probability Density

The **Central Limit Theorem** is of fundamental importance to probability and statistics. This theorem simply states that if a random variable can be represented as a sum of any finite number of independent random variables, then under most conditions, it is normally distributed.

More formally, the **Central Limit Theorem** is as follows:

Let $X_1, X_2, \ldots, X_n, \ldots$ be a sequence of independent random variables each with mean u_i and variance s_i^2, $i = 1, 2, \ldots$. Let $Y = X_1 + X_2 + \ldots + X_n$. Under some general conditions that are beyond the scope of this development, an approximation to the standard normal distribution is given by

$$Z = (Y - U_n)/S_n$$

where $U_n = u_1 + u_2 + \ldots + u_n$

and $S_n = (s_1^2 + s_2^2 + \ldots + s_n^2)^{0.5}$

Note that as n (the number of random variables that are summed) increases, the distribution of Z more closely approximates $N(0, 1)$. This theorem emphasizes the great importance of the normal distribution, since it gives a mathematical justification of why the normal distribution works well in so many statistical applications.

Approximating Normal Probabilities.   The use of the normal distribution usually requires consulting tables of values. Although this may be adequate for a few hand calculations, these tables are cumbersome and are difficult to incorporate into computer programs. The smooth nature of the normal curve allows a closed form alternative to tables via polynomial approximation.

Consider again the following equation of the standard normal distribution:

$$P(X \le x0) = \frac{1}{\sqrt{2\pi}} \int_{-\infty}^{x0} e^{-x^2/2} dx$$

There are two approximations concerning this equation that are useful. The first approximates the cumulative normal probability $p0$ for a given x_0. The second approximation determines the specific value of x_0 required for a given cumulative probability $p0$. Note that the second case is really the inverse of the first.

Approximating the Normal Distribution.   The goal is to estimate the cumulative normal probability $P(X \le x0)$ for a given $x0$.

For $x0 \ge 0$, the probability is approximated with the following truncated power series:

$$P(X \le x0) = 1 - f(x0) * (b_1 t + b_2 t^2 + b_3 t^3 + b_4 t^4 + b_5 t^5)$$

where $t = (1 + 0.2316419 * x0)^{-1}$ and $f(x0) = 0.39894228 * e^{-x0^2/2}$
and $\quad b_1 = 0.31938153, \qquad b_2 = -0.356563782$
$\qquad b_3 = 1.781477937, \qquad b_4 = -1.821255978$
$\qquad b_5 = 1.330274429$

For $x0 < 0$, the complementary power series is used:

$$P(X \le x0) = f(x0) * (b_1 t + b_2 t^2 + b_3 t^3 + b_4 t^4 + b_5 t^5)$$

The error of these approximations is less than $7.5e - 8$.

See Functions: nprob(), normal().

Approximating the Inverse Normal Distribution.   This is the inverse of the previous approximation. The goal is to estimate the value of x_0 for a given cumulative normal probability, $p0 = P(X \le x_0)$.
For $p0 \le 0$, the value is estimated with the following truncated power series:

$$x_0 = -t + (c_0 t + c_1 t^2 + c_2 t^3)/(1 + d_1 t + d_2 t^2 + d_3 t^3)$$

where $t = (-2 * \ln(p0))^{0.5}$
and $\quad c_0 = 2.515517, \qquad d_1 = 1.432788$
$\qquad c_1 = 0.802853, \qquad d_2 = 0.189269$
$\qquad c_2 = 0.010328, \qquad d_3 = 0.001308$

For $p0 > 0.5$, the following complementary power series is used:

$$x_0 = t - (c_0 t + c_1 t^2 + c_2 t^3)/(1 + d_1 t + d_2 t^2 + d_3 t^3)$$

where $t = (-2 * \ln(1 - p0))^{0.5}$.

The error of this approximation is less than $4.5e - 4$.

See Functions: invprob(), normal().

4.3.2 Uniform Distribution

The **uniform** random variable is one of the most important random variables in probability and statistics. Many probability distributions can be generated with computations that involve the uniform distribution. Like the normal distribution, the uniform distribution has a standard form:

$$u(x) = 1 \qquad \text{for } 0 < x < 1$$
$$= 0 \qquad \text{otherwise}$$

It can be shown that the mean of the standard uniform density is 1/2 and the variance is 1/12. However, the standard form can easily be transformed to a uniform random variable with arbitrary mean u and variance s^2:

$$X' = (b - a)(X - 1/2) + (a + b)/2$$

where X is the standard uniform random variable and X' is the uniform random variable with mean u and variance s^2:

$$u = (a + b)/2$$
$$s^2 = (b - a)^2/12$$

This transformation can be used to generate uniform white noise samples of any mean and variance.

The uniform probability density function for the random variable X' is plotted in Figure 4.5.

UNIFORM DENSITY

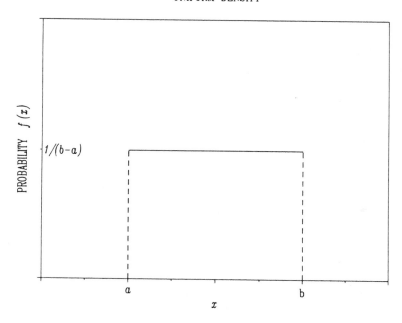

Figure 4.5   Uniform Probability Density

A useful shorthand notation in referring to uniform variables is to say that X' is $U(a, b)$ and X is $U(0, 1)$.

See Function: urand().

Normal Random Variable from a Uniform Distribution.   A normal random variable of arbitrary mean and variance can be approximated with the uniform distribution. This approximation is very useful because of the widespread availability of uniform random number generators.

The method uses a special case of the previously described Central Limit Theorem. Consider the sum of a sequence of 12 independent random variables X_i, which are uniformly distributed over the interval $(0, 1)$:

$$Y = X_1 + X_2 + X_3 + \ldots + X_{12}$$

Recall that the mean and variance of the standard uniform distribution were given by

$$u_x = 1/2, \quad \text{and} \quad s_x^2 = 1/12$$

It can be shown that the mean and variance of the random variable Y are given by

$$u_y = 12 * (1/2) = 6$$
$$s_y^2 = 12 * (1/12) = 1$$

From the Central Limit Theorem, the standard normal random variable, Z, with zero mean and unity variance can be approximated by:

$$Z = (Y - 6)/s_y = Y - 6$$

Note that since the variance of Y is unity, the division by the standard deviation (also unity) in the above formula is avoided. This is the reason for summing exactly 12 uniform random variables. Summing more than 12 uniform random variables would improve the statistical approximation, but would require a division by the standard deviation, s_y.

The desired normal random variable, W, with mean u_w and standard deviation s_w is obtained with the simple transformation

$$W = Z * s_w + u_w = (Y - 6) * s_w + u_w$$

Thus, normal random numbers of any mean and variance can easily be approximated with a uniform random number generator.

See Functions: normal(), urand().

Poisson Random Variable from a Uniform Distribution.   The Poisson distribution is very useful in modeling the number of discrete events that occur in a fixed time interval. The time, t, between events is exponentially distributed:

$$f(t) = ae^{-at} \quad \text{for } t > 0 \text{ and } a > 0$$
$$= 0 \quad \text{otherwise}$$

The mean and variance of the exponential random variable are

$$u_t = 1/a$$
$$s_t^2 = 1/a^2$$

The uniform distribution can be used to generate an exponential random variable of arbitrary mean and variance. In turn, the exponential approximation can be used to generate the desired Poisson random variable.

The most straightforward method of generating a Poisson random integer is to sum a sequence of K independent random variables Y_i, which are exponentially distributed with unity mean and variance (i. e., $a = 1$). K is incremented until the sum is equal to, or exceeds, the desired Poisson mean, $p0$. At this point, the value of K is the desired Poisson random integer. The sequence of exponentially distributed random variables, Y_i, is easily constructed from any standard uniform random number sequence X_i according to:

$$Y_i = -\log_e(X_i)$$

Note that the random variable Y_i/a is exponentially distributed with parameter a.
    Consider the sum of K of these random variables:

$$Y = Y_1 + Y_2 + Y_3 + \ldots + Y_K$$

It is easy to show that the above sum of exponential random variables, Y, has a mean and variance of K, where K is approximately the desired Poisson parameter $p0$. This follows directly since:

1. The mean of the sum is the sum of the means.
2. The variance of the sum is the sum of the variances.

Recall that for the Poisson distribution, the variance must equal the mean. Thus, Y is the desired Poisson random variable since:

$$s_y^2 = u_y = K = p0$$

A slightly more efficient method of generating Poisson random variables uses a variation of the technique described above. This approach generates the Poisson random variable directly from the uniform distribution. K standard uniform random variables, X_i, are multiplied together. As K is incremented, the product declines until it is less than or equal to e^{-p0}. At this point, the value of K is the desired Poisson random integer. This approach results from raising both sides of the above expression for Y to the base of e:

$$e^{-Y} = X_1 * X_2 * X_3 * \ldots * X_K \geq e^{-p0}$$

where $Y_i = -\log_e(X_i)$ and $\exp(-Y_i) = X_i$.
    Note that this approach avoids the natural logarithm and the need to generate the exponential random sequence.

See Functions: poisson(), urand().

5

Statistics

Statistics evolved from the need to analyze social and political trends. Despite its origin, statistics has developed into a highly sophisticated numerical science. Modern statistics goes beyond data reduction, forming the basis of all applied probability. Statistics is the foundation of many new and diverse scientific fields such as reliability theory, decision theory, and epidemiology.

This chapter attempts to cover those areas of statistics that are more frequently used by engineers and scientists, rather than those that are found in advanced statistics books.

5.1 EXPECTATION AND RANDOM VARIABLES

In the chapter on probability, the concept of the **mean** and **variance** of a random variable was introduced. Recall that a **random variable** is any function that maps all the events of a probabilistic experiment (the sample space) to the x-axis. As discussed previously, each random variable has its own characteristic probability density function and probability distribution function.

This section defines the **expectation** of a random variable. The mean and variance are discussed in more detail and are shown to be special cases of the more general concept of expectation.

5.1.1 Expectation of a Function

If X is a random variable and if $Y = H(X)$ is a function of X, then Y is also a random variable with its own probability density function. The **expectation** of the random variable Y will be denoted as $E(Y)$ and is defined below.

If X is a discrete random variable, the expectation of a function of X, $H(X)$, is defined by:

$$E(Y) = E(H(X)) = \sum_{j=1}^{\infty} H(x_j)p(x_j)$$

where $p(x_j)$ is the discrete probability density of X.

Similarly, if X is a continuous variable, the expectation of a function of X, $H(X)$, is defined by

$$E(Y) = E(H(X)) = \int_{-\infty}^{+\infty} H(x)f(x)\,dx$$

where $f(x)$ is the continuous probability density of X.

Properties of Expectation.   There are a number of important properties of the expected value that hold for both discrete and continuous random variables. They are listed here without proof:

1. If C is a constant, and $X = C$, then $E(X) = C$.
2. If C is a constant, then $E(CX) = CE(X)$.
3. Given two random variables, X and Y, $E(X + Y) = E(X) + E(Y)$.
4. If X and Y are independent, then $E(XY) = E(X)E(Y)$.

5.1.2 Mean and Variance of a Random Variable

In the previous section, the expectation of a function of a random variable was defined. This definition may be used to formally define the mean and variance of a random variable.

The Mean of a Random Variable.   The **mean** u of a random variable X is simply the expected value of X, $E(X)$. Thus, if X is a discrete random variable, its mean is defined by

$$E(X) = \sum_{i=1}^{\infty} x_i p(x_i)$$

where $p(x_i)$ is the discrete probability density of X.

Similarly, if X is a continuous variable, its mean is defined by

$$E(X) = \int_{-\infty}^{+\infty} xf(x)\,dx$$

where $f(x)$ is the continuous probability density of X.

The mean represents the average value of the probability density function.

The Variance of a Random Variable.   The **variance** s^2 of a random variable X is the expected value of $(X - u)^2$, and is denoted VAR(X).

Properties of Variance.   There are several important properties of the variance of discrete and continuous random variables:

1. VAR(X) = $E(X - E(X))^2 = E(X^2) - E^2(X)$, where $E(X) = u$.
2. If C is a constant, then VAR($X + C$) = VAR(X).
3. If C is a constant, then VAR(CX) = C^2VAR(X).
4. If X and Y are independent, then VAR($X + Y$) = VAR(X) + VAR(Y).

The variance quantifies the amount of spread that a random variable has about its mean value.

Sample Variance.   In statistical sampling, the above definition of variance is often referred to as the **population variance**. In practice, the entire population can rarely be observed and if the variance is desired, it must be estimated from a small sample of the population. To ensure that the mean of estimate equals the true variance (i. e., an unbiased estimate), the above variance definition must be modified slightly. The **sample variance** of a sample of size m is given by

$$s_s^2 = [(X_1 - u_x)^2 + (X_2 - u_x)^2 + \ldots + (X_m - u_x)^2]/(m - 1)$$

Similarly, the **sample standard deviation** is s_s. Note that this variance expression is similar to the true variance definition except for the factor of $m-1$ in the denominator.

   See Function: stat().

5.1.3 Chebyshev's Theorem

The concepts of mean and variance are very useful in predicting the outcomes of random events. Chebyshev's theorem relates the mean and variance of any arbitrary random variable. This famous inequality is an explicit mathematical bound which shows how the variance directly measures statistical variability. **Chebyshev's theorem** is as follows:

If a random variable has a probability density with a mean u and a standard deviation s, the probability that the variable deviates from the mean by at least k standard deviations is at most $1/k^2$. Mathematically,

$$P(|X - u| \geq ks) \leq 1/k^2$$

Chebyshev's inequality is a very powerful statistical statement. For example, this theorem proves that the probability that a random variable will achieve a value that is two standard deviations away from the mean is at most 1/4. Likewise, the probability that a random variable will achieve a value that is 10 standard deviations away from the mean is at most 1/100.

Example:

> Consider 1600 flips of a fair coin. The number of heads is a binomial random variable with mean $u = 1600/2 = 800$ and standard deviation $s = (1600/4)^{0.5} = 20$. For $k = 5$, Chebyshev's theorem states that there is a probability of at least 0.96 that there will be between 700 and 900 heads. Equivalently, the *proportion* of heads will fall between 0.4375 and 0.5625.

5.2 REGRESSION AND LEAST-SQUARES

Errors are inherent in the measurement of any physical quantity. In making measurements, usually the goal is to determine the true values from the observations as accurately as possible. **Regression** is a statistical technique that reduces random errors by effectively smoothing multiple measurements.

Few numerical techniques have had a more pervasive effect on applied mathematics than the field of least-squares analysis. The contributions of least-squares methods are most obvious in the areas of:

1. Data smoothing
2. Statistical trend analysis (regression)
3. Fourier analysis
4. Functional approximation and minimization.

It is interesting to note that the mathematician who first discovered least-squares (Gauss) considered his findings trivial and did not report them until a decade after his discovery.

5.2.1 Generalized Least-Squares

One of the most powerful curve-fitting techniques is the method of **generalized least-squares**. For any set of one-dimensional functions, this technique determines the linear combination that best fits the data.

The hypothesis is that a set of M observations (x_i, y_i) can be represented by the equation:

$$y(x) = a_1 f_1(x) + a_2 f_2(x) + a_3 f_3(x) + \ldots + a_N f_N(x)$$

The goal is to find the coefficients a_j of the best-fitting linear combination of the N one-dimensional basis functions $f_j(x)$. For a solution to exist, the number of observations must be greater than the number of basis functions, (i. e., $M \geq N$). Solving for the coefficients a_j is equivalent to determining the composite function $y(x)$ that minimizes the **mean-square error** across the entire data interval:

$$\text{MSE} = (y(x_1) - y_1)^2 + (y(x_2) - y_2)^2 + \ldots + (y(x_M) - y_M)^2$$

Note that if y_i exactly equals each $y(x_i)$, then the error of the approximation is zero. This is generally more than one can expect with real data. In practice, there is usually some minimal nonzero error.

Before proceeding it will be helpful to introduce a shorthand notation. Let Σ denote the finite sum over the M data points. For example, Σf_j means the sum of the jth basis function across all the data points:

$$\Sigma f_j = f_j(x_1) + f_j(x_2) + \ldots + f_j(x_M)$$

Similarly, $\Sigma y f_j$ is defined as:

$$\Sigma y f_j = y_1 f_j(x_1) + y_2 f_j(x_2) + \ldots + y_M f_j(x_M)$$

and the expression $\Sigma f_k f_j$ denotes:

$$\Sigma f_k f_j = f_k(x_1) f_j(x_1) + f_k(x_2) f_j(x_2) + \ldots + f_k(x_M) f_j(x_M)$$

where $1 \leq k \leq N$ and $1 \leq j \leq N$.

The minimum MSE solution to the generalized least-squares problem can be found by solving:

$$
\begin{bmatrix}
\Sigma f_1 f_1 & \Sigma f_1 f_2 & \Sigma f_1 f_3 & \cdots & \Sigma f_1 f_N \\
\Sigma f_2 f_1 & \Sigma f_2 f_2 & \Sigma f_2 f_3 & \cdots & \Sigma f_2 f_N \\
\Sigma f_3 f_1 & \Sigma f_3 f_2 & \Sigma f_3 f_3 & \cdots & \Sigma f_3 f_N \\
\vdots & \vdots & \vdots & & \vdots \\
\Sigma f_N f_1 & \Sigma f_N f_2 & \Sigma f_N f_3 & \cdots & \Sigma f_N f_N
\end{bmatrix}
\begin{bmatrix}
a_1 \\ a_2 \\ a_3 \\ \vdots \\ a_N
\end{bmatrix}
=
\begin{bmatrix}
\Sigma y f_1 \\ \Sigma y f_2 \\ \Sigma y f_3 \\ \vdots \\ \Sigma y f_N
\end{bmatrix}
$$

These are called the **normal** equations. Note that the normal equations are symmetric. This property can be exploited in determining least-squares solutions.

See Function: least_sq().

5.2.2 Polynomial Regression

One of the more effective fitting functions is the Nth order polynomial:

$$y = a_1 + a_2 x + a_3 x^2 + \ldots + a_N x^N$$

The normal equations for a polynomial fit are

$$
\begin{bmatrix}
M & \Sigma x & \Sigma x^2 & \cdots & \Sigma x^N \\
\Sigma x & \Sigma x^2 & \Sigma x^3 & \cdots & \Sigma x^{N+1} \\
\Sigma x^2 & \Sigma x^3 & \Sigma x^4 & \cdots & \Sigma x^{N+2} \\
\vdots & \vdots & \vdots & & \vdots \\
\Sigma x^N & \Sigma x^{N+1} & \Sigma x^{N+2} & \cdots & \Sigma x^{2N}
\end{bmatrix}
\begin{bmatrix}
a_1 \\ a_2 \\ a_3 \\ \vdots \\ a_N
\end{bmatrix}
=
\begin{bmatrix}
\Sigma y \\ \Sigma y x \\ \Sigma y x^2 \\ \vdots \\ \Sigma y x^N
\end{bmatrix}
$$

These equations can be solved with a variety of computer approaches such as LU decomposition or Gaussian elimination. It should be noted that computer precision usually limits the order of the polynomial to less than ten.

See Function: curvreg().

5.2.3 Linear Regression

The simplest type of approximating curve is a line. The constants a and b must be found which give the closest agreement between the empirical data and the equation:

$$y = bx + a$$

This is just a special case of the polynomial regression described above. The solution can be found by using Cramer's Rule and the following normal equations:

$$\begin{bmatrix} M & \Sigma\,x \\ \Sigma\,x & \Sigma\,x^2 \end{bmatrix} \begin{bmatrix} a \\ b \end{bmatrix} = \begin{bmatrix} \Sigma\,y \\ \Sigma\,yx \end{bmatrix}$$

Using Σ to denote summation over all the observations, the solution is given by:

$$b = \frac{\Sigma\,yx - \Sigma\,x\Sigma\,y/M}{\Sigma\,x^2 - (\Sigma\,x)^2/M}$$

$$a = [\Sigma\,y - b\Sigma\,x]/M.$$

The **coefficient of determination** r^2 indicates how closely a function approximates a data set. Sometimes called the **correlation coefficient**, the value of r^2 is always between zero and one. The closer r^2 is to one, the better the approximation. For a linear approximation, the coefficient is given by:

$$r^2 = \frac{[\Sigma\,yx - \Sigma\,x\Sigma\,y/M]^2}{[\Sigma\,x^2 - (\Sigma\,x)^2/M][\Sigma\,y^2 - (\Sigma\,y)^2/M]}$$

   See Function: linreg().

5.2.4 Logarithmic Regression

Another useful approximating function is the natural logarithm:

$$y = a + b\ln(x)$$

where $x_i > 0$.
   The normal equations for the logarithmic fit are

$$\begin{bmatrix} M & \Sigma\,\ln(x) \\ \Sigma\,\ln(x) & \Sigma\,\ln^2(x) \end{bmatrix} \begin{bmatrix} a \\ b \end{bmatrix} = \begin{bmatrix} \Sigma\,y \\ \Sigma\,y\ln(x) \end{bmatrix}.$$

Solving for the logarithmic coefficients yields

$$b = \frac{\Sigma\,y\ln(x) - \Sigma y\,\Sigma\ln(x)/M}{\Sigma\,\ln^2(x) - (\Sigma\,\ln(x))^2/M}$$

$$a = [\Sigma\,y - b\Sigma\,\ln(x)]/M$$

The correlation coefficient for the logarithmic fit is given by

$$r^2 = \frac{[\Sigma\,\ln(x)y - \Sigma\,y\Sigma\,\ln(x)/M]^2}{[\Sigma\,y^2 - (\Sigma\,y)^2/M][\Sigma\,\ln^2(x) - (\Sigma\,\ln(x))^2/M]}$$

5.2.5 Fourier Regression

Any data set can be represented by a combination of a finite sine and cosine series:

$$y = a_0 + a_1 \cos(wx) + a_2 \cos(2wx) + \ldots + a_N \cos(Nwx)$$
$$+ b_1 \sin(wx) + b_2 \sin(2wx) + \ldots + b_N \sin(Nwx)$$

The normal equations for a finite sinusoidal series fit are

$$
\begin{bmatrix}
M & \Sigma \cos(wx) & \Sigma \sin(wx) & \cdots \\
\Sigma \cos(wx) & \Sigma \cos^2(wx) & \Sigma \sin(wx)\cos(wx) & \cdots \\
\Sigma \sin(wx) & \Sigma \sin(wx)\cos(wx) & \Sigma \sin^2(wx) & \cdots \\
\vdots & \vdots & \vdots & \ddots \\
\Sigma \cos(Nwx) & \Sigma \cos(Nwx)\cos(wx) & \Sigma \cos(Nwx)\sin(wx) & \cdots \\
\Sigma \sin(Nwx) & \Sigma \sin(Nwx)\cos(wx) & \Sigma \sin(Nwx)\sin(wx) & \cdots
\end{bmatrix}
\begin{bmatrix}
a_0 \\
a_1 \\
b_1 \\
\vdots \\
a_N \\
b_N
\end{bmatrix}
$$

$$
=
\begin{bmatrix}
\Sigma\, y \\
\Sigma\, y \cos(wx) \\
\Sigma\, y \sin(wx) \\
\vdots \\
\Sigma\, y \cos(Nwx) \\
\Sigma\, y \sin(Nwx)
\end{bmatrix}
$$

The Fourier coefficients can be determined by any number of approaches for solving linear systems of equations. However, it should be mentioned that this is neither the most efficient nor the most accurate approach to sinusoidal analysis. This is especially true as the number of harmonics N increases. For large N ($N > 10$), the FFT algorithm described in the Digital Signal Processing chapter is a far more efficient and accurate method. The FFT method is also a least-squares fit and has almost no limit on the order (N) of the sinusoidal series.

5.2.6 Nonlinear Regression

This section describes a few nonlinear forms of regression. For generalized least-squares, the fitting function can be expressed as a linear combination of arbitrary basis functions. In this context the term nonlinear does **not** mean "not a line," but refers to functions that cannot be represented by a linear combination of basis functions. Although these regression types are not currently implemented in this package, you could easily implement them yourself.

Exponential Regression.   The exponential function is one of the most commonly used curve-fitting functions. A linear combination of weighted exponentials can be a very effective regression formula:

$$y = c_1 + c_2 e^x + \ldots$$

This is the exponential form that is implemented in the GRAFIX program.

One of the simplest nonlinear approximating functions is the exponential curve:

$$y = a \, e^{bx}$$

where $a > 0$, and $y_i > 0$.

This problem can be linearized by taking the logarithm of both sides of the equation:

$$\ln(y) = \ln(a) + bx$$

The normal equations for the exponential fit are

$$\begin{bmatrix} M & \Sigma\, x \\ \Sigma\, x & \Sigma\, x^2 \end{bmatrix} \begin{bmatrix} \ln(a) \\ b \end{bmatrix} = \begin{bmatrix} \Sigma\, \ln(y) \\ \Sigma\, x \ln(y) \end{bmatrix}$$

Solving for the coefficients of the exponential function yields

$$b = \frac{\Sigma\, x \ln(y) - \Sigma\, x \Sigma\, \ln(y)/M}{\Sigma\, x^2 - (\Sigma\, x)^2/M}$$

$$a = \exp[\Sigma\, \ln(y)/M - b\Sigma\, x/M]$$

The correlation coefficient of the exponential fit is given by

$$r^2 = \frac{[\Sigma\, \ln(y)x - \Sigma\, x \Sigma\, \ln(y)/M]^2}{[\Sigma\, x^2 - (\Sigma\, x)^2/M][\Sigma\, \ln^2(y) - (\Sigma\, \ln(y))^2/M]}$$

Geometric Power Curve Regression.   Like the exponential curve, the geometric power curve can be useful in approximating data:

$$y = ax^b$$

where $a > 0$, $x_i > 0$, and $y_i > 0$.

This problem can be linearized by taking the logarithm of both sides of the equation:

$$\ln(y) = \ln(a) + b\ln(x)$$

The normal equations for the geometric power fit are

$$\begin{bmatrix} M & \Sigma\, \ln(x) \\ \Sigma\, \ln(x) & \Sigma\, \ln^2(x) \end{bmatrix} \begin{bmatrix} \ln(a) \\ b \end{bmatrix} = \begin{bmatrix} \Sigma\, \ln(y) \\ \Sigma\, \ln(y) \ln(x) \end{bmatrix}$$

Solving for the coefficients of the geometric power function yields

$$b = \frac{\Sigma\, \ln(x) \ln(y) - \Sigma\, \ln(x) \Sigma\, \ln(y)/M}{\Sigma\, \ln^2(x) - (\Sigma\, \ln(x))^2/M}$$

$$a = \exp[\Sigma\, \ln(y) - b\Sigma\, \ln(x)]/M$$

For the geometric power fit the correlation coefficient is given by

$$r^2 = \frac{[\Sigma\, \ln(y) \ln(x) - \Sigma\, \ln(x) \Sigma\, \ln(y)/M]^2}{[\Sigma\, \ln^2(x) - (\Sigma\, \ln(x))^2/M][\Sigma\, \ln^2(y) - (\Sigma\, \ln(y))^2/M]}$$

Hyperbolic Regression.   In speech analysis and synthesis, the generalized hyperbolic curve is a useful approximating function:

$$y = \frac{1}{a_1 + a_2 x + a_3 x^2 + \ldots + a_N x^N}$$

where $y_i > 0$.

The normal equations for the hyperbolic fit are

$$
\begin{bmatrix}
M & \Sigma\, x & \Sigma\, x^2 & \ldots & \Sigma\, x^N \\
\Sigma\, x & \Sigma\, x^2 & \Sigma\, x^3 & \ldots & \Sigma\, x^{N+1} \\
\Sigma\, x^2 & \Sigma\, x^3 & \Sigma\, x^4 & \ldots & \Sigma\, x^{N+2} \\
\vdots & \vdots & \vdots & & \vdots \\
\Sigma\, x^N & \Sigma\, x^{N+1} & \Sigma\, x^{N+2} & \ldots & \Sigma\, x^{2N}
\end{bmatrix}
\begin{bmatrix}
a_1 \\ a_2 \\ a_3 \\ \vdots \\ a_N
\end{bmatrix}
=
\begin{bmatrix}
\Sigma\, 1/y \\
\Sigma\, x/y \\
\Sigma\, x^2/y \\
\vdots \\
\Sigma\, x^N/y
\end{bmatrix}
$$

Again, a variety of computer approaches can be used to determine the best-fitting coefficients. To avoid computer precision problems, the order of the denominator polynomial should be limited to 10.

6

Linear Algebra

Linear algebra is a branch of mathematics that is concerned with the theory and application of systems of linear equations. The mathematics of matrices and vectors are fundamental to the theory of linear algebra. Those who are unfamilar with matrix notation may at first find it a little confusing. However, matrices provide a convenient method of dealing with systems of many variables. Often in working with linear equations, only the coefficients of the variables are important, not the variables themselves. Matrices are just a mathematical "shorthand" that maintains the respective positions of each of the coefficients. Computer libraries of matrix functions allow the analyst to quickly and efficiently solve large sets of linear equations that would be much too tedious by hand.

6.1 MATRICES

A matrix is a rectangular array of numbers. Each element of a matrix is a scalar quantity (a real, or a complex number). For example, the matrix M has two rows and three columns, and a total of $2 \times 3 = 6$ elements:

$$M = \begin{bmatrix} 1 & 2 & 4 \\ 3 & 7 & 9 \end{bmatrix}$$

6.1.1 Matrix Addition and Subtraction

Matrices can be added and subtracted only if they have the same number of rows and columns (i. e., the same dimensions). To add or subtract two matrices, simply add or subtract the corresponding elements and enter the answers into the same places of the resulting matrix. The ollowing is an example of the addition of 2×2 matrices:

$$\begin{bmatrix} 1 & 2 \\ 3 & 7 \end{bmatrix} + \begin{bmatrix} 7 & 2 \\ 3 & 4 \end{bmatrix} = \begin{bmatrix} 1+7 & 2+2 \\ 3+3 & 7+4 \end{bmatrix} = \begin{bmatrix} 8 & 4 \\ 6 & 11 \end{bmatrix}$$

Subtracting the above matrices yields:

$$\begin{bmatrix} 1 & 2 \\ 3 & 7 \end{bmatrix} - \begin{bmatrix} 7 & 2 \\ 3 & 4 \end{bmatrix} = \begin{bmatrix} 1-7 & 2-2 \\ 3-3 & 7-4 \end{bmatrix} = \begin{bmatrix} -6 & 0 \\ 0 & 3 \end{bmatrix}$$

See Functions: mxadd(), Cmxadd(), mxsub(), Cmxsub(), vadd(), Cvadd(), vsub(), Cvsub().

6.1.2 Scaling a Matrix

Sometimes it is necessary to multiply a matrix by a scalar quantity (i. e., a real or a complex number). The following is an example of scalar multiplication:

$$3 \times \begin{bmatrix} 1 & 2 \\ 3 & 7 \end{bmatrix} = \begin{bmatrix} 3 \times 1 & 3 \times 2 \\ 3 \times 3 & 3 \times 7 \end{bmatrix} = \begin{bmatrix} 3 & 6 \\ 9 & 21 \end{bmatrix}$$

See Functions: mxscale(), CmxscalR(), CmxscalC(), vscale(), CvscalR(), CvscalC().

6.1.3 Matrix Multiplication

Two matrices can be multiplied only if the number of columns in the first matrix is equal to the number of rows in the second matrix. The element in the ith row and jth column of the product matrix is formed by taking the dot product (see dot product in the Vectors section 6.3.1) between the ith row of the first matrix and the jth column of the second matrix. The following is an example of multiplication of a 2×3 matrix by a 3×2 matrix:

$$\begin{bmatrix} 1 & 2 & 9 \\ 3 & 7 & 5 \end{bmatrix} \begin{bmatrix} 1 & 7 \\ 2 & 3 \\ 4 & 1 \end{bmatrix} = \begin{bmatrix} 1 \times 1 + 2 \times 2 + 9 \times 4 & 1 \times 7 + 2 \times 3 + 9 \times 1 \\ 3 \times 1 + 7 \times 2 + 5 \times 4 & 3 \times 7 + 7 \times 3 + 5 \times 1 \end{bmatrix}$$
$$= \begin{bmatrix} 16 & 22 \\ 37 & 47 \end{bmatrix}$$

Unlike scalar arithmetic, the order of matrix multiplication is important since, in general, $A * B$ is not equal to $B * A$. Thus, matrix multiplication is not commutative.

See Functions: mxmul(), mxmul1(), mxmul2(), Cmxmul(), Cmxmul1(), Cmxmul2().

6.1.4 Matrix Transpose

The **transpose** of a matrix is obtained by interchanging the row and column indices of the elements of the matrix. For example, if

$$
A = \begin{bmatrix}
a_{11} & a_{12} & a_{13} & \cdots & a_{1n} \\
a_{21} & a_{22} & a_{23} & \cdots & a_{2n} \\
a_{31} & a_{32} & a_{33} & \cdots & a_{3n} \\
\vdots & \vdots & \vdots & & \vdots \\
a_{m1} & a_{m2} & a_{m3} & \cdots & a_{mn}
\end{bmatrix}
$$

then

$$
A^T = \begin{bmatrix}
a_{11} & a_{21} & a_{31} & \cdots & a_{m1} \\
a_{12} & a_{22} & a_{32} & \cdots & a_{m2} \\
a_{13} & a_{23} & a_{33} & \cdots & a_{m3} \\
\vdots & \vdots & \vdots & & \vdots \\
a_{1n} & a_{2n} & a_{3n} & \cdots & a_{mn}
\end{bmatrix}
$$

Note that since A is an $m \times n$ matrix, A^T must be an $n \times m$ matrix. Some useful properties of matrix transposition are

1. $(ABC\ldots)^T = \ldots C^T B^T A^T$
2. $(A + B)^T = A^T + B^T$
3. $(cA)^T = cA^T$, where c is a scalar
4. $(A^T)^T = A$

See Functions: mxtransp(), Cmxtrans().

6.1.5 Some Special Square Matrices

Square matrices are the most important class of matrices, and their algebra has some special properties. The redundancy in special square matrices can be exploited to reduce the number of computations required to solve sets of simultaneous linear equations.

For square matrices it is possible to define an operation that is somewhat analogous to scalar division called matrix inversion (see matrix inversion in section 6.1.9).

For scalar arithmetic, multiplication by 1 (unity) results in the original scalar value. A similar operation also exists for matrices. The **identity matrix** I is a square matrix with all of the diagonal elements equal to one and all other elements equal to zero, for example

$$
I = \begin{bmatrix}
1 & 0 & 0 & 0 & 0 \\
0 & 1 & 0 & 0 & 0 \\
0 & 0 & 1 & 0 & 0 \\
0 & 0 & 0 & 1 & 0 \\
0 & 0 & 0 & 0 & 1
\end{bmatrix}
$$

Multiplication of a matrix with the identity matrix produces the original matrix.

A matrix U is said to be **upper diagonal** if all of its elements below the diagonal are zero. Upper diagonal matrices have the following form:

$$U = \begin{bmatrix} u_{11} & u_{12} & u_{13} & u_{1n} \\ 0 & u_{22} & u_{23} & u_{2n} \\ 0 & 0 & u_{33} & u_{3n} \\ 0 & 0 & 0 & u_{nn} \end{bmatrix}$$

There are, in general, only $n(n+1)/2$ nonzero elements in an upper diagonal matrix.

A matrix L is **lower diagonal** if all of its elements above the diagonal are zero. Lower diagonal matrices have the following form:

$$L = \begin{bmatrix} l_{11} & 0 & 0 & 0 \\ l_{21} & l_{22} & 0 & 0 \\ l_{31} & l_{32} & l_{33} & 0 \\ l_{n1} & l_{n2} & l_{n3} & l_{nn} \end{bmatrix}$$

Like the upper diagonal matrix, there are at most only $n(n+1)/2$ nonzero elements in a lower diagonal matrix.

A matrix T is **tridiagonal** if it is of the form

$$T = \begin{bmatrix} t_{11} & t_{12} & 0 & 0 \\ t_{21} & t_{22} & t_{23} & 0 \\ 0 & t_{32} & t_{33} & t_{mn} \\ 0 & 0 & t_{nm} & t_{nn} \end{bmatrix} \quad \text{where } m = n - 1$$

There are, in general, only $3n - 2$ nonzero elements in a tridiagonal matrix.

The elements of a **symmetric** matrix have the following symmetry around the main diagonal:

$$a_{ij} = a_{ji} \quad \text{where } 0 \le i, j \le n - 1$$

The transpose of a symmetric matrix is the original matrix. Householder developed an efficient method solving linear systems of real-symmetric matrices.

A **Hermitian matrix** has complex elements that exhibit complex conjugate symmetry around the main diagonal. The following is an example of a 3×3 Hermitian matrix:

$$M = \begin{vmatrix} a + i0 & b - ic & c - id \\ b + ic & e + i0 & f - ig \\ c + id & f + ig & h + i0 \end{vmatrix} \quad \text{where } i^2 = -1$$

Note that the elements along the main diagonal of the Hermitian matrix must be real. Thus, a real symmetric matrix is also Hermitian, since the imaginary parts of its elements are all zero.

A **Toeplitz** matrix M is a real symmetric matrix with the additional row and column symmetry shown below:

$$M = \begin{bmatrix} m_1 & m_2 & m_3 & \cdots & m_n \\ m_2 & m_1 & m_2 & \cdots & m_{n-1} \\ m_3 & m_2 & m_1 & \cdots & m_{n-2} \\ \vdots & \vdots & \vdots & & \vdots \\ m_n & m_{n-1} & m_{n-2} & \cdots & m_1 \end{bmatrix}$$

Note that the entire Toeplitz matrix is defined by any one of its row or column vectors. Levinson developed an efficient method for solving linear systems of Toeplitz matrices.

Sets of equations that can be represented by upper diagonal, lower diagonal, tridiagonal, or Toeplitz matrices are much easier to solve than those represented by more general matrices.

See Functions: mxident(), levinson().

6.1.6 Matrix Trace

The **trace** is defined for a square matrix as the sum of all its diagonal elements.

Two useful properties of the trace of a matrix are

1. $\text{Tr}(A + B) = \text{Tr}(A) + \text{Tr}(B)$
2. $\text{Tr}(ABC) = \text{Tr}(BCA) = \text{Tr}(CAB)$

For example, the trace of the 3×3 identity matrix is **3**. Note that, in general, the trace of the product of two matrices is not equal to the product of the traces of each of the matrices.

See Functions: mxtrace(), Cmxtrace().

6.1.7 Matrix Determinant

The **determinant** of a matrix, $|A|$, is a scalar value that indicates whether or not a system of linear equations has a unique solution. If the determinant of the system matrix is nonzero, then the set of equations has a unique solution. The determinant is of fundamental importance to the concepts of **matrix rank**, **linear independence**, and **matrix singularity**.

There is a common formula that can be followed to find the determinant of any square matrix. For two-dimensional matrices, the determinant is given by

$$|A| = a_{11}a_{22} - a_{12}a_{21}$$

For three-dimensional matrices, the determinant is given by

$$|A| = a_{11}a_{22}a_{33} + a_{12}a_{23}a_{31} + a_{13}a_{21}a_{32} - a_{13}a_{22}a_{31} - a_{11}a_{21}a_{33}$$

where a_{ij} is the element of the ith row and jth column of matrix A.

Determinants of real matrices have an interesting geometrical interpretation. If A is a two-dimensional square matrix, the absolute value of its determinant is equal to the area of the parallelogram formed by its row (or column) vectors. Similarly, if A is a three-dimensional matrix, the absolute value of its determinant is equal to the volume of the parallelepiped formed by its row (or column) vectors.

If two rows (or columns) of a matrix are interchanged, then the determinant changes its sign.

See Functions: mxdeterm(), Cmxdeter().

6.1.8 Matrix Singularity and Rank

A square matrix is **singular** if its determinant is zero. Conversely, if the determinant of a matrix is nonzero, then it is nonsingular. This is important because algebraic equations that have a nonsingular system matrix have a unique solution.

The **rank** of a square matrix is defined as the size of the largest nonzero determinant that can be formed from the matrix. If A is an $n \times n$ matrix, then the maximum rank of A is n. Furthermore, if A is of maximal rank, then A is said to be nonsingular. Equivalently, if A is maximal rank, then it has a nonzero determinant and the rows (and columns) of A are a **linear independent basis set**.

6.1.9 Matrix Inverse

A square matrix A has an **inverse**, A^{-1}, if and only if

$$A^{-1}A = AA^{-1} = I$$

where I is the identity matrix.

There is just one condition required for the existence of an inverse matrix: the original matrix must be nonsingular. That is, the matrix must be of maximal rank, or equivalently, its determinant must be nonzero. Matrix inversion is similar to scalar division and trying to invert a singular matrix is analogous to a scalar divide by zero.

See Functions: mxinv(), mxinv22(), mxinv33(), Cmxinv(), Cmxinv22(), Cmxinv33().

6.2 SIMULTANEOUS LINEAR EQUATIONS

Any system of simultaneous equations can be expressed in matrix form. For example,

$$3x + 2y = 2$$
$$5x + 5y = 1$$

in matrix form is

$$\begin{bmatrix} 3 & 2 \\ 5 & 5 \end{bmatrix} \begin{bmatrix} x \\ y \end{bmatrix} = \begin{bmatrix} 2 \\ 1 \end{bmatrix}$$

Note that the matrix representation eliminates the need for repeating the *"plus"* signs, the *"equals"* signs, and the x and y characters.

There are a variety of applied techniques available for solving this set of linear equations. It is beyond the scope of this section to describe them all. It is more important to know when to use each technique. The reader may verify that the correct answer is $x = 1.6$, $y = -1.2$.

Cramer's rule is useful for hand calculations with low order systems such as this. This formula is easily generalized for linear systems of two equations. Consider the following two equations:

$$\begin{bmatrix} a_{11} & a_{12} \\ a_{21} & a_{22} \end{bmatrix} \begin{bmatrix} x \\ y \end{bmatrix} = \begin{bmatrix} b_1 \\ b_2 \end{bmatrix}$$

The solution to this system of equations can be calculated as the ratio of two determinants. Solving for x yields

$$x = \frac{\begin{vmatrix} b_1 & a_{12} \\ b_2 & a_{22} \end{vmatrix}}{\begin{vmatrix} a_{11} & a_{12} \\ a_{21} & a_{22} \end{vmatrix}} = \frac{b_1 a_{22} - a_{12} b_2}{a_{11} a_{22} - a_{21} a_{12}}$$

and similarly, the solution for y is

$$y = \frac{\begin{vmatrix} a_{11} & b_1 \\ a_{21} & b_2 \end{vmatrix}}{\begin{vmatrix} a_{11} & a_{12} \\ a_{21} & a_{22} \end{vmatrix}} = \frac{a_{11} b_2 - b_1 a_{21}}{a_{11} a_{22} - a_{21} a_{12}}$$

Note that the determinant of the system matrix is in the denominator in both of the above expressions, and must be nonzero for there to be a solution (i. e., the system matrix must be nonsingular).

When the number of equations is large, Cramer's rule becomes cumbersome. For these cases, a variety of computer solutions are more appropriate. The "brute-force" approach is to employ matrix inversion. Although matrix inversion is a convenient method for solving systems of linear equations, it is not the most efficient, nor the most accurate. Direct matrix reduction techniques such as LU decomposition and Gaussian elimination often provide superior speed and accuracy. This is especially true for large matrices that are nearly singular. However, for small matrices, matrix inversion is usually adequate.

See Functions: lineqn(), Clineqn().

6.2.1 Pseudo-Inverse

Pseudo-inverse approaches are very useful in matching simple geometric shapes to real-world objects. These techniques have also proved to be very useful in calibration and coordinate alignment for robotic vision. These approaches are very powerful and are optimal in the sense that they minimize mean-square error (see Chapter 5, Statistics).

Suppose a data set is known to belong to a linear process. We may wish to estimate the parameters that characterize this process but we find that the data is corrupted with random noise (e. g., measurement errors). For example, the points on a wall all lie in a plane. With no noise, only three points are needed to characterize the plane's equation. However, in measuring the distance to the wall, random measurement noise is added. This noise degrades the estimate of the plane's true equation. One way to reduce these random effects is to take several additional data points (more than three) and overconstrain the system. Pseudo-inverse methods can be used to solve this augmented set of equations.

Consider the following set of overconstrained linear equations:

$$D_{MN} v_N = z_M$$

where M is the number of equations and N is the number of unknowns ($M \geq N$).

Solving for v_N would be straightforward if D_{MN} were square (i. e., $M = N$) and invertible. Nonetheless, the above equation can be easily transformed into an equivalent square system by premultiplying both sides of the equation by the transpose matrix, D_{NM}^T:

$$[D_{NM}^T D_{MN}]v_N = D_{NM}^T z_M$$

Observe that this transforms the overconstrained system of equations into a set that can be easily solved.

Let

$$A_{NN} = [D_{NM}^T D_{MN}] \quad \text{and} \quad w_N = D_{NM}^T z_M$$

then v_N can be determined from the equivalent system of equations:

$$A_{NN}v_N = w_N$$
$$v_N = A_{NN}^{-1}w_N$$

If A_{NN} is nonsingular, then the **pseudo-inverse matrix** A_{NN}^{-1} exists and the solution v_N is minimum mean-square. The pseudo-inverse matrix is a real-symmetric matrix.

See Function: pseudinv().

6.3 VECTORS

A vector is simply a matrix with one of its dimensions equal to one. A row vector is a $1 \times n$ matrix, and a column vector is an $n \times 1$ matrix. All of the algebra of matrices also applies to vectors. For vectors with real elements, the **dot product, cross product,** and **magnitude** have useful geometrical applications.

6.3.1 Dot Product

The **dot product** (or inner product) has an interesting geometrical interpretation. If θ is the angle between the two vectors v_1 and v_2, then:

$$v_1 * v_2 = |v_1| |v_2| \cos(\theta)$$

When the dot product is normalized by the magnitude of the larger vector, it represents the projection of the smaller vector onto the larger. If the dot product between two vectors is zero, then the vectors are perpendicular (i. e., the angle) between them is 90 degrees).

See Function: vdot().

6.3.2 Cross Product

The **cross product** (sometimes called the vector product) between two three-dimensional vectors is given by the following determinant:

$$v_1 \times v_2 = \begin{vmatrix} i & j & k \\ x1 & y1 & z1 \\ x2 & y2 & z2 \end{vmatrix}$$

where i, j and k are the unit vectors in the x, y, and z directions. The variables $x1$, $y1$, and $z1$ are the components of v_1. The variables $x2$, $y2$, and $z2$ are the components of v_2.

The cross product has an interesting geometrical interpretation. If θ is the angle between the two vectors, then:

$$v_1 \times v_2 = u|v_1|\,|v_2|\sin(\theta)$$

where u is the unit vector perpendicular to the plane of v_1 and v_2, and so directed that a right-handed screw driven in the direction of u would carry v_1 into v_2.

The cross product is not commutative, since

$$v_1 \times v_2 = -v_2 \times v_1$$

If the cross product between two vectors is zero, then the vectors are colinear (i. e., the angle between them is 0 degrees).

See Function: vcross().

6.3.3 Vector Magnitude

The **magnitude** of a real vector, v, is the square root of the sum of the squares of its elements:

$$|v| = [v_1^2 + v_2^2 + v_3^2 + \ldots + v_n^2]^{0.5}$$

The magnitude of a vector is the same as its length.

See Functions: vmag(), Cvmag().

6.4 EIGENANALYSIS

Eigensystems play a crucial part in the theory of electrical and mechanical resonance and in the theory of statics as well. Eigenvalues and eigenvectors can be used to determine the stability of feedback systems and to control the convergence of associated tracking algorithms. The concepts of similarity transforms and matrix diagonalization are of fundamental importance to eigenanalysis because the eigenvalues are invariant to such transforms. The general eigenvalue problem is one of determining the similarity of a matrix to a diagonal form. This section deals with eigenvalues of **real symmetric** and **complex Hermitian** matrices.

6.4.1 Real symmetric matrices

A nonsingular $n \times n$ real symmetric matrix has n real eigenvalues and n distinct eigenvectors. The most common eigenanalysis technique for real symmetric matrices is Jacobi's method. The **cyclic Jacobi** method consists of a sequence of similarity transforms designed to diagonalize a real symmetric matrix **A**. This technique derives an orthogonal transformation matrix **P** whose columns are the desired n eigenvectors. The n eigenvalues are the diagonal elements of the diagonalized matrix

$$C = P^{-1}AP$$

This diagonalization is sometimes referred to as a **similarity transformation**.

The Jacobi method obtains the orthogonal matrix P as an infinite product of rotation matrices P_l where the operative elements are of the form:

$$P_l = \begin{bmatrix} \cos \omega & -\sin \omega \\ \sin \omega & \cos \omega \end{bmatrix}$$

and all of the other elements are identical to the identity matrix. The cosine terms are constrained to intersect the main diagonal; the sine terms can be anywhere off the main diagonal. Let the above four entries represent the elements in positions (i, i), (i, k), (k, i), and (k, k). Then the corresponding elements of $P_l^{-1} A P_l$ are computed as:

$$b_{ij} = a_{ii} \cos^2 \omega + 2a_{ik} \sin \omega \cos \omega + a_{kk} \sin^2 \omega$$
$$b_{ki} = b_{ik} = (a_{kk} - a_{ii}) \sin \omega \cos \omega + a_{ik}(\cos^2 \omega - \sin^2 \omega)$$
$$b_{kk} = a_{ii} \sin^2 \omega - 2a_{ik} \sin \omega \cos \omega + a_{kk} \cos^2 \omega$$

The angle ω is chosen such that

$$\tan(2\omega) = 2a_{ik}/(a_{ii} - a_{kk})$$

which then makes $b_{ki} = b_{ik} = 0$. At each step, the Jacobi algorithm annihilates a pair of off-diagonal elements. Unfortunately, successive steps introduce nonzero contributions to positions that were formerly zero. Nevertheless, the iterative product $P_0 P_1 P_2 \ldots$ produces the desired orthogonal matrix P.

Similarly, the desired diagonal matrix C is obtained from

$$C = \ldots P_2^{-1} P_1^{-1} P_0^{-1} A P_0 P_1 P_2 \ldots$$

The cyclic Jacobi method is practically foolproof for all real symmetric matrices.

See Function: mxeigen().

6.4.2 Complex Hermitian matrices

A nonsingular $n \times n$ complex Hermitian matrix has n real eigenvalues and n distinct eigenvectors. A Hermitian matrix exhibits complex conjugate symmetry around its main diagonal, which also means that the elements along the main diagonal must be real. Thus, a real symmetric matrix is also Hermitian, since the imaginary parts of its elements are all zero.

The eigenanalysis of any $n \times n$ Hermitian matrix can be transformed into an equivalent problem that involves a $2n \times 2n$ real symmetric matrix. The real symmetric eigensystem can then be solved with Jacobi transformations (see W. H. Press, et al., pp. 381–382).

The method is as follows:

1. The complex matrix m is separated into its real and imaginary parts:

$$m[n][n] = A[n][n] + iB[n][n]$$

2. Likewise, the complex eigenvector, v, is separated into its real and imaginary components:

$$v[n] = a[n] + ib[n]$$

3. The following complex eigenvalue equation transforms into one real equation of twice the size according to:

$$(A + iB)\lambda(x + iy) = \lambda(a + ib)$$

$$\begin{bmatrix} A & -B \\ B & A \end{bmatrix} \begin{bmatrix} a \\ b \end{bmatrix} = \lambda \begin{bmatrix} a \\ b \end{bmatrix}$$

4. The real symmetric eigensystem is solved with Jacobi transformations. The eigenvalues of the real system are ordered as redundant pairs:

$$\text{eigenvalues} == \lambda_1, \lambda_1, \lambda_2, \lambda_2, \ldots, \lambda_n, \lambda_2$$

$$\text{eigenvectors} == \begin{bmatrix} a1 \\ b1 \end{bmatrix}, \begin{bmatrix} -b1 \\ a1 \end{bmatrix}, \begin{bmatrix} a2 \\ b2 \end{bmatrix}, \begin{bmatrix} -b2 \\ a2 \end{bmatrix}, \ldots, \begin{bmatrix} an \\ bn \end{bmatrix}, \begin{bmatrix} -bn \\ an \end{bmatrix}$$

5. The complex eigensystem solution is then found by taking every other eigenvalue and eigenvector from the real system:

$$\text{eigenvalues} == \lambda_1, \lambda_2, \ldots, \lambda_n$$

$$\text{eigenvectors} == \begin{bmatrix} a1 \\ b1 \end{bmatrix}, \begin{bmatrix} a2 \\ b2 \end{bmatrix}, \ldots, \begin{bmatrix} an \\ bn \end{bmatrix}$$

See Function: Cmxeigen().

7

Numerical Analysis

Numerical analysis deals with the computational issues of applied mathematics. The focus of this field is on "results-oriented" mathematics. That an expression can be represented as an infinite recursion or sum is only the beginning for numerical analysis. Numerical mathematics must also deal with practical issues, like the number of terms that are needed for a given accuracy, and the speed compared to alternative techniques.

It is often necessary to estimate the values of data sets in between samples. The data points may not be equally spaced and may not be represented by a convenient function. The goal of **interpolation** is to produce a smooth curve that passes through all the data points.

Finite sums and differences are the tools that the numerical mathematician needs to perform integration and differentiation. Numerical calculus is valuable because it facilitates the evaluation of even the most difficult mathematical expressions. Numerical **integration** and **differentiation** can be performed on tables of data as well as analytical equations.

This chapter deals with the following numerical techniques:

1. Interpolation
2. Integration
3. Differentiation
4. Function minimization

7.1 INTERPOLATION

Interpolation is something of an art. In practice, it is often impossible to objectively evaluate the performance of any interpolation approach, since it depends on the true (and, in general, unknown) values of the data in between the known points. Interpolation performance is often judged subjectively. The question is asked: Does the interpolating function produce a curve that is pleasing to the eye? The answer is usually **yes**, as long as the data is oversampled and not corrupted with noise.

7.1.1 Piecewise Interpolation

For any arbitrary set of $n+1$ data points, an nth order polynomial can always be found that passes through all of the points. For even medium size data sets, numerical precision can limit the polynomial order that can be chosen. Over 100 years ago, Runge noted that the problems of high order polynomial interpolation are compounded by large oscillations in between the data points.

One solution is to divide the data interval into pieces and interpolate each piece independently. For polynomial interpolation of sparsely sampled data, this tends to produce sharp cusps at the edges of each subinterval. Spline approaches eliminate these cusps by matching the slope at the edges of each segment.

7.1.2 Linear Interpolation

Linear interpolation is the most straightforward interpolation method. Data points are simply connected with straight line segments. Each interpolated value lies along an appropriate straight line segment. For any two points, the interpolation at x is given by

$$y = [y_1(x - x_2) - y_2(x - x_1)]/(x_1 - x_2)$$

where (x_1, y_1) and (x_2, y_2) are the known data points.

Linear interpolation is a special case of the Lagrange formula described in the following section.

7.1.3 Lagrange n-Point Interpolation

Lagrange interpolation is an extension of the linear interpolation formula. This is one of the most versatile interpolation approaches. At each data point, this technique computes the $(n - 1)$th order polynomial that passes through the ordinates of each group of n neighboring points. If the data are samples from an mth order polynomial, and $m < n$, the interpolation is exact everywhere. The interpolation is always exact at the input data samples (many interpolation techniques do not have this property).

The one problem with Lagrange interpolation is that the derivative of the polynomial is totally unconstrained. If the data points are far apart, this causes the interpolation function to have sharp cusps at many points. Also, for large orders, the polynomial functions have oscillations in between the data points.

See Functions: interp(), interp1().

7.1.4 Cubic Spline Interpolation

Some of the drawbacks of polynomial interpolation are overcome with spline functions. A spline is a low order interpolating polynomial with slope and curvature constraints. The cubic spline interpolates between data points with the following third order equation:

$$f_i(x) = b_{0i} + b_{1i}x + b_{2i}x^2 + b_{3i}x^3$$

where $x_i \le x \le x_{i+1}$.

The smooth appearance of the spline results from matching the slope and the curvature of the cubic formulas in adjacent intervals. Noting that the second derivative of a cubic is a line, we can develop a tridiagonal system of equations to constrain the slope and curvature of the cubic formulas across adjacent intervals of the data. All that is needed to solve this set of equations is the specification of the second derivative at the endpoints of the data. Often it is convenient to assume that the second derivative at the endpoints of the data is zero. This condition corresponds to a natural spline interpolant.

Interpolating with a cubic spline is similar to interpolating with a second order Lagrange polynomial. This is because the slope constraints of the spline are analogous to an order constraint for the Lagrange polynomial. The major difference is that, in general, the second order Lagrange polynomial will not be smooth at all the data points. However, cubic spline interpolation is a little more difficult to use than polynomial interpolation since the second derivatives must be computed over the entire interval of interest (and specified at the endpoints).

See Function: spline().

7.1.5 Coordinates of a Centroid

It frequently happens that a variable or function is given in tabular form and that the abscissa values are of more interest than the ordinates:

1. In spectral analysis, one may wish to determine the frequency of an inharmonic sinusoid.
2. In image processing, the position of a light source that straddles several camera pixels may be of interest (e. g., a pixel tracker).
3. In one-dimensional correlation procedures, estimates of fractional time lags are often required to determine the pure delay between two similar time series.
4. In two-dimensional correlation procedures, determining the amount of spatial misregistration between two similar images may be important.

For these cases, the **coordinates of the centroid** are more informative than the actual value of the centroid. Centroiding techniques are ideal for these applications. Because of the focus on the coordinate values, centroiding is sometimes referred to as **inverse interpolation**.

For sampled data, the one-dimensional equation for the coordinates of a centroid is

$$c_x = \frac{\displaystyle\sum_i i f(i)}{\displaystyle\sum_i f(i)}$$

This equation is simple and theoretically correct given infinite digital resolution and infinite signal-to-noise ratio. However, for real data with low signal-to-noise ratios, straightforward application of this formula can lead to more than half a resolution bin of error. **This amount of error defeats the entire purpose of the centroid.**

The reader can easily verify the inaccuracy of this formula by applying it to the frequency spectrum of a low amplitude sinusoid plus white noise. The sinusoid can be set to a known (inharmonic) frequency. The coordinates of the centroid will appear to fluctuate, depending on the number of bins included in the computation of the centroid. This is because the inclusion of noise bins biases the estimate of the centroid as much as half a resolution bin.

One solution is to decide which bins are "valid" (signal) and to exclude the "invalid" (noise) bins from the centroid calculation. If the signal or function of interest is known to be **unimodal**, then a simple threshold can be set. The assumption is that the data is a quantized version of a function that has a single maximum and is monotonically decreasing below this value.

Let u_f be the mean of the absolute value of the data computed over all the bins of the centroid. Assuming an equally likely distribution of the noise (i. e., white noise) throughout the bins, a simple decision strategy results:

> bin i is "valid" if $|f(i)| \geq u_f$
> bin i is "invalid" otherwise

The centroid is then computed over all the "valid" bins. For low signal-to-noise ratios, this strategy can lead to an order of magnitude improvement in the estimate of the centroid coordinates compared to the original quantization resolution.

This centroiding strategy is easily extended to two dimensions and yields comparable results.

See Functions: mxcentro(), v_centro().

7.2 INTEGRATION

Numerical integration provides quick and accurate solutions to difficult integrals or integrals without analytical solutions. Specifically, these approaches are used to estimate definite integrals of the following form:

$$y(x) = \int_{x_0}^{x_1} f(x)\, dx$$

where x_0 and x_1 are the endpoints of the interval of interest and $f(x)$ is any one-dimensional function of x.

Numerical integration techniques involve fitting simple curves to data over subintervals of the domain of interest. The simple curves are easily integrated and can be summed together to approximate the total integral.

7.2.1 Rectangular Rule

The simplest integration formula is the Rectangular Rule. This rule divides the integration interval into rectangular panels. Each panel has a height equal to the value of the function at a data point and a width equal to the distance between data points. If the distance between data points is small, then the approximation to the integral is good. For data that is equally spaced, the integral I is

$$I = dx(y_0 + y_1 + y_2 + \ldots + y_{n-1})$$

where dx is the distance between data points and y_i is the value of the function at x_i.

7.2.2 Trapezoidal Rule

The Trapezoidal Rule connects data points with a straight line. The area of each panel is the average height times the panel width. For equally spaced data, the integral I is

$$I = dx[(y_0 + y_1)/2 + (y_1 + y_2)/2 + \ldots + (y_{n-2} + y_{n-1})/2]$$

where dx is the distance between data points and y_i is the value of the function at x_i.

7.2.3 Romberg Integration

Romberg integration successively applies the Trapezoidal Rule to efficiently estimate the integral of a function. To demonstrate the approach, the first two integral approximations are

$$I_0 = 2dx[(y_0 + y_2)/2 + (y_3 + y_5)/2 + \ldots + (y_{n-3} + y_{n-1})/2]$$
$$I_1 = dx[(y_0 + y_1)/2 + (y_1 + y_2)/2 + \ldots + (y_{n-2} + y_{n-1})/2]$$

Using Richardson's improvement formula (see the Differentiation section 7.3.4), a more accurate estimate of the integral can be obtained by combining the I_0 and I_1:

$$I = I_1 + (I_1 - I_0)/3$$

If the difference between I_0 and I_1 is small, then the approximation to the integral is good. Otherwise, the interval between data points is halved and the procedure is repeated.

One disadvantage of Romberg's approach is in the integration of periodic functions over an integer number of periods. If the grid locations of the integration panels nearly align with the zero crossings of the function, the Romberg approach may terminate prematurely with an incorrect answer. These situations can easily be avoided by segmenting the integration interval into lengths that are not integer multiples of the period.

See Function: romberg().

7.2.4 Simpson's Rule

Simpson's integration technique uses parabolic arcs to approximate the data. The data interval must be divided into an even number of equal panels of width dx. The most common form is the **one-third** rule:

$$I = dx[(y_0 + 4y_1 + 2y_2 + 4y_3 + \ldots + 2y_{n-3} + 4y_{n-2} + y_{n-1})]/3$$

It is interesting to note that if the input data is parabolic, then Simpson's formula provides the exact integral.

See Function: integrat().

7.3 DIFFERENTIATION

Difference formulas are used to approximate the derivatives of tabulated data as well as difficult analytical expressions. Although a multiplicity of formulas is available, for a given application some formulas are better than others. Central difference formulas are the most accurate. The forward and backward differences are also useful in evaluating the derivatives at the edges of tables of data.

7.3.1 First Order Differences

The first derivative of the function $f(x)$ can be approximated by

$$f'(x) = [f(x + dx) - f(x)]/dx$$

if dx is small. The truncation error of this approximation is proportional to the size of the increment dx (i. e., a first order approximation, $O(dx)$). If this equation is used as an estimate of the derivative at x, then it is called a **backward difference**. However, if it is used as an estimate of the derivative at $x + dx$, then it is called a **forward difference**. Forward differences estimate the derivative at a point with data that precedes the point. Similarly, backward differences approximate the derivative with data that follow the point.

A more accurate approximation to the derivative is given by the following equation:

$$f'(x) = [f(x + dx) - f(x - dx)]/(2dx)$$

where again it is assumed that dx is small. The truncation error of this approximation is proportional to dx^2 (i. e., a second order approximation, $O(dx^2)$). This two-sided equation is the basis of the **central difference** formulas. Central differences use an equal number of data points before and after the point of interest.

7.3.2 Higher Order Differences

Difference formulas allow the approximation of the nth derivative of a function. For tabulated data, it is assumed that the function passes through each discrete point. **In computing the derivative of tables of data, it is necessary that the data points be equally spaced.** The difference approximations consist of weighted local averages of the data.

The difference coefficients are related to Pascal's triangle and the binomial coefficients, as shown in the following table. Note that the alternating sign of these coefficients also depends on the order of the derivative. In the forward and backward difference formulas, the ith coefficient for the nth derivative is given by

$$C_{ni} = (-1)^{n+i} n!/[i!(n - i)!]$$

where $i = 0, 1, \ldots, n$.

The coefficients of the difference formulas are the same for the forward and backward equations. The only real distinction between these formulas is in their use, not in the coefficients themselves. In fact, for even order derivatives, the central difference coefficients are the same as the forward and backward coefficients. For odd order derivatives, the central difference coefficients are not the same as the forward and backward coefficients. For these cases, the middle coefficient is always zero.

nth Derivative n	Forward and Backward Difference Coefficients	Central Difference Coefficients
1	$-1 \quad 1$	$-1 \quad 0 \quad 1$
2	$1 \quad -2 \quad 1$	$1 \quad -2 \quad 1$
3	$-1 \quad 3 \quad -3 \quad 1$	$-1 \quad 2 \quad 0 \quad -2 \quad 1$
4	$1 \quad -4 \quad 6 \quad -4 \quad 1$	$1 \quad -4 \quad 6 \quad -4 \quad 1$
5	$-1 \quad 5 \quad -10 \quad 10 \quad -5 \quad 1$	$-1 \quad 4 \quad -6 \quad 0 \quad 6 \quad -4 \quad 1$

7.3.3 Examples of Difference Calculations

Before the difference (binomial) coefficients can be used to compute derivatives, they must be normalized by the sum of the absolute values of all the coefficients:

$$C_n = |C_{n0}| + |C_{n1}| + |C_{n2}| + \ldots + |C_{nn}|$$

The coefficients are applied to the appropriate data points as a weighted local average about the point of interest.

For example, the second derivative at the point x_i is approximated by

$$y''(x_1) = [C_{20}y(x_0) + C_{21}y(x_1) + C_{22}y(x_2)]/(C_2 dx)$$
$$= [y(x_0) - 2y(x_1) + y(x_2)]/(4\ dx)$$

When the coefficients are applied in this way to successive data points, the process is called a **convolution**. Convolution is discussed in more detail in Chapter 8 on Digital Signal Processing.

Consider the following table of x, y data:

x	1.1	1.2	1.3	1.4	1.5	1.6	1.7
y	-2	4	-1	6	8	4	3

We will use this table of data for all the following examples to evaluate derivatives at specific points of interest.

Example 1:

Approximate the first derivative at $x = 1.4$.
We will use the central first difference formula normalized by $C_1 = 2$:

$$y'(1.4) = [C_{10}y(1.3) + C_{11}y(1.5)]/(C_1 dx)$$
$$= [-1 * y(1.3) + 1 * y(1.5)]/(2 * 0.1)$$
$$= (1 + 8)/0.2 = 45$$

Example 2:

Approximate the first derivative at $x = 1.1$.

Since $x = 1.1$ is on the left edge of the table, we must use the backward first difference formula. The normalization factor is $C_1 = 2$.

$$y'(1.1) = [C_{10}y(1.1) + C_{11}y(1.2)]/(C_1 dx)$$
$$= [-1 * y(1.1) + 1 * y(1.2)]/(2 * 0.1)$$
$$= (2 + 4)/0.2 = 30$$

Example 3:

Approximate the first derivative at $x = 1.7$.

Since $x = 1.7$ is on the right edge of the table, we must use the forward first difference formula. The normalization factor is $C_1 = 2$.

$$y'(1.7) = [C_{10}y(1.6) + C_{11}y(1.7)]/(C_1 dx)$$
$$= [-1 * y(1.6) + 1 * y(1.7)]/(2 * 0.1)$$
$$= (-4 + 3)/0.2 = -5$$

7.3.4 Improving the Difference Estimates

For the lower order differences, there are several other formulas that can provide slightly better estimates of the derivatives. These formulas operate across more data points than those previously described. For m-points, the formulas represent the derivatives of an $m - 1$ order polynomial that passes through the points. If the polynomial matches the data well, then so does the derivative.

For example, a five-point central difference formula for estimating the first derivative is shown in the following table:

offset	-2	-1	0	1	2
coefficient	1/12	$-2/3$	0	2/3	$-1/12$

For data that is sufficiently oversampled, this five-point formula usually provides a better estimate of the first derivative than the three-point formula described previously.

Finally, it should be mentioned that all the central difference formulas can be modified to interpolate between samples as they differentiate. This is useful if the derivative is needed in between data points. The derivative estimates are best at the central data points.

Richardson's Improvement Theorem. Richardson's improvement theorem is of fundamental importance to numerical approximations with truncated series. This approach can be used to improve estimates of the derivatives of any differentiable function. Consider two estimates of the derivative, each made with a different step size. Essentially, Richardson's formula computes the relative weighting of each of these estimates needed to form a more accurate approximation of the derivative.

This theorem is as follows:

Suppose that an approximation is of a known truncation order, $O(h^n)$. Suppose also that this approximation is evaluated for two arbitrary step sizes, h and rh, where $r > 1$. If $F(h)$ is the evaluation for step size h, and $F(rh)$ is the evaluation for step size rh, then an improved estimate $F_1(h)$ may be formed according to:

$$F_1(h) = [r^n F(h) - F(rh)]/(r^n - 1)]$$

This new estimate will have a smaller truncation error than either $F(h)$ or $F(rh)$. That is, the truncation order of $F_1(h)$ will be larger than n.

This improvement is sometimes called **Romberg extrapolation**, since it is also used in connection with the trapezoidal rule of integration. Since the trapezoidal rule is an $O(h^2)$ approximation, $n = 2$. The ratio of the step sizes r is 2, since Romberg's integration approach changes its step size by factors of two. Recall that the weighting on the integral approximation with the smaller step size was $4/3 = r^n/(r^n - 1)$, whereas the weighting on the integral approximation with the larger step size was $-1/3 = -1/(r^n - 1)$.

See Functions: deriv(), deriv1().

7.4 FUNCTION MINIMIZATION

Even the simplest nonlinear equation can have no analytical solution. This section describes the numerical minimization of differentiable functions. There are several approaches, and the performance of each is highly dependent on the function that is minimized.

In curve fitting, the function to be minimized is often nonnegative, such as mean-squared error. For these applications, root-finding is equivalent to function minimization. The **Newton–Raphson** one-dimensional root-finding approach is described next.

For multidimensional smooth functions, gradient minimization techniques are very robust. The method of **conjugate gradients** is discussed in this section, since it is one of the more effective approaches.

7.4.1 Finding Roots of One-Dimensional Functions

One of the most common and difficult problems in mathematical analysis is the solution of nonlinear equations. For example, consider the following equation:

$$\sin(x) = 3 * x + 0.1$$

Although this is a very simple equation, there is no analytical method for solving for x. All attempts at a straightforward solution are of no avail. Numerical root-finding algorithms are ideal for problems such as these.

As indicated by the term "root-finding," much of the early work in numerical minimization dealt with polynomials. However, most of these techniques work just as well with any differentiable function.

Newton–Raphson Method.   The Newton–Raphson technique is a simple iterative approach for finding the value of x where $f(x) = 0$. The formula results from the first two terms of the Taylor series expansion at x_i:

$$f(x) = f(x_i) + f'(x_i) * (x - x_i) + \dots$$

Setting $f(x) = 0$ and solving for $x = x_{i+1}$ yields

$$x_{i+1} = x_i - f(x_i)/f'(x_i)$$

The value of the function is determined at x_{i+1}, and if it is not close enough to zero, the iteration is repeated.

One very nice feature of the Newton–Raphson technique is that the number of required iterations is usually insensitive to the initial guess. This is because the step size is weighted by the derivative of the function. The worse the initial guess, the larger the initial weighting. The number of iterations depends more on the behavior of $f'(x)$ as $f(x)$ approaches zero. For example, finding the zeros of $\sin(x)$ requires very few iterations, since the slope of the sine function is maximum at the zero-crossings. However, finding the zeros of $\cos(x) - 1$ requires comparatively more iterations since its derivative (and the weighted step size) vanishes at the zero-crossings. In either case, the number of required iterations is fairly independent of the initial guess.

In this form, the Newton–Raphson algorithm explicitly requires the derivative $f'(x)$. It is often more convenient to approximate the derivative with the two-point difference formula:

$$f'(x_i) = [f(x_i) - f(x_i + \delta)]/\delta$$

where a good choice for δ is

$$\delta = x_i * e^{-5}.$$

Incorporating this approximation, the new algorithm becomes

$$x_{i+1} = x_i - \delta * [-1 + f(x_i + \delta)/f(x_i)]^{-1}$$

This approximation alleviates the need to specify the derivative of the function, yet it maintains the convergence speed of the original Newton–Raphson approach.

Example:

Find the square root of 36. That is, find the zeros of

$$f(x) = 36 - x^2$$

For this case, the Newton-Raphson algorithm simplifies to

$$x_{i+1} = x_i + (36 - x_i^2)/(2x_i + \delta)$$

Choosing $\delta = 1.e - 5$ and a first guess of $x = 15$, the results of each iteration are as follows:

$$x_0 = 15$$
$$x_1 = 8.700003$$
$$x_2 = 6.4189677$$
$$x_3 = 6.0136734$$
$$x_4 = 6.0000157$$
$$x_5 = 6.0000000$$

Thus, for this example, the Newton–Raphson algorithm converges to within eight significant digits of the correct answer in five steps.

See Function: newton().

7.4.2 Finding Roots of Polynomials

The roots of polynomials have been of great interest to mathematicians over the centuries. The nth order polynomial of the form

$$f(x) = \text{coef}[0] + x * \text{coef}[1] + x^2 * \text{coef}[2] + \ldots + x^{n-1} * \text{coef}[n-1]$$

has, in general, n complex roots. Yet closed form solutions exist only for polynomials less than fifth order, and the solutions for third and fourth order polynomials are very cumbersome.

The Newton–Raphson method is a useful tool for the iterative evaluation of roots of differentiable functions (e. g., polynomials). Polynomials with real coefficients are analytic functions and are differentiable everywhere in the complex plane. When used in conjunction with the Cauchy–Riemann equations, the Newton–Raphson technique can be extended to evaluate complex roots of high order polynomials.

See Function: p_roots().

7.4.3 Minimization of Multidimensional Functions

This section describes some numerical approaches to solving nonlinear equations that may have no analytical solution. The simplest multidimensional minimization techniques are the unconstrained gradient-search approaches. The negated gradient at a point on the surface points in the direction of greatest decrease. If the gradient exists, and if unimodality can be assumed, the minimum of the function can be found by successively following the surface in the direction of its steepest descent. For multimodal surfaces, multiple starting points are often required.

Gradient approaches work best when the scales of the independent variables are chosen so that the components of the typical gradient vector are of comparable magnitude. These search procedures can be very inefficient if a few components of the gradient are consistently much smaller than the rest.

Method of Steepest Descent.   The method of steepest descent is most easily described in three dimensions. We wish to find the solution of the equation $f(x, y, z) = 0$. Let $f(x_0, y_0, z_0)$ be a first guess of the minimum. If this is not the minimum, the guess can be improved by considering the gradient of the function (f_x, f_y, f_z) at the guess:

$$x_1 = x_0 - h * f_x(x_0, y_0, z_0)$$
$$y_1 = y_0 - h * f_y(x_0, y_0, z_0)$$
$$z_1 = z_0 - h * f_z(x_0, y_0, z_0)$$

where h is a small value normalized by the magitude of the gradient vector. The function is evaluated at (x_1, y_1, z_1) to verify that this estimate is closer to zero and the process is iterated until the desired accuracy is reached. If the new estimate is not closer to zero, the step size h is halved until a smaller error is found. If the surface is smooth and unimodal (e. g., quadratic), this procedure is guaranteed to find the minimum.

The Conjugate Gradient Approach.   Although the method of steepest descent usually converges to the desired minimum, it is not the fastest gradient approach. This technique tends to zigzag down even the smoothest surface, since each new direction of descent tends to be orthogonal (but not conjugate) to the last direction. Thus, this algorithm tends to either overshoot or undershoot the surface fall line.

The **conjugate gradient** approach offers a solution to this problem. The n-dimensional approach searches along "conjugate directions," which are weighted averages of the current gradient with the gradients of the last n steps. This averaging tends to smooth the search path, resulting in a much more rapid convergence than the method of steepest descent. For a more detailed description of the conjugate gradient approach, the reader should refer to the original work of Fletcher and Reeves (1964).

See Function: conjgrad().

8

Digital Signal Processing

Signal processing deals with the enhancement, extraction, and representation of information for communication or analysis. Many different fields of engineering rely on signal processing technology. Acoustics, telephony, radio, television, seismology, and radar are some examples.

Initially, signal processing systems were implemented exclusively with analog hardware. However, recent advances in high-speed digital technology have made discrete signal processing systems more popular. Digital systems have an advantage over analog systems in that they can process signals with an extraordinary degree of precision. Unlike the resistive and capacitive networks of analog systems, digital systems can be built numerically with the simple operations of addition and multiplication.

Digital signal processing is a field of numerical mathematics that is concerned with the processing of discrete signals. This area of mathematics deals with the principles that underly all digital systems.

8.1 BACKGROUND

The following sections present some concepts that are important to the understanding of digital signal processing (**DSP**). Although these sections present many DSP fundamentals, they are in no way intended to be a complete description of the field. It is recommended that the interested reader consult the excellent references cited at the end of this book.

8.1.1 The Sampling Theorem

Shannon's **sampling theorem** is of fundamental importance to digital signal processing. This theorem is as follows:

> Suppose that $x(t)$ is a continuous time signal that is bandlimited with a maximum frequency of f_M. Then $x(t)$ can be uniquely represented by a sequence of time samples:
>
> $$x(mT) = \ldots, x(-2T), x(-T), x(0), x(T), x(2T), \ldots$$
>
> where T is the **sampling interval**, and $1/T = 2f_M$ is called the **Nyquist rate**. The maximum frequency of the signal f_M is called the **Nyquist frequency**.

In essence, the sampling theorem establishes a minimum rate at which an analog signal must be sampled. This sampling frequency is $F_s = 1/T$. Sampling at any rate faster than the Nyquist rate (i. e., making T shorter than $2f_M$) also produces a valid representation of $x(t)$.

In digital sampling, frequencies are often specified normalized to F_s. For example, $F_s/2$, the Nyquist frequency, is the highest frequency that can be represented and has a normalized value of 1/2.

The above theorem is very important, since in practice all signals can be considered to be bandlimited over the frequency range of interest. For example, musical sounds with frequency components higher than 20 KHz cannot be heard by most humans. For listening purposes, the sound only needs to be sampled at the Nyquist rate $2f_M = 40$ KHz.

8.1.2 The Discrete Fourier Transform

The discrete Fourier transform (**DFT**) plays an important role in digital signal analysis. The DFT serves as a gateway between the time representation and the frequency representation of a signal. The **forward DFT** transforms a time signal into the frequency domain and is defined as:

$$X(k) = [x(0) + x(1)e^{-j\omega k} + \ldots x(m)e^{-jm\omega k} \ldots + x(n-1)e^{-j(n-1)\omega k}]$$

for $k = 0, 1, 2, \ldots, n-1$
where $\omega = 2 * \pi/n$ and $\pi = 3.141592654\ldots$.

The **inverse DFT** transforms the frequency representation of a signal back into the time domain and is defined as:

$$x(m) = [X(0) + X(1)e^{j\omega m} + \ldots X(k)e^{jk\omega m} \ldots + X(n-1)e^{j(n-1)\omega m}]/n$$

for $m = 0, 1, 2, \ldots, n-1$
where again $\omega = 2 * \pi/n$ and $\pi = 3.141592654\ldots$.

The time sequence $\{x(0), x(1), \ldots, x(n-1)\}$ and the frequency sequence $\{X(0), X(1), \ldots, X(n-1)\}$ are Fourier transform pairs. It should be noted that both these sequences are equally spaced. The interval between the samples of the time sequence is T. The interval between the samples of the frequency sequence is $1/(nT)$.

Example 1

Find the frequency resolution of a 128-point FFT for a sample rate of 100 Hz.
The interval between the samples of the time sequence is:

$$T = 1/100 = 0.01 \text{ second}$$

This yields a frequency resolution of

$$1/(128 * T) = 0.78125 Hz$$

Throughout this development, it is assumed that the Nyquist condition of the sampling theorem has been satisfied. The factor of T, found in some DFT definitions, is redundant and is eliminated here, since it in no way affects the above transforms.

Example 2

The frequency composition of time domain signals can be analyzed with the DFT. The DFT of the complex sinusoid $A * [\cos(\omega*k*i) + j*\sin(\omega*k*i)]$ consists of one frequency component at the kth harmonic. The DFT of the complex sinusoid

$$A * [\cos(\omega * k * i) - j * \sin(\omega * k * i)]$$

is also one frequency component at the $(n-k)$th harmonic $k = 1, 2, \ldots, n/2$ and $i = 0, 1, 2, \ldots, n-1$. Both components are of magnitude $A * n$.

Two-Dimensional DFT. The **two-dimensional DFT** is a simple extension of the one-dimensional formulas. The **forward 2-D discrete Fourier transform** $X(k, l)$ for a square array of $N \times N$ elements is

$$X(k, l) = \sum_{m=0}^{N-1} \sum_{n=0}^{N-1} x(m, n) e^{-j2\pi(km+ln)/N}; \quad k, l = 0, 1, \ldots, N-1$$

The **inverse 2-D discrete Fourier transform** $x(m, n)$ is given by

$$x(m, n) = \frac{1}{N^2} \sum_{k=0}^{N-1} \sum_{l=0}^{N-1} X(k, l) e^{j2\pi(km+ln)/N}$$

The two-dimensional DFT can be efficiently computed by decomposing the 2-D transform into separable 1-D row and column FFTs.

See Functions: fft2d(), fft2d_r().

8.1.3 The Discrete Cosine Transform

The discrete cosine transform (**DCT**) has become a useful tool in image compression. The 8×8 2-D DCT has been adopted by the JPEG committee as part of its standard for compressing still images. The DCT is closely related to the discrete Fourier transform.

One-Dimensional Discrete Cosine Transform.   The Fourier transform of any real symmetric function can be represented by a cosine series. Fewer edge effects result if the cosine series implements a half-sample shift of the following form:

$$X(k) = \sqrt{\frac{2}{N}}\; a(k) \sum_{n=0}^{N-1} x(n) \cos \frac{(2n+1)k\pi}{2N}; \quad k = 0, 1, \ldots, N-1$$

$$a(0) = 1/\sqrt{2} \quad \text{and} \quad a(k) = 1; \quad k \neq 0$$

$$x(n) = \sqrt{\frac{2}{N}} \sum_{k=0}^{N-1} X(k)a(k) \cos \frac{(2n+1)k\pi}{2N}; \quad n = 0, 1, \ldots, N-1$$

Note that, except for a constant scaling factor, the forward transform $X(k)$ is identical to the inverse transform $x(n)$.

Example

If the real sinusoid $A * \cos(\omega * k * (2 * i + 1))$ is used to generate the x vector, the transformed vector will consist of one frequency component at the kth harmonic, where $k = 0, 1, 2, \ldots, n-1$, $i = 0, 1, 2, \ldots, n-1$, $\omega = \pi/(2*n)$, and $\pi = 3.141592654\ldots$. If k is greater than zero, the magnitude of the component is $A(n/2)^{1/2}$.

See Function: dct().

Two-Dimensional Discrete Cosine Transform.   Due to separability, the 2-D **DCT** can be calculated with consecutive 1-D row and column cosine transforms. However, a fast block matrix approach is more efficient [see P. M. Embree, et al. (1991)].

The 2-D DCT is of the form

$$X(k, l) = \frac{2}{N}\; a(k)a(l) \sum_{m=0}^{N-1} \sum_{n=0}^{N-1} x(m, n) \cos \frac{\pi k(2m+1)}{2N} \cos \frac{\pi l(2n+1)}{2N}$$

$$a(0) = 1/\sqrt{2}; \quad a(j) = 1; \quad j \neq 0$$

$$x(m, n) = \frac{2}{N} \sum_{k=0}^{N-1} \sum_{l=0}^{N-1} a(k)a(l)X(k, l) \cos \frac{\pi k(2m+1)}{2N} \cos \frac{\pi l(2n+1)}{2N}$$

Note that, as in the one-dimensional case, the forward transform $X(k, l)$ is very similar to the inverse transform $x(m, n)$.

Example

If the real 2-D sinusoid $A * \cos(\omega * k * (2 * i + 1)) * \cos(\omega * l * (2 * j + 1))$ is used to generate the matrix m, the transformed matrix will consist of one frequency component at the kth row and the lth column, where $k = 0, 1, 2, \ldots, n-1$, $l = 0, 1, 2, \ldots, n-1$, $i = 0, 1, 2, \ldots, n-1$, $j = 0, 1, 2, \ldots, n-1$, $\omega = \pi/(2*n)$, and $\pi = 3.141592654\ldots$. If k and l are greater than zero, the magnitude of the component is $A * n/2$.

See Function: dct2d().

8.1.4 Convolution

The **convolution theorem** is very important to digital signal processing. This theorem is as follows:

Suppose that $x(n)$ and $h(n)$ are time signals with discrete Fourier transforms $X(k)$ and $H(k)$. The **convolution** of $x(n)$ and $h(n)$ is defined as the sequence $y(n)$, where:

$$y(n) = \sum_{k=-\infty}^{\infty} x(k)\,h(n-k)$$

Then the discrete Fourier transform $y(n)$ is given by

$$Y(k) = X(k)\,H(k)$$

It is assumed that $X(k)$ and $H(k)$ are Fourier transforms of time domain sequences that have been properly augmented with zeros so as to implement linear, rather than circular, convolution.

Thus convolution in the time domain corresponds to multiplication in the frequency domain. This theorem can be used to prove Shannon's sampling theorem. Convolution is also very important to digital filter theory.

Note also that multiplication in the time domain corresponds to convolution in the frequency domain. This property can be exploited both in spectral estimation and in designing FIR filters with window methods.

Example

Find the convolution of two identical square pulses $x(n)$ and $h(n)$ of unit height, each with a duration of three samples. The answer is the triangular pulse $y(n)$ with a duration of five samples, as shown in the following table.

n	-3	-2	-1	0	1	2	3
$x(n)$	0	0	1	1	1	0	0
$h(n)$	0	0	1	1	1	0	0
$y(n)$	0	1	2	3	2	1	0

Convolution can be efficiently implemented with the FFT.

See Functions: convolve(), convofft().

Two-Dimensional Convolution.   The one-dimensional definition of convolution is easily extended to two dimensions. The **2-D convolution** of $x(k_1, k_2)$ with $h(k_1, k_2)$ is $y(n_1, n_2)$, where

$$y(n_1, n_2) = \sum_{k_1-0}^{N_1-1} \sum_{k_2=0}^{N_2-1} x(k_1, k_2)h(n_1 - k_1, n_2 - k_2)$$

Two-dimensional convolution can be efficiently implemented with the 2-D FFT.

See Function: conv2dft().

8.1.5 Windowing

Window functions are frequently used in digital signal processing. The most common applications are in spectral analysis, antenna design, and digital filtering. Data windows are usually applied with simple multiplication in the time domain.

The time-domain weights of a 16-point Hanning window are shown in Figure 8.1. Note the tapered nature of this curve. Most data windows have this basic bell shape. Windows can also be implemented in the frequency domain with convolution (see the Convolution Theorem). The frequency response of a 16-point Hanning window is shown in Figure 8.2.

HANNING WINDOW IN TIME DOMAIN

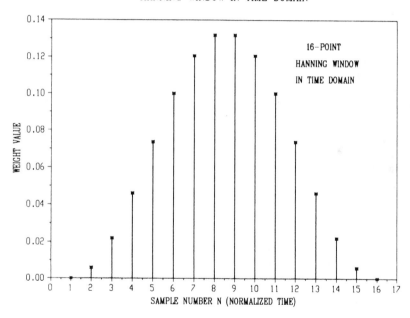

Figure 8.1   16-Point Hanning Window in Time Domain

Note that the first, second, and third sidelobes of the frequency response are approximately −30, −40, and −50 dB, respectively. This is the characteristic shape of all Hanning windows. The definition of a **dB** is given in the Digital Filtering section 8.2 that follows. Several other commonly used windows are

1. Hamming window
2. Kaiser-Bessel window
3. Blackman window
4. Approximate Blackman window
5. Blackman-Harris window

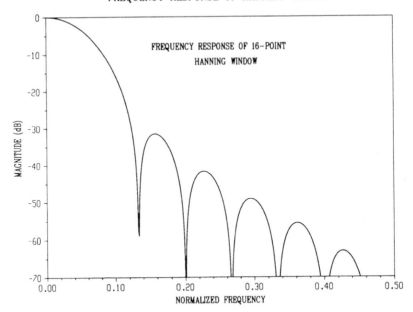

Figure 8.2   Frequency Response of a Hanning Window

It is beyond the scope of this section to describe the subtle differences between these various windows. The effects of these various windows are qualitatively the same. For more information about the advantages of data windowing, the interested reader is advised to consult the excellent reference by F. J. Harris (1978).

Most window functions have a fixed windowing effect. However, the Kaiser–Bessel window provides a selectable amount of windowing by varying the value of b in the equation below. This makes the Kaiser–Bessel window useful for a variety of digital filter design applications.

A Kaiser–Bessel window of length $M + 1$ is of the form

$$w[n] = I_0[b(1 - [(n - a)/a]^2)^{1/2}]/I_0(b), \qquad 0 \le n \le M$$
$$= 0.0, \qquad\qquad\qquad\qquad\qquad \text{otherwise}$$

where I_0 is the zeroth-order modified Bessel function of the first kind and $a = M/2$ and b is the shaping parameter.

See Function: tdwindow().

8.2 DIGITAL FILTERING

The purpose of most filtering applications is either to enhance desirable signal frequencies, or to reject undesirable frequencies (e. g., noise). Digital filters are useful in many engineering and data processing applications. Not only are their frequency responses often sharper than analog filters, but their phase responses can be made to be almost exactly linear. Many recent developments in high fidelity audio technology are due to digital signal processing techniques. For example, the improved dynamic range of compact disc players is mostly due to digital encoding and filtering methods.

This section deals with the following two kinds of digital filters:

1. Finite impulse response (**FIR**) filters
2. Infinite impulse response (**IIR**) filters

8.2.1 Finite Impulse Response (FIR) Filters

The basic form of all FIR filters is shown in Figure 8.3. Each stage of z^{-1} corresponds to one sample of delay. The triangular symbols signify multiplication by the appropriate filter weight $h(i)$. The sequence of filter weights $h(n)$ is often referred to as the **impulse response** of the filter. The frequency content of a signal is altered by convolving the signal with the impulse response of the filter.

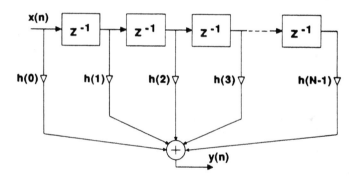

Figure 8.3   Finite Impulse Response Filter

The most commonly used filters are **lowpass**, **highpass**, and **bandpass** designs. Idealizations of these filter types are shown in Figure 8.4. Lowpass filters reject high frequencies. Highpass filters reject low frequencies. Bandpass filters reject high and low frequencies.

The extremely sharp rolloffs of ideal filters can only be realized with an infinite number of time delays. The impulse response of an ideal lowpass filter with cutoff frequency F_c is given by

$$h_L(n) = \frac{\sin(2\pi F_c n)}{\pi n} \qquad -\infty < n < \infty$$

The impulse response of an ideal highpass filter with cutoff frequency F_c is given by

$$h_H(n) = 1 - \frac{\sin(2\pi F_c n)}{\pi n} \qquad -\infty < n < \infty$$

The impulse response of an ideal bandpass filter with upper cutoff frequency F_c and lower cutoff frequency F_1 is given by

IDEAL FILTER TYPES

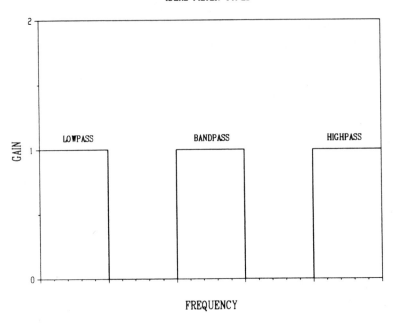

Figure 8.4   Ideal Filter Types

$$h_B(n) = \frac{\sin(2\pi F_c n) - \sin(2\pi F_1 n)}{\pi n} \qquad -\infty < n < \infty$$

Unfortunately, the finite nature of most digital data dictates that the digital filter must also be finite. In general, the duration of the filter's impulse response should be less than one-tenth of the duration of the data that is to be filtered. Truncating the number of filter weights produces undesirable oscillations at the corner frequencies of the filter's response. These oscillations are often called the **Gibbs phenomenon**. Window design approaches reduce the Gibbs effect by placing a tapered window across the sequence of truncated filter weights.

It is beyond the scope of these sections to provide all the mathematical development necessary to understand digital filter theory. For more information about digital filtering, the interested reader is advised to consult the excellent references on the subject. Instead, the Kaiser window method of filter design is described. The following sections deal exclusively with Kaiser window design approaches.

Kaiser Window Filter Design.   The basic approach of window design methods is to truncate the infinite length impulse response of an ideal frequency filter by multiplying by a time-domain window (in this case, a Kaiser–Bessel window). Time-domain windows are bell-shaped, which tapers the filter impulse response and reduces the Gibbs effect. The Kaiser–Bessel window is useful because of the flexibility it lends to filter designs. Recall that a Kaiser–Bessel window of length $N = M + 1$ is of the form

$$\omega[n] = I_0[b(1 - [(n-a)/a]^2)^{1/2}]/I_0(b), \qquad 0 \le n \le M$$
$$= 0.0, \qquad\qquad\qquad\qquad\qquad\qquad \text{otherwise}$$

where I_0 is the zeroth-order modified Bessel function of the first kind and $a = M/2$ and b is the shaping parameter.

In the frequency domain, the passband of a filter is separated from its stopband by the transition band. The gain in the passband is reasonably close to one. Frequencies in the stopband are attenuated by some specified amount d_s (i. e., the gain in the stopband is less than $d_s < 1$). Frequently the gain in the stopband is specified in *decibels* (dB):

$$D_s \text{ dB} = 20 \log_{10}(d_s)$$

The stopband attenuation of Kaiser window filters is controlled by varying the shaping parameter, b. Increasing b increases the stopband attenuation. Kaiser determined that the stopband attenuation, $D_s = -A$ was empirically related to b:

$$\begin{aligned} b &= 0.1102(A - 8.7), & &\text{if } A > 50 \\ b &= 0.5842(A - 21)^{0.4} + 0.07886(A - 21) & &\text{if } 21 \le A \le 50 \\ b &= 0.0, & &\text{if } A < 21 \end{aligned}$$

Values of b for some typical stopband attenuations, A, are given in the following table.

b	2.120	3.384	4.538	5.658	6.764	7.865	8.960	10.056
A	30	40	50	60	70	80	90	100

Let F_t be the width of the transition band of the filter, normalized by the sampling rate (i. e., $0 \le F_t \le 0.5$) For lowpass and highpass filters, the length of the filter, N, needed to achieve the specified values of A and F_t is given by

$$N = (A - 8)/(14.357 F_t)$$

For similar design specifications, symmetric bandpass filters are about twice as long as lowpass and highpass filters.

Lowpass Filtering.   The simplest digital lowpass filter is an **n-point averager**. Every output sample of this filter is a local average of n neighboring input samples. For a comparison, the impulse response of a 4-point averager is shown in Figure 8.5 along with the impulse response of a 10-point Kaiser–Bessel lowpass filter.

Digital lowpass filters are characterized by a passband, $(0, F_1)$, over which the frequency response is reasonably close to unity gain. They also often have a stopband, $(F_2, F_s/2)$ over which frequencies are attenuated by some specified amount. The maximum attenuation is at the highest frequency, $F_s/2$.

The frequency parameter, fc, sets the nominal passband (-6 dB) of the Kaiser lowpass filter, since it is the average of F_1 and F_2:

$$fc = (F_1 + F_2)/2$$

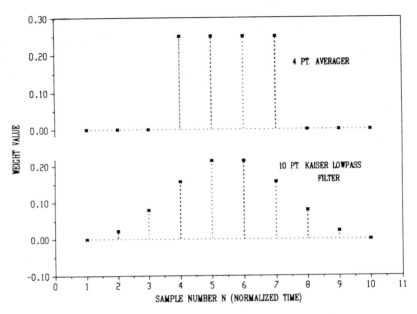

Figure 8.5   Impulse Response of Lowpass Filters

FIR LOWPASS FILTER RESPONSE

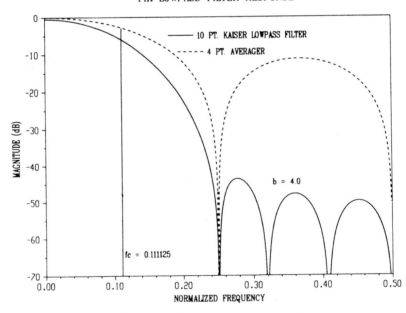

Figure 8.6   Frequency Response of Lowpass Filters

The transition band, F_t, is the difference between the passband and the stopband frequencies:

$$F_t = F_2 - F_1$$

As described previously, these parameters can be used to design a lowpass filter with any desired frequency response. The 3 dB cutoff of a filter, f_0, is the frequency where the response is at half power (0.707107). The (3 dB) cutoff of an "n-point" averager is

$$f_0 = 0.4445/n$$

The frequency responses of these two lowpass filters are shown in Figure 8.6. The stopband parameter, $b = 4$, sets the stopband attenuation of the Kaiser lowpass filter to about -45 dB. The cutoff frequency, $f_c = 0.111125$, of the Kaiser filter was set equal to the 3 dB cutoff of the 4-point averager, $f_0 = 0.4445/4$. The frequency response of the Kaiser lowpass filter at fc is attenuated by approximately 6 dB. Note that the 10-point Kaiser filter has a much better lowpass response than the 4-point averager.

See Function: lowpass().

Highpass Filtering.   The "2-point" difference filter is the simplest highpass filter. This filter is an antisymmetric (odd) function. It is implemented by replacing each data point with the difference of itself and its nearest neighbor. For a comparison, the impulse response of a 2-point difference filter is shown in Figure 8.7 along with the impulse response of a 25-point Kaiser–Bessel highpass filter. The upper plot is the response of a "2-point" difference filter, and the lower plot is the response of a Kaiser highpass filter.

Digital highpass filters are characterized by a passband, $(F_2, F_s/2)$, over which the filter's response is reasonably close to unity gain. They also often have a stopband, $(0, F_1)$, over which frequencies are attenuated by some specified amount (i. e., the gain in the stopband is less than $d_s < 1$). The maximum attenuation is at zero frequency (DC).

The frequency parameter, fc, sets the nominal passband (-6 dB) of the Kaiser lowpass filter, since it is the average of F_1 and F_2:

$$fc = (F_1 + F_2)/2$$

The transition band, F_t, is the difference between the passband and the stopband frequencies:

$$F_t = F_2 - F_1$$

As described previously, these parameters can be used to design a highpass filter with any desired frequency response.

It is instructive to compare the frequency response of the "2-point" difference filter with the Kaiser highpass filter. The frequency responses are plotted in Figure 8.8. The 3 dB cutoff of a "2-point" difference filter is:

$$f_0 = \sin^{-1}(0.707107)/3.14159265 = 0.25$$

The 6 dB cutoff frequency of the Kaiser filter is $fc = 0.10$. The stopband parameter, $b = 2.16$, sets the stopband attenuation of the Kaiser lowpass filter to about -32 dB. Note that the 25-point Kaiser filter has a much better highpass response than the 2-point difference filter.

See Function: highpass().

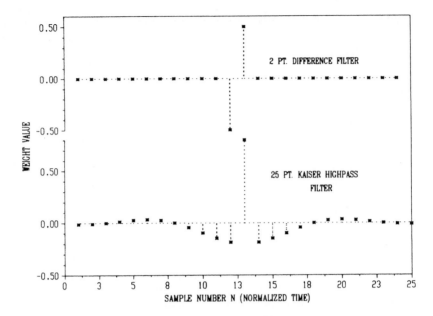

Figure 8.7   Impulse Response of Highpass Filters

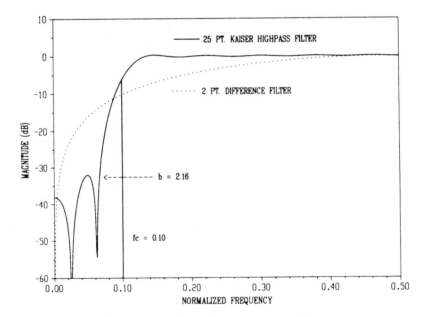

Figure 8.8   Frequency Response of Highpass Filters

Bandpass Filtering.   Digital bandpass filters are characterized by one pass-band, (F_1, F_2), over which the filter's response is reasonably close to unity gain. The impulse response of a 31-point Kaiser bandpass filter is shown in Figure 8.9. Bandpass

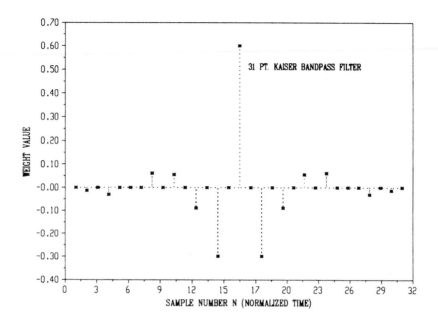

Figure 8.9   Impulse Response of a Bandpass Filter

filters have two stopbands, $(0, F_{s1})$, and $(F_{s2}, F_s/2)$, over which frequencies are attenuated by some specified amount (i. e., the gain in the stopband is less than $d_s < 1$).

The cutoff frequencies, fl and fh, set the nominal passband (and stopband) of the Kaiser window filter, since they are averages of the "corner" frequencies:

$$fl = (F_{s1} + F_1)/2$$
$$fh = (F_2 + F_{s2})/2$$

The response of a Kaiser bandpass filter at the cutoff frequencies, fl and fh, is attenuated by approximately 6 dB.

The transition band, F_t, is the difference between the passband and the stopband frequencies:

$$F_t = F_{s2} - F_2 = F_1 - F_{s1}$$

Frequently, the gain in the stopband is specified in *decibels* (dB):

$$D_s dB = 20 \log_{10}(d_s)$$

As described previously, these parameters can be used to design filters with any desired frequency response.

The length of the bandpass filter, N, needed to achieve the specified values of A and F_t is given by

$$N = (A - 8)/(7.1785F_t)$$

Note that bandpass filters are about twice as long as highpass and lowpass designs for comparable transition bandwidths and stopband attenuations.

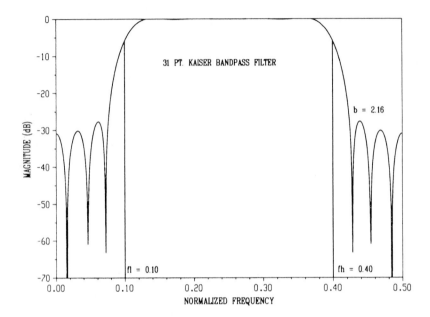

Figure 8.10   Bandpass Frequency Response

The frequency response of the Kaiser bandpass filter has the symmetric shape shown in Figure 8.10. The -6 dB cutoff frequencies for the filter are $fl = 0.10$ and $fh = 0.40$. The stopband attenuation is about -27 dB ($b = 2.16$).

See Function: bandpass().

8.2.2 Infinite Impulse Response Filters

An **infinite impulse response (IIR)** filter is a digital filter that theoretically "rings" forever when excited by an impulse. However, in practice the responses of these filters can be considered finite after some nominal time interval. The general IIR filter is described by the following difference equation:

$$y(n) = \sum_{r=1}^{N} b_r y(n-r) + \sum_{k=0}^{M} a_k x(n-k)$$

The z-transform of this difference equation is given by

$$H(z) = \frac{Y(z)}{X(z)} = \frac{\displaystyle\sum_{k=0}^{M} a_k z^{-k}}{1 - \displaystyle\sum_{r=1}^{N} b_r z^{-r}}$$

The **frequency response** of this filter is determined by evaluating the above z-transform on the unit circle $z = e^{j\omega t_s}$:

$$H(e^{j\omega t_s}) = \frac{\displaystyle\sum_{k=0}^{M} a_k e^{jk\omega t_s}}{1 - \displaystyle\sum_{r=1}^{N} b_r e^{jr\omega t_s}}$$

IIR filters are often implemented as a cascade of second order **biquad** sections, as shown in Figure 8.11. The transfer function of each biquad stage can be expressed as a ratio of second order polynomials:

$$H(z) = \frac{Y(z)}{X(z)} = \frac{a_0 + a_1 z^{-1} - a_2 z^{-2}}{1 - b_1 z^{-1} - b_2 z^{-2}}$$

Choosing the poles in each stage to be complex conjugate pairs reduces the errors due to coefficient truncation.

The classical techniques of analog filter design can also be used to design IIR filters. Several standard techniques exist for transforming analog filter designs into IIR designs. **Impulse invariance** and **bilinear transformation** are the most popular.

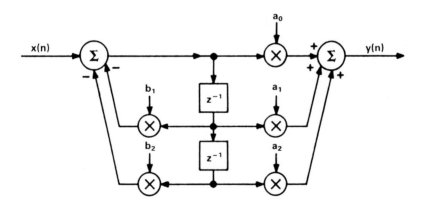

Figure 8.11   Biquad IIR Filter

See Function: iirfiltr().

The Bilinear Transform Method.   The **bilinear transform (BLT)** is the most popular technique for translating analog filter designs into IIR designs. The following substitution defines the bilinear transformation from the s-plane to the z-plane:

$$s = \frac{z - 1}{z + 1}$$

This mapping can be used to translate analog filter transfer functions into digital IIR filter transfer functions. However, the critical frequencies of the analog design must be prewarped to compensate for the nonlinear mapping of the analog frequency f_A axis onto the normalized digital frequency f_D axis:

$$f_A = \tan(\pi * f_D)$$

where $0 < f_D < 1/2$ and $\pi = 3.14159\ldots$

Example:

Given the following analog lowpass filter, use the BLT to design an equivalent IIR lowpass filter, assuming a sampling rate of 500 radians/sec.

$$H(s) = \frac{100}{s + 100}$$

Solution: The cutoff frequency (100 radians/sec) must be prewarped to

$$\omega_A = \tan(\pi * 100/500) = 0.726542528$$

The resulting frequency-scaled Laplace transfer function is

$$\hat{H}(s) = \frac{0.726542528}{s + 0.726542528}$$

Letting $s = (z - 1)/(z + 1)$, the desired IIR filter is given by

$$H(z) = \frac{0.726542528 + 0.726542528z}{(0.726542528 - 1) + (0.726542528 + 1)z}$$

$$= \frac{0.726542528 + 0.726542528z^{-1}}{1.726542528 - 0.273457472z^{-1}}$$

See Function: bilinear().

8.2.3 Data Smoothing

Data smoothing is a lowpass filtering process that reduces the high frequency noise, thus enhancing the low frequency signals. The noise typically consists of both high and low frequencies, while it is assumed that the signal has mainly low frequencies.

The design concerns for smoothing filters are quite similar to those of decimation filters. It could be said that a smoothing filter is simply a decimation filter, where the data rate has not been reduced.

Each new signal and noise combination must be considered separately in the smoothing filter design. It is impossible to design a smoothing filter that will work for all situations.

All smoothing processes remove noise at the expense of introducing some errors in the signal. For example, if the signal has high frequency information, some of this energy will be lost. Special care must be taken to ensure that the noise reduction process does not lead to excessive signal loss. In smoothing by large amounts, it is best that the signals of interest are sufficiently oversampled or they will be filtered out with the noise.

See Function: smooth().

8.2.4 Median Filtering

Median filtering is a nonlinear technique that is ideal for removing shot-noise or noise from digital dropout sometimes found in images. This technique has also been used in speech processing for smoothing estimates of zero crossings and pitch contours. In its simplest form the median filter is a sliding window that extends over an odd number of data points. At each new window position, the central data point is replaced by the median of the window. The effect of this processing is to filter out large (high frequency) noise spikes without significantly degrading the desired data. A fast sorting algorithm (e. g., Heapsort) can be used to speed up the filtering. Median smoothing is ideal for correcting isolated errors due to random data loss.

See Function: median().

8.3 SAMPLE RATE CONVERSION

Frequently in digital signal processing, it is necesssary to match systems that are sampled at different rates. **Sample rate conversion** is the process of converting a signal sampled at one rate to any other rate. When the new sampling rate is lower, this process is called **decimation**. When the new sampling rate is higher, this process is called **interpolation**. The following sections present some of the fundamentals of sample rate conversion.

8.3.1 Decimation

Originally, the term **decimation** was used in reference to a 10 percent reduction in sampling rate. Currently this signal processing term is used for reductions of arbitrary amounts. Lowering the sampling rate of a data set can be very useful in many data processing applications. A common case occurs with graphics packages, which are often designed to handle some maximum number of data points. One may wish to plot a much larger number of points, and may effectively do so by using decimation techniques.

When the sampling rate is reduced by an integer amount, the resampling is called **integer decimation**. The output sequence is formed by "hopping" a lowpass filter across the input samples in integral steps that correspond to the rate reduction factor. It should be emphasized that the output sequence represents all of the low-frequency information of the original input sequence, and is not simply a subset formed by deleting the intermediate input points.

When decimating by large factors, one must take care that the signals of interest are sufficiently oversampled, or they will be filtered out with the noise. It is important to note that even if the input data are appropriately oversampled, all resampling approaches introduce some errors (resampling ambiguities). These errors can be controlled by adjusting the stopband attenuation, d_s, of the lowpass filter to match the accuracy that is needed to represent the signal. For example, if the signal needs to be represented to only three-place accuracy, the stopband attenuation of the lowpass filter should be set to $d_s = 0.001$ (or $20_* \log(d_s) = -60dB$). Unnecessarily high accuracies should be avoided since they lead to resampling filters with long impulse responses, which exaggerate filter edge effects.

See Function: resample(), downsamp().

8.3.2 Interpolation

Interpolation is the opposite of the decimation process just described (i. e., the sampling rate is increased). The most common form of interpolation is encountered in the interpretation of tabulated data. Before the advent of the calculator and the computer, scientists and mathematicians had to rely on volumes of tables for computations that involved special functions. Thus, interpolation approaches evolved over hundreds of years from the need to generate and evaluate the points "in between" mathematical tables. The famous Lagrange interpolation has already been described in the chapter on numerical analysis.

The interpolation of digital signal processing is similar to the classical interpolation methods. For an excellent comparison of these approaches, the reader is encouraged to consult the paper in the references by R. W. Schafer and L. R. Rabiner (1973). There are some basic differences in the digital signal processing approach to interpolation:

1. The n-point interpolation is implemented by inserting $n - 1$ zeros between the original time samples and filtering the result with an FIR lowpass filter.

2. The interpolated output sequence is, in general, equally spaced.

3. The interpolation at the input data samples is not exact. In Lagrange interpolation, the appropriate output samples are always equal to the original input values.

4. The DSP techniques cannot interpolate at the edges of data sets.

5. For a sufficiently high sampling rate, the accuracy of the interpolation is determined by the design of the lowpass filter. Often this allows the interpolation to be arbitrarily accurate.

See Functions: resample(), interp1(), interp(), spline().

8.4 SPECTRAL ANALYSIS

Spectral analysis is the determination of the frequency content of a signal. In low-noise environments, the harmonic content of signals with constant amplitude and phase is well characterized by the **fast Fourier transform** (FFT). The FFT is simply an efficient approach to evaluating the previously described discrete Fourier transform.

Power spectral analysis provides a valuable solution to the problem of estimating the harmonic content of random signals. Most real-world signals are not well represented by constant amplitude and constant phase sinusoids. For example, speech signals often resemble noise, more than sinusoids. Due to their inherent randomness, such signals are better characterized statistically by their mean (average value) and variance (average power) over the observation interval of interest.

8.4.1 The Fast Fourier Transform

The discrete Fourier transform (DFT) plays an important role in digital signal analysis. Yet it was not until the development of the fast Fourier transform (FFT) that the DFT became computationally feasible [J. W. Cooley and J. W. Tukey (1965)]. Although the FFT is most commonly used for spectral estimation, its efficiency can be exploited to perform a variety of signal processing tasks such as convolution and correlation. This makes the FFT one of the most valuable tools in digital signal processing.

With most FFT approaches, all the calculations can be done "in place." This means that the desired transform is written over the original input data. Many FFT algorithms require that the length of the transform, N, must be a "power of two":

$$N = 2^M$$

where M is some integer.

Direct evaluation of the DFT requires a number of computations that is proportional to N^2, where N is the length of the transform. Yet with FFT approaches, the number of required computations is only proportional to $N * \log_2(N) = NM$. Thus, the relative increase in efficiency is proportional to

$$N / \log_2(N) = N/M$$

which becomes quite large as N increases.

"Radix-2" FFT algorithms decompose the full transform into M stages of $N/2$ 2-point transforms. "Radix-4" FFT algorithms are implemented with $M/2$ stages of $N/4$ 4-point transforms. "Radix-4 + 2" algorithms break the transform down into as many "radix-4" stages as possible, and finish with a "radix-2" stage if necessary.

The increase in efficiency that results from using an FFT can be quantified by counting the total number of required operations (real adds and multiplies). The following table compares the total number of operations required for "radix-2" and "radix-4" FFT algorithms with direct DFT evaluation.

N	Direct DFT Computation	Radix-2 FFT	Radix-4 FFT
16	$1.860e + 003$	$2.300e + 002$	$1.820e + 002$
64	$3.200e + 004$	$1.542e + 003$	$1.254e + 003$
256	$5.212e + 005$	$8.710e + 003$	$7.174e + 003$
1024	$8.376e + 006$	$4.506e + 004$	$3.738e + 004$
4096	$1.342e + 008$	$2.212e + 005$	$1.843e + 005$
16384	$2.147e + 009$	$1.049e + 006$	$8.765e + 005$

Note the tremendous reduction in number of operations required by the FFT algorithms for all of the transform lengths (N) shown. This is especially true as N increases. Note also that the "radix-4" algorithm requires on the average about 18 percent fewer operations than the "radix-2" FFT.

See Functions: fftrad2(), fft42(), fftreal(), fftr_inv().

8.4.2 The Power Spectrum

A power spectrum is a display of the average power (variance) of a signal as a function of frequency. While there are several methods of computing the power spectrum, the periodogram approach is the most common. Most modern spectrum analyzers use this approach. Periodogram analysis exploits the computational efficiency of the fast Fourier transform (FFT). The signal length is divided into equal subintervals (possibly overlapping) where the subinterval length is a "power of two." Each segment is Fourier transformed, and all of the transforms are averaged, yielding the desired power spectrum.

Welch (1970) found that the maximum benefit of overlapping occurs when the overlapped length is half the segment length. The variance of the spectral estimate is about a factor of two less than it is for the nonoverlapping case. However, twice as much computation is required.

Periodograms allow a convenient tradeoff between the frequency resolution and the variance of the spectral estimate. If more frequency resolution is desired (and more noise variance is tolerable), simply divide the signal length into fewer segments and perform longer Fourier transforms.

Time domain windows can be used to help reduce degradations in spectral estimates caused by finite data length. Multiplying the time data by tapered windows lowers the sidelobes that cause "spectral-leakage" in the frequency domain.

See Functions: powspec(), Cpowspec().

8.4.3 Correlation and Covariance

In many DSP applications it is necessary to compare two random signals. It may be suspected that signals are similar, though not identical. **Correlation** procedures provide a quantitative measure of waveform similarity and are particularly useful in random noise environments. Correlation is important in matched filtering applications and power spectral estimation.

Autocorrelation (sometimes called autocovariance) is a useful tool in random signal and power spectral analysis. The autocorrelation function and the power spectrum are Fourier transform pairs.

The autocorrelation function $R_{xx}(m)$ of the **zero-mean** real sequence $x(n)$ is given by:

$$R_{xx}(m) = \frac{1}{N - |m|} \sum_{n=0}^{N-|m|-1} x(n)x(n+m) \quad \text{where } |m| < N$$

Although the autocorrelation is defined for $2 * N - 1$ lags, the central 10 percent are the most accurate estimates of the true autocorrelation. The autocorrelation function is symmetric for real data.

The **crosscorrelation** $R_{xy}(m)$ (sometimes called covariance) between two real **zero-mean** sequences $x(n)$ and $y(n)$ is

$$R_{xy}(m) = \frac{1}{N-m} \sum_{n=0}^{N-m-1} x(n)y(n+m) \qquad 0 \le m < N$$

$$R_{xy}(-m) = \frac{1}{N-m} \sum_{n=0}^{N-m-1} x(n+m)y(n) \qquad 0 \le m < N$$

Again, the correlation estimate is most accurate for the central (e. g., 10 percent) lags. Unlike autocorrelation, the **crosscorrelation** function is not generally symmetric.

Both autocorrelation and crosscorrelation can be efficiently implemented with the FFT.

See Functions: crosscor(), autocor(), autofft(), crossfft().

Two-Dimensional Correlation.   Two-dimensional correlation can be used to measure the spatial misregistration between two similar images. The definitions for autocorrelation and crosscorrelation can be directly extended to two dimensions. The crosscorrelation $R_{xy}(n_1, n_2)$ between two real **zero-mean** 2-D sequences $x(n_1, n_2)$ and $y(n_1, n_2)$ has the following four quadrant structure:

$$R_{xy}(n_1, n_2) = \frac{1}{N_1-n_1} \frac{1}{N_2-n_2} \sum_{k_1=0}^{N_1-n_1-1} \sum_{k_2=0}^{N_2-n_2-1} x(n_1, n_2)y(n_1+k_1, n_2+k_2)$$

$$R_{xy}(n_1, -n_2) = \frac{1}{N_1-n_1} \frac{1}{N_2-n_2} \sum_{k_1=0}^{N_1-n_1-1} \sum_{k_2=0}^{N_2-n_2-1} x(n_1, n_2+k_2)y(n_1+k_1, n_2)$$

$$R_{xy}(-n_1, n_2) = \frac{1}{N_1-n_1} \frac{1}{N_2-n_2} \sum_{k_1=0}^{N_1-n_1-1} \sum_{k_2=0}^{N_2-n_2-1} x(n_1+k_1, n_2)y(n_1, n_2+k_2)$$

$$R_{xy}(-n_1, -n_2) = \frac{1}{N_1-n_1} \frac{1}{N_2-n_2} \sum_{k_1=0}^{N_1-n_1-1} \sum_{k_2=0}^{N_2-n_2-1} x(n_1+k_1, n_2+k_2)y(n_1, n_2)$$

where $0 \le n_1 < N_1$ and $0 \le n_2 < N_2$.

The autocorrelation formulas are a trivial modification of the crosscorrelation definitions. To save space, the autocorrelation formulas are not given. However, it may be worth noting an additional symmetry that results for the 2-D autocorrelation function:

$$R_{xx}(-n_1, -n_2) = R_{xx}(n_1, n_2)$$

As in the one-dimensional case, the estimates of the central lags are the most accurate.

The 2-D FFT can be used to efficiently correlate moderately large sets of 2-D data.

See Functions: auto2dft(), cros2dft().

8.5 ADAPTIVE SIGNAL PROCESSING

Adaptive filters automatically change their coefficients based on the signal and noise statistics to enhance the signal. The adaptive filters presented in this section are all of the **transversal** (FIR) type, as shown in Figure 8.12. The circular symbols with arrows are meant to indicate that the filter weights $h(i)$ are not fixed, but adapt depending on the signal and noise environment. There are two basic adaptive modes of operation: **open-loop** and **closed-loop**. Batch filtering processes like the Wiener filter are often used for the open-loop mode. The closed-loop mode of operation is efficiently implemented with the **LMS** (least mean square) algorithm developed by Widrow. In the steady state, both modes of operation approach the Wiener solution.

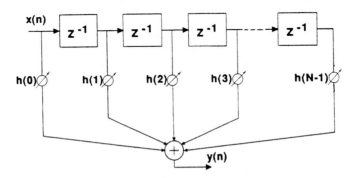

Figure 8.12 Adaptive Transversal Filter

8.5.1 Wiener Filtering

Norbert Wiener's work in adaptive filtering originally appeared in a book of limited circulation in February 1942. This work detailed an intricate mathematical procedure for solving the transcendental problem needed to derive optimal prediction filters. Wiener's results sparked several related military reports that dealt with gunnery prediction and fire control. However, Wiener is perhaps remembered most for developing an approximate (and more computationally feasible) solution to the optimal filter problem. This solution is the famous **Wiener–Hopf** equation expressed in matrix form:

$$
\begin{bmatrix}
r_{xx}(0) & r_{xx}(1) & r_{xx}(2) & \cdots & r_{xx}(N-1) \\
r_{xx}(1) & r_{xx}(0) & r_{xx}(1) & \cdots & r_{xx}(N-2) \\
r_{xx}(2) & r_{xx}(1) & r_{xx}(0) & \cdots & r_{xx}(N-3) \\
\vdots & \vdots & \vdots & & \vdots \\
r_{xx}(N-1) & r_{xx}(N-2) & r_{xx}(N-3) & \cdots & r_{xx}(0)
\end{bmatrix}
\begin{bmatrix}
h_0 \\
h_1 \\
h_2 \\
\vdots \\
h_{N-1}
\end{bmatrix}
$$

$$
=
\begin{bmatrix}
r_{xx}(0) \\
r_{xx}(1) \\
r_{xx}(2) \\
\vdots \\
r_{xx}(N-1)
\end{bmatrix}
$$

where
$$R_{xx}(m) = [r_{xx}(0), r_{xx}(1), \ldots, r_{xx}(N-1)]$$
is the autocorrelation function of the signal of interest $X(n)$ and
$$H_{LMS}(m) = [h_0, h_1, \ldots, h_{N-1}]^T$$
is the desired **Wiener filter**. The Wiener filter is a transversal matched filter. Under the proper assumptions, the Wiener solution can be shown to be minimum mean square.

The Wiener–Hopf equation can be solved directly using standard matrix techniques. However, Levinson developed a much more efficient approach that exploits the Toeplitz symmetry of the autocorrelation matrix.

See Function: levinson().

The Wiener Prediction Filter.   Prediction theory has many applications ranging from digital data compression to military gunfire control. The Wiener–Hopf equation can be extended to include the prediction of future values of a signal. Suppose we are interested in estimating the future value of a signal at time $t + k/f_s$, where f_s is the sampling rate. The **Wiener prediction filter** is obtained by solving the following modified set of normal equations:

$$
\begin{bmatrix}
r_{xx}(0) & r_{xx}(1) & r_{xx}(2) & \cdots & r_{xx}(N-1) \\
r_{xx}(1) & r_{xx}(0) & r_{xx}(1) & \cdots & r_{xx}(N-2) \\
r_{xx}(2) & r_{xx}(1) & r_{xx}(0) & \cdots & r_{xx}(N-3) \\
\vdots & \vdots & \vdots & & \vdots \\
r_{xx}(N-1) & r_{xx}(N-2) & r_{xx}(N-3) & \cdots & r_{xx}(0)
\end{bmatrix}
\begin{bmatrix}
h_0 \\ h_1 \\ h_2 \\ \vdots \\ h_{N-1}
\end{bmatrix}
$$

$$
=
\begin{bmatrix}
r_{xx}(0+k) \\
r_{xx}(1+k) \\
r_{xx}(2+k) \\
\vdots \\
r_{xx}(N-1+k)
\end{bmatrix}
$$

Note the only change is in the constraining vector $R_{xx}(m+k)$, and that for $k = 0$, the above equation simplifies to the original Wiener filter problem.

Example Program: Design a Wiener prediction filter for a sine wave.

```
#include <stdio.h>
#include <stdlib.h>
#include <math.h>
#include "mathlib.h"
main()
{
Real_Vector signal,sig_noise,filter,sig_covar,covar_ahead,y;
int i,ny, predict_time = 4;  /* desired positive prediction time in samples */
```

(continued on next page)

(continued)

```
double noise, amplitude = 50.0;
unsigned n=128,nfilt=16;
Real w[16];

/* main program designs a Wiener prediction filter to estimate future values of
   a signal in background noise */

        signal = valloc(NULL, n);
        sig_noise = valloc(NULL, n);
        covar_ahead = valloc(NULL, nfilt);
        for (i=0; i<n; i++) {
           signal[i] = amplitude*sin(twopi_*i*0.14);
           noise = ((double) rand()/RAND_MAX - 0.5); /* uniform +0.5, -0.5 */
           sig_noise[i] = signal[i] + noise;
        }
        sig_covar = autocor(signal, n, nfilt+predict_time);
        for(i=0; i<nfilt; i++) covar_ahead[i] = sig_covar[i+predict_time];
        filter = levinson(sig_covar, covar_ahead, nfilt);
        for (i=0; i<nfilt; i++) w[i] = filter[i];
        y = convolve(sig_noise, n, w, nfilt, 1, 1, 0, &ny);
        printf(" true signal      Wiener predicted signal, %d samples ahead\n",
               predict_time);
        for (i=predict_time; i<predict_time+17; i++) {
           printf("%2d  %8.4f          %8.4f\n", i, signal[i-predict_time],y[i]);
        }
        return 0;
}
```

Program Output:

	true signal	Wiener predicted signal, 4 samples ahead
4	0.0000	-0.4050
5	38.5257	38.9296
6	49.1144	50.0329
7	24.0877	23.9791
8	-18.4062	-18.3570
9	-47.5528	-48.0411
10	-42.2164	-42.1333
11	-6.2667	-6.6494
12	34.2274	33.8501
13	49.9013	50.6411
14	29.3893	30.3072
15	-12.4345	-12.1562
16	-45.2414	-45.3712
17	-45.2414	-44.7656
18	-12.4345	-12.6516
19	29.3893	30.2071
20	49.9013	49.9910

See Function: levinson().

8.5.2 The LMS Algorithm

The goal of all adaptive filtering is to approximate the Wiener–Hopf equation as closely as possible. Even with Levison's recursion and exact correlation estimates, the direct matrix approach can be computationally intensive. Furthermore, for pseudo-stationary signals, the elements of the correlation matrix are slowly time-varying and cannot be estimated perfectly. The **LMS** (least mean square) algorithm has several advantages over other adaptive estimation approaches:

1. The iterative LMS minimization formula is much simpler than that of gradient approaches.

2. The LMS method does not require explicit calculation of the correlation matrices.

3. The LMS technique does not require a matrix inversion.

The details of the derivation of the LMS method are not presented here. It is recommended that the interested reader consult the excellent references listed at the end of this section.

The LMS algorithm uses a gradient estimate that is based on the instantaneous error

$$e_k = d_k - y_k$$

where d_k is the desired response and y_k is the output of the adaptive filter. The filter update equation is

$$H_{k+1} = H_k + 2u_k e_k X_k$$

where H_k is the vector of N filter coefficients at the kth iteration and X_k is the vector of input signal values

$$X_k = [x_k, x_{k-1}, \ldots, x_{k-N+1}]^T$$

The convergence parameter u_k is included to smooth the noise problems associated with the instantaneous error estimation. The stable range for u_k varies with the input signal power s^2

$$u_k = u/(N * s^2), \qquad 0 < u < 1$$

where N is the number of filter coefficients. The normalized convergence parameter u controls the speed and accuracy of convergence. The goal is to select u such that the algorithm quickly converges to the Wiener–Hopf solution. Experimentation is often required to avoid choosing u to be too small or too large.

If the signal power is changing, it is sometimes convenient to allow the signal power estimate to adapt according to

$$\hat{s}_k^2 = a * x_k^2 + (1 - a) * \hat{s}_{k-1}^2$$

where $0 \le a < 1$.

The following sections discuss the application of the LMS algorithm to the noise canceling.

References.   B. Widrow and S. D. Stearns (1985); S. D. Stearns and R. A. David (1988).

See Function: lmsadapt().

8.5.3 Adaptive Interference Canceling

The general block diagram for **adaptive interference canceling** is shown in Figure 8.13. It is assumed that the signal has been corrupted with noise known to be correlated with a given noise source. Provided the signal is uncorrelated with the noise source, the LMS algorithm may be used to enhance the desired signal. Note that the output of the adaptive filter is a best estimate of the noise that corrupts the signal of interest. A program that demonstrates the use of the LMS algorithm for interference canceling is given along with the description of the *lmsadapt* function.

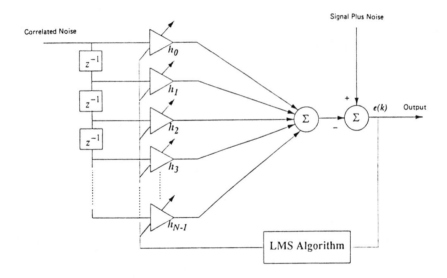

Figure 8.13   Adaptive Interference Cancelling

See Function: lmsadapt().

8.5.4 Adaptive Line Enhancement

Unlike interference canceling, the adaptive filter output is an estimate of the desired signal for **adaptive line enhancement**. The general approach to adaptive line enhancement is illustrated in Figure 8.14. The desired signal may be wideband, and is shown in this figure to be a finite sum of sinusoids. A constant delay is chosen so that the noise before the delay block is uncorrelated with the noise that is input to the adaptive filter. A one-sample delay is adequate for white noise. For bandlimited noise, a delay of at least the reciprocal of the bandwidth (i. e., at least two samples) is required.

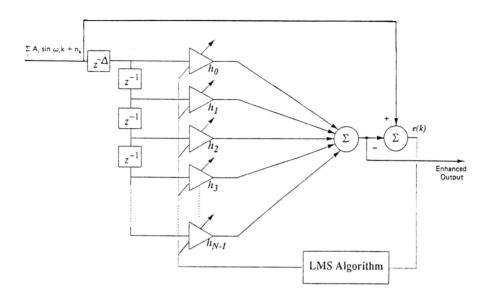

Figure 8.14   Adaptive Line Enhancement

Example Program: Design an Adaptive Line Enhancer for a sine wave.

```c
#include <stdio.h>
#include <stdlib.h>
#include <math.h>
#include "mathlib.h"
main()
{
Real_Vector desired, source, corrupted, error;
int k, ndelays=20, ndata=250;
double err_before, err_after, freq, tmp, rms_desired, noise;
double mu = 0.01, sig_power = 0.5, rho = 0.0;

/* Adaptive Line Enhancement with the LMS algorithm : Given a sinusoid in
   unknown (white) noise, cancel the unknown noise and enhance the sinusoid   */

    source = valloc(NULL, ndata);
    corrupted = valloc(NULL, ndata);
    desired = valloc(NULL, ndata);

    freq = 2*pi_/30.; /* for this example, the desired signal is a sinusoid  */
    source[0] = 0.0;
    for(k = 0; k<ndata-1; k++) {
        noise = 4.0*((double) rand()/RAND_MAX - 0.5);
        desired[k] = sin(freq*k);
        corrupted[k] = desired[k] + noise;
        source[k+1] = corrupted[k];
```

(continued on next page)

(continued)

```
    }
/* a delay of one sample is enough for white noise to decorrelate          */
/* at least a two sample delay would be needed for bandlimited noise        */
    noise = 4.0*((double) rand()/RAND_MAX - 0.5);
    desired[k] = sin(freq*(ndata-1));
    corrupted[k] = desired[ndata-1] + noise;

        /* enhance the sinusoid with the LMS algorithm: */

    error = lmsadapt(source, corrupted, ndata, ndelays, mu, rho, sig_power);

    err_after = err_before = rms_desired = 0.0;
    for(k = 0; k<ndata; k++) {
        tmp = desired[k] - corrupted[k];
        err_before += tmp*tmp;
        tmp = desired[k] - source[k];
        err_after += tmp*tmp;
        rms_desired += desired[k]*desired[k];
    }
    printf("error before LMS filtering = %7.4f\n", err_before/rms_desired);
    printf(" error after LMS filtering = %7.4f\n", err_after/rms_desired);
    mathfree();
    return 0;
}
```

Program Output:

```
error before LMS filtering =   2.6136
 error after LMS filtering =   0.4735
```

See Function: lmsadapt().

9

Functions

This chapter describes all of the functions in the C/Math Toolchest. The functions are listed in alphabetical order. Each function begins with a prototype, as defined in the MATHLIB.H header file. The prototype specifies the type of the function return value as well as the type of each argument. Following the prototype is a description of the function and its return values, along with a program example. In some cases, a list of related functions is also provided, as well as references to books that offer a more detailed discussion of the topic. A complete list of references is provided in Appendix B. The appendix also lists the library functions by category.

The listings of the example programs in this section are provided to illustrate how each function is used. These example programs are also supplied on the C/Math Toolchest disks. If you installed these files to a hard disk, you will find the example programs in the subdirectory named EXAMPLES. All the program files have a C extension. The file names are the same as the names of the library functions.

acosh—atanh

```
#include "mathlib.h"
double acosh(double x);
double asinh(double x);
double atanh(double x);
```

DESCRIPTION

The *acosh* function calculates the inverse hyperbolic cosine of x. The value of x must be greater than or equal to 1. Like the square root function, the inverse hyperbolic cosine is double-valued $\pm$. Only the positive value is returned by this routine.

The *asinh* function calculates the inverse hyperbolic sine of x, which is defined for all x.

The *atanh* function calculates the inverse hyperbolic tangent of x. The value of x must be in the range $(-1, 1)$.

The *acosh*, *asinh*, and the *atanh* functions are plotted in Figure 9.1.

INVERSE HYPERBOLIC FUNCTIONS

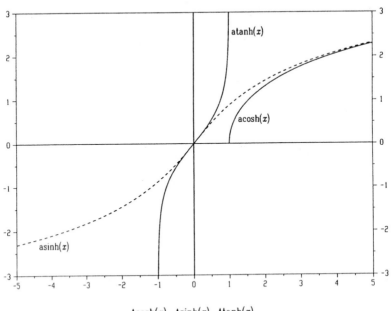

Acosh(x), Asinh(x), Atanh(x).

Figure 9.1   Inverse Hyperbolic Functions

RETURNS

acosh: inverse hyperbolic cosine of x, if $x \geq 1$. If $x < 1$, the following error message is generated: "Argument is outside the domain of the function" (math_errno = E_DOMERR), and -HUGE_VAL is returned.

asinh: inverse hyperbolic sine of x, for all x.

atanh: inverse hyperbolic tangent of x, if x is in the range $(-1, 1)$. Otherwise, the following error message is generated: "Argument is outside the domain of the function" (math_errno = E_DOMERR). -HUGE_VAL is returned if $x \leq -1$. +HUGE_VAL is returned if $x \geq +1$.

```c
#include "mathlib.h"
#include <stdio.h>

main()
{
double dx,x;
int i,n=20;

    dx = 4.0/(double) n;
    x=1;
    printf(" evaluate:  acosh(1) ..... acosh(4)\n");
    for(i=0; i<=n; ++) {
        printf("%11.7f %11.7f\n", x, acosh(x));
        x += dx;
    }

    dx = 10.0/(double) n;
    x=-5;
    printf(" evaluate:  asinh(-5) ..... asinh(5)\n");
    for(i=0; i<=n; i++) {
        printf("%11.7f %11.7f\n", x, asinh(x));
        x += dx;
    }

    dx = 2.0/(double) n;
    x=-1.0+dx;
    printf(" evaluate:  atanh(-0.9) ..... atanh(0.9)\n");
    for(i=0; i<=n; i++) {
        printf("%11.7f %11.7f\n", x, atanh(x));
        x += dx;
        if(x>=1.0-dx) break;
    }
    return 0;
}
```

auto2dft

```
#include "mathlib.h"
Real_Matrix auto2dft(Real_Matrix a, unsigned *rows, unsigned *cols);
```

DESCRIPTION

Autocorrelation (sometimes called autocovariance) is a useful tool in random signal and power spectral analysis. The 2-D FFT can be used to efficiently correlate moderately large sets of 2-D data (i. e., rows > 64 and columns > 64).

The *auto2dft* function uses a 2-D FFT approach to compute the autocorrelation of the real matrix *a*.

The *rows* and *cols* arguments specify the number of rows and columns in the input matrix *a*. Notice that these arguments are passed as pointers, since they are also used to return the number of rows and columns in the resulting autocorrelation matrix.

The zero lag value is always the centermost element of the resulting autocorrelation matrix x (e. g., $x[rows/2][cols/2]$). For best results, the correlation lags should be within 10 percent of the central lag point. This will usually ensure that the estimate of each of the lags is accurate. For example, if matrix *a* has 100 rows and columns, the crosscorrelation matrix x will have 99 rows and columns and the central lag will be at $x[49][49]$. The most accurate lags are in the rectangle bounded by the two points $x[44][44]$ and $x[54][54]$.

The autocorrelation output matrix is the return value of this function and is automatically allocated by the *auto2dft* function. The numbers of rows and columns in the output matrix are returned by the *rows* and *cols* arguments. This resulting output autocorrelation matrix should be freed when it is no longer needed.

The *auto2dft* function calls the *fft2d* function.

RETURNS

A *rows* by *cols* real matrix representing the 2-D autocorrelation of the real matrix *a* if successful, or NULL if an error occurs.

SEE ALSO

autofft(), autocor(), crosscor(), crossfft(), cros2dft().

REFERENCES

R. B. Blackman and J. W. Tukey (1959); E. O. Brigham (1974); A. V. Oppenheim and R. W. Schafer (p. 539, 1975).

```
#include <stdio.h>
#include <stdlib.h>
#include <math.h>
#include "mathlib.h"
main()
{
unsigned i, j, auto_rows, auto_cols, rows=9, cols=9;
double yval0, xval0, sigma=0.5, r0=4.0, c0=4.0;
Real_Matrix m0, autocorrelation;

/* Find the 2-D autocorrelation of a 2-D Gaussian pulse centered at (r0,c0)  */
    m0 = mxalloc(NULL, rows, cols);
    for(i=0; i<rows; i++) {      /* create one 2-dimensional Gaussian pulse  */
        xval0 = (i-r0)*(i-r0);   /* with center  at (r0,c0) and  a variance  */
        for(j=0; j<cols; j++) { /* of sigma                                  */
            yval0 = (j-c0)*(j-c0);
            m0[i][j] = exp(-(xval0+yval0)/(2*sigma));
        }
    }
    auto_rows = rows;
    auto_cols = cols;
    autocorrelation = auto2dft(m0, &auto_rows, &auto_cols);
    for(i=0; i<auto_rows; i++) {
        for(j=0; j<auto_cols; j++) {
            printf("%4.0f",fabs(90*autocorrelation[i][j]));
        }
        printf("\n");
    }
    mathfree();
    return 0;
}
```

PROGRAM OUTPUT

```
0   0   0   0   0   0   0   0   0
0   0   0   0   0   0   0   0   0
0   0   0   0   0   0   0   0   0
0   0   0   0   1   0   0   0   0
0   0   0   1   2   1   0   0   0
0   0   0   0   1   1   0   0   0
0   0   0   0   0   0   0   0   0
0   0   0   0   0   0   0   0   0
0   0   0   0   0   0   0   0   0
```

autocor

#include "mathlib.h"
Real__Vector autocor(Real__Vector *a*, unsigned *na*, unsigned *nlags*);

DESCRIPTION

Autocorrelation (sometimes called autocovariance) is a useful tool in random signal and power spectral analysis. The autocorrelation function is symmetric for real data. For best results, the number of desired correlation lags should be less than 10 percent of the number of input data samples. This will usually assure that the estimate of each of the lags is accurate.

The *autocor* function computes the autocorrelation of the input vector *a* of length *na*.

The resulting autocorrelation vector is returned as a vector of length *nlags*. This vector represents half of the symmetric autocorrelation function, where the first element is the central lag. The autocorrelation vector should be freed when it is no longer needed.

RETURNS

A real vector of length *nlags* representing the autocorrelation of the input vector *a* if successful, or NULL if an error occurs. If *na* < *nlags*, the error message "Not enough input data" is generated (math__errno = E__NOTENOUGH).

SEE ALSO

autofft(), auto2dft(), crosscor(), crossfft(), cros2dft().

REFERENCES

R. B. Blackman and J. W. Tukey (1959); A. V. Oppenheim and R. W. Schafer (p. 539, 1975).

```c
#include <stdio.h>
#include <stdlib.h>
#include <math.h>
#include "mathlib.h"
main()
{
Real_Vector signal,Rss;
int i;
unsigned n=128,nlags=13;

/* main program to autocorrelate a vector of data  */

        signal = valloc(NULL, n);
        for (i=0; i<n; i++) signal[i] = 0.0;
        for (i=n/4; i<3*n/4; i++) {
            signal[i] = 1.0;
        }
        Rss = autocor(signal, n, nlags);
        for (i=0; i<nlags; i++) {
            printf("%2d %f\n", i, Rss[i]);
        }
        vfree(signal);
        vfree(Rss);
        return 0;
}
```

PROGRAM OUTPUT

```
 0 0.250000
 1 0.242126
 2 0.234127
 3 0.226000
 4 0.217742
 5 0.209350
 6 0.200820
 7 0.192149
 8 0.183333
 9 0.174370
10 0.165254
11 0.155983
12 0.146552
```

autofft

#include "mathlib.h"
Real__Vector autofft(Real__Vector *a*, unsigned *na*, unsigned *nlags*);

DESCRIPTION
Autocorrelation (sometimes called autocovariance) is a useful tool in random signal and power spectral analysis.

The *autofft* function uses an FFT to compute the autocorrelation of vector *a* of length *na*.

The autocorrelation vector is returned as a vector of length *nlags*. This vector represents half of the symmetric autocorrelation function, where the first element is the central lag. The autocorrelation function is symmetric for real data. For best results, the number of desired correlation lags should be less than 10 percent of the number of input data samples (i. e., *nlags* < .10 * *na*). This will usually ensure that the estimate of each of the lags is accurate.

The FFT can be used to efficiently correlate moderately large data sets (i. e., *na* > 512 and *nlags* > 256). For small data sets, the *autocor* function is more efficient.

The *autofft* function calls the *fftr_inv* and the *fftreal* functions.

RETURNS
A real vector of length *nlags* representing the autocorrelation of the vector *a* if successful, or NULL if an error occurs. If *na* < *nlags*, the error message "Not enough input data" is generated (math_errno = E_NOTENOUGH).

SEE ALSO
autocor(), auto2dft(), crosscor(), crossfft(), cros2dft().

REFERENCES
R. B. Blackman and J. W. Tukey (1959); E. O. Brigham (1974); A. V. Oppenheim and R. W. Schafer (p. 539, 1975).

```c
#include <stdio.h>
#include <stdlib.h>
#include <math.h>
#include "mathlib.h"
main()
{
Real_Vector signal,Rss;
int i;
unsigned n=128,nlags=13;

/* main program to autocorrelate a vector of data  */

        signal = valloc(NULL, n);
        for (i=0; i<n; i++) signal[i] = 0.0;
        for (i=n/4; i<3*n/4; i++) {
            signal[i] = 1.0;
        }
        Rss = autofft(signal, n, nlags);
        for (i=0; i<nlags; i++) {
            printf("%2d %f\n", i, Rss[i]);
        }
        vfree(signal);
        vfree(Rss);
        return 0;
}
```

PROGRAM OUTPUT

```
 0 0.250000
 1 0.242126
 2 0.234127
 3 0.226000
 4 0.217742
 5 0.209350
 6 0.200820
 7 0.192149
 8 0.183333
 9 0.174370
10 0.165254
11 0.155983
12 0.146552
```

bandpass

#include "mathlib.h"
void bandpass(Real *weights* [ ], int *nweights*, double *fh*, double *fl*, double * *dB*, int *half*);

DESCRIPTION

The *bandpass* function designs a digital bandpass filter which can then be implemented with the *convolve* function. The basic approach of window design methods is to truncate the infinite length impulse response of an ideal frequency filter by multiplying it by a time-domain window (in this case, a Kaiser–Bessel window). Time-domain windows are bell-shaped, (see the *tdwindow* function), which tapers the filter impulse response.

It is assumed that the data to be filtered is sampled at uniform intervals of time, T. The sampling frequency is $F_s = 1/T$. In digital sampling, frequencies are often specified normalized to F_s. For example, $F_s/2$, the Nyquist frequency, is the highest frequency that can be represented and has a normalized value of 1/2.

Digital bandpass filters are characterized by one passband, (F_1, F_2), over which the filter's response is reasonably close to unity gain. These filters also often have two stopbands, $(0, F_{s1})$, and $(F_{s2}, F_s/2)$, over which frequencies are attenuated by some specified amount (i. e., the gain in the stopband is less than $d_s < 1$).

The fl and fh arguments specify the low and high frequency cutoffs, which set the nominal passband (and stopband) of the Kaiser window filter, since they are averages of the "corner" frequencies:

$$fl = (F_{s1} + F_1)/2$$
$$fh = (F_2 + F_{s2})/2$$

The two cutoff frequencies are the frequencies at which the magnitude of the filter response is approximately -6 dB.

The transition band, F_t, is the difference between the passband and the stopband frequencies:

$$F_t = F_{s2} - F_2 = F_1 - F_{s1}$$

The dB argument specifies the desired stopband attenuation in *decibels*:

$$dB = 20 \log_{10}(G_s)$$

where G_s is the gain in the stopband.

The *nweights* argument specifies the length of the filter. The length of the filter needed to achieve the desired values of dB and F_t is given by:

$$nweights = (|dB| - 8)/(7.1785F_t)$$

The *nweights* filter weights representing the impulse response are returned via the *weights* array. The *weights* array should be declared as an array of type **Real** with at least *nweights* elements.

The *half* argument specifies whether to return all the filter weights, or just half of them. Since the Kaiser–Bessel bandpass filter is a symmetric (even) function, the second half of the filter weights is a mirror image of the first half. This symmetry can be exploited to reduce by a factor of 2, the number of multiplications (and array storage locations) required to implement the filter. If *half* is zero, then all the *nweights* coefficients are returned through the *weights* array. However, if *half* is nonzero, then only the first *nweights*/2 coefficients are returned. The remaining half of the coefficients could be filled into the *weights* array using the following C statement:

```
for (i = 0; i < nweights/2; i++)
    weights[nweights-1-i] = weights[i];
```

If the filter is symmetric, the *convolve* function needs only the first half of the coefficients.

It should be mentioned that the above design formulas are approximations. If an application requires an exact frequency response, then the filter design should be checked to see that it meets the desired specification. It may be necessary to slightly change the filter design parameters to achieve a more exact frequency response.

The response of a 31-point Kaiser bandpass filter is shown in the bandpass filter section. The -6 dB cutoff frequencies for this filter are $fl = 0.10$ and $fh = 0.40$. The stopband attenuation is about -27 dB. In the example below, an interference tone is placed at the -27 dB peak in the stopband of the bandpass filter at $f = 0.06$. The signal is placed in the passband at $f = 0.20$. The 31-point Kaiser bandpass filter reduces the interference from 0.5000 (per data point) to 0.001566 (per data point).

RETURNS

The *nweights* (*nweights*/2 if *half* is nonzero) coefficients representing the impulse response of a bandpass filter with cutoff frequencies fl and fh, and stopband attenuation dB is returned via the *weights* array.

SEE ALSO

lowpass(), highpass(), convolve(), downsamp(), deriv(), interp(), tdwindow().

REFERENCE

A. V. Oppenheim and R. W. Schafer (1989).

```
#include "mathlib.h"
#include <stdio.h>
#include <stdlib.h>
#include <math.h>

#define NWEIGHTS 31
#define NDATA    100

main()
{
Real_Vector data, y;
Real weights[NWEIGHTS];
int i, ny, ndec=1, itype=1, isym=-1;
double truth, interfer, off, init_error=0.0, residual_error=0.0;

    data=valloc(NULL, NDATA);/* allocate a vector for data */
    for(i=0; i<NDATA; i++) {/* generate data, one cycle of a sinusoid+noise */
        truth = sin(twopi_*i*0.20);    /*put signal in passband of filter & */
        interfer = sin(twopi_*i*0.06);/*interference at the peak in stopband*/
        data[i] = truth + interfer;
        init_error += interfer*interfer/(double) NDATA;
    }
    bandpass(weights, NWEIGHTS, 0.40, 0.10, 27.0, 0); /* design bandpass    */

    y = convolve(data, NDATA, (Real_Vector) weights, NWEIGHTS,
                 ndec, itype, isym, &ny);
    off = (double) (NWEIGHTS-1)/2.; /* align truth with filtered data      */
    for(i=0; i<ny; i++) { /* compute the error after filtering            */
        truth = sin(twopi_*0.20*(i+off));
        residual_error += (truth-y[i])*(truth-y[i])/(double) ny;
    }/* The Kaiser bandpass should remove most of the interference:       */
    printf("the error before filtering is %f\n", init_error); /*  = 0.500000*/
    printf("the error after filtering is %f\n", residual_error);/*= 0.001566*/
    return 0;
}
```

besi0 - beskn

#include "mathlib.h"
double besi0(double x);
double besi1(double x);
double besin(int n, double x);
double besk0(double x);
double besk1(double x);
double beskn(int n, double x);

DESCRIPTION

The modified bessel functions have an exponential behavior shown in Figure 9.2.

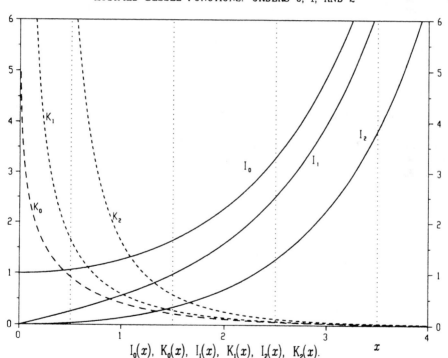

MODIFIED BESSEL FUNCTIONS: ORDERS 0, 1, AND 2

$I_0(x)$, $K_0(x)$, $I_1(x)$, $K_1(x)$, $I_2(x)$, $K_2(x)$.

Figure 9.2   Modified Bessel Functions

The *besi0* function calculates the zero order modified bessel function $i0(x)$. The $i0(x)$ function is an even function of x, that is, $i0(x) = i0(-x)$.

The *besi1* function calculates the first order modified bessel function $i1(x)$. The $i1(x)$ function is an odd function of x, specifically, $i1(x) = -i1(x)$.

The *besin* function calculates the nth order modified bessel function $in(x)$. If n is even, $in(x)$ is an even function of x. If n is odd, $in(x)$ is an odd function of x.

The *besk0*, *besk1*, and *beskn* functions calculate the modified bessel functions $k0(x)$, $k1(x)$, and $kn(x)$, respectively. These functions are nonnegative and are defined only for positive values of x. If $x \leq 0$, the following error message is generated: "Argument is outside the domain of the function" (math_errno = E_DOMERR), and +HUGE_VAL is returned.

RETURNS

besi0: the zero order modified bessel function, $i0(x)$, for any value of x.

besi1: the first order modified bessel function, $i1(x)$, for any value of x.

besin: the nth order modified bessel function, $in(x)$, for any value of x.

besk0: the zero order modified bessel function, $k0(x)$, if $x > 0$; and HUGE_VAL, if $x \leq 0$.

besk1: the first order modified bessel function, $k1(x)$, if $x > 0$; and HUGE_VAL, if $x \leq 0$.

beskn: the nth order modified bessel function, $kn(x)$, if $x > 0$; and HUGE_VAL, if $x \leq 0$.

NOTES

The relative accuracy for the *besi0*, *besi1*, and *besin* functions is 14 decimal places for all x tested.

The relative accuracy for the *besk0*, *besk1*, and *beskn* functions is 14 decimal places for $x < 8$; and 9 decimal places for $x \geq 8$.

The *besi0*, *besi1*, *besk0*, and *besk1* functions call the *chebser* function. The *besin* function calls the *besi0* and *besi1* functions. The *beskn* function calls the *besk0* and *besk1* functions.

REFERENCES

C. W. Clenshaw (1963, pp. 34–36); Abramowitz and Stegun (1964, pp. 358–453).

```
#include "mathlib.h"
#include <stdio.h>
main()
{
double dx,x;
int i,n=20;

        dx = 10.0/(double) n;
        x=1;
        printf("x=1 ... 10           I0(x)           I1(x)\n");
        for(i=0; i<n-1; i++) {
                printf("%10.7f %15.7f %15.7f\n", x, besi0(x),besi1(x));
                x += dx;
        }
        x=1;
        printf("x=1 ... 10           K0(x)           K1(x)\n");
        for(i=0; i<n-1; i++) {
                printf("%10.7f %15.7f %15.7f\n", x, besk0(x),besk1(x));
                x += dx;
        }
        x=1;
        printf("x=1 ... 10           I2(x)           K2(x)\n");
        for(i=0; i<n-1; i++) {
                printf("%10.7f %15.7f %15.7f\n", x, besin(2,x),beskn(2,x));
                x += dx;
        }
        return 0;
}
```

besj0 - besyn

```
#include "mathlib.h"
double besj0(double x);
double besj1(double x);
double besjn(int n, double x);
double besy0(double x);
double besy1(double x);
double besyn(int n, double x);
```

DESCRIPTION

The bessel functions have the exponentially damped oscillatory behavior, as shown in Figure 9.3.

BESSEL FUNCTIONS: ZERO ORDER AND FIRST ORDER

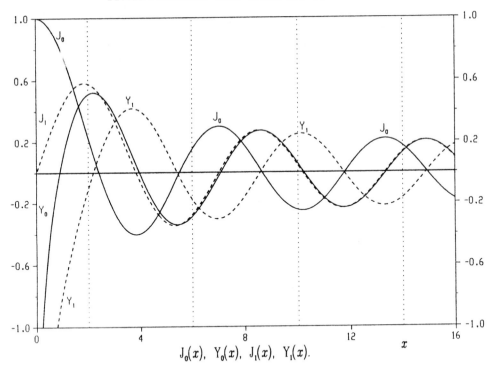

Figure 9.3   Bessel Functions: Orders 0 and 1

The *besj0* function calculates the zero order bessel function $j0(x)$. The $j0(x)$ function is an even function of x; that is, $j0(x) = j0(-x)$.

The *besj1* function calculates the first order bessel function $j1(x)$. The $j1(x)$ function is an odd function of x, specifically, $j1(x) = -j1(x)$.

The *besjn* function calculates the nth order bessel function $jn(x)$. If n is even, $jn(x)$ is an even function of x. If n is odd, $jn(x)$ is an odd function of x.

The *besy0, besy1,* and *besyn* functions calculate the bessel functions $y0(x)$, $y1(x)$, and $yn(x)$, respectively. These functions are defined only for positive values of x. If $x \leq 0$, the following error message is generated: "Argument is outside the domain of the function" (math_errno = E_DOMERR), and -HUGE_VAL is returned.

RETURNS

besj0: the zero order bessel function, $j0(x)$, for any value of x.

besj1: the first order bessel function, $j1(x)$, for any value of x.

besjn: the nth order bessel function, $jn(x)$, for any value of x.

besy0: the zero order bessel function, $y0(x)$, if $x > 0$; and -HUGE_VAL), if $x \leq 0$).

besy1: the first order bessel function, $y1(x)$, if $x > 0$; and -HUGE_VAL), if $x \leq 0$.

besyn: the nth order bessel function, $yn(x)$, if $x > 0$; and -HUGE_VAL), if $x \leq 0$.

NOTES

The relative accuracy for the *besj0, besj1,* and *besjn* functions is 14 decimal places for all x tested.

The relative accuracy for the *besy0, besy1,* and *besyn* functions is also 14 decimal places for all x tested.

The *besj0, besj1, besy0,* and *besy1* functions call the *pseries* function. The *besjn* function calls the *besj0* and *besj1* functions. The *besyn* function calls the *besy0* and *besy1* functions.

REFERENCES

John F. Hart (1968, pp. 141–149); Abramowitz and Stegun, (1964, pp. 358–453).

```
#include "mathlib.h"
#include <stdio.h>
main()
{
double dx,x;
int i,n=20;

        dx = 10.0/(double) n;
        x=dx;
        printf("x=1/2 ... 10        J0(x)              J1(x)\n");
        for(i=0; i<n; i++) {
                printf("%10.7f %15.7f %15.7f\n", x, besj0(x),besj1(x));
                x += dx;
        }
        x=dx;
        printf("x=1/2 ... 10        Y0(x)              Y1(x)\n");
        for(i=0; i<n; i++) {
                printf("%10.7f %15.7f %15.7f\n", x, besy0(x),besy1(x));
                x += dx;
        }
        x=dx;
        printf("x=1/2 ... 10        J2(x)              Y2(x)\n");
        for(i=0; i<n; i++) {
                printf("%10.7f %15.7f %15.7f\n", x, besjn(2,x),besyn(2,x));
                x += dx;
        }
        return 0;
}
```

bilinear

#include "mathlib.h"
int bilinear(Real as[ ], Real bs[ ], Real az[ ], Real bz[ ], unsigned nn, unsigned nd);

DESCRIPTION

The classical techniques of analog filter design can also be used to design **infinite impulse response (IIR)** filters. An IIR filter is a digital filter that theoretically "rings" forever when excited by an impulse. However, in practice the responses of these filters can be considered finite after some nominal time interval. The **bilinear transformation** is the most popular technique for transforming analog filter designs into IIR designs. The following substitution defines the bilinear transformation from the s-plane to the z-plane:

$$s = \frac{z-1}{z+1}$$

This mapping can be used to translate analog filter transfer functions into digital IIR filter transfer functions. However, the critical frequencies of the analog design must be prewarped to compensate for the nonlinear mapping of the analog frequency f_A axis onto the normalized digital frequency f_D axis:

$$f_A = \tan(\pi * f_D)$$

where $0 < f_D < 1/2$ and $\pi = 3.14159\ldots\ldots$

The *bilinear* function transforms a general s-plane transfer function into the z-plane via the bilinear transform. This routine **does not** prewarp the critical frequencies of the analog design.

The input array that contains the numerator s-plane polynomial $as[\,]$ is of length nn. The input array that contains the denominator s-plane polynomial $bs[\,]$ is of length nd. The output array that contains the numerator z-plane polynomial $az[\,]$ is of length nd. The output array that contains the denominator z-plane polynomial $bz[\,]$ is of length nd.

The s-plane polynomial arrays are ordered in ascending powers of s:

$$H(s) = \frac{as[0] * s^0 + as[1] * s^1 + \ldots + as[nn-1] * s^{nn-1}}{bs[0] * s^0 + bs[1] * s^1 + \ldots + bs[nd-1] * s^{nd-1}}$$

The z-plane polynomial arrays are ordered in ascending powers of z^{-1}.

$$H(z) = \frac{az[0] * z^0 + az[1] * z^{-1} + \ldots + az[nn-1] * z^{-(nd-1)}}{bz[0] * z^0 + bz[1] * z^{-1} + \ldots + bz[nd-1] * z^{-(nd-1)}}$$

RETURNS

0 if successful or -1 if an error occurs. If successful, the bilinear transform of the s-plane arrays as and bs are returned in the z-plane arrays, az and bz. If an error occurs, **math_errno** is set to E_TRANSFER if $nd < nn$, or E_MALLOC if unable to allocate enough temporary storage space.

SEE ALSO
iir_read(), iirfiltr().

REFERENCE
A. V. Oppenheim and R. W. Schafer (1975).

```
#include <stdio.h>
#include <stdlib.h>
#include <math.h>
#include "mathlib.h"
#define NCOEFF 3
main()
{
Real as[NCOEFF], bs[NCOEFF], az[NCOEFF], bz[NCOEFF];
unsigned nn = NCOEFF, nd = NCOEFF;

/*  2nd order example: GIVEN H(s),    FIND      H(z)   WHERE:

        as[0]*s^2 + as[1]*s + as[0]                az[0]+az[1]*z^-1+az[0]*z^-2
 H(s) = _____  <===> H(z) = _____

        bs[0]*s^2 + bs[1]*s + bs[0]                bz[0]+bz[1]*z^-1+bz[0]*z^-2
*/
                /* THE ANSWER IS:                      */
                /*       2 +  0*z^(-1) +  2*z^(-2)     */
                /* H(z) = _____    */
                /*                                     */
                /*       3 +  0*z^(-1) +  1*z^(-2)     */

    as[0] = 1; as[1] = 0; as[2] = 1;   bs[0] = 1; bs[1] = 1; bs[2] = 1;

    bilinear(as, bs, az, bz, nn, nd);  /* call the bilinear transform routine */

    printf("      %2.0f + %2.0f*z^(-1) + %2.0f*z^(-2)\n", az[0], az[1], az[2]);
    printf("H(z) = _____\n\n");
    printf("      %2.0f + %2.0f*z^(-1) + %2.0f*z^(-2)\n", bz[0], bz[1], bz[2]);
    return 0;
}
```

PROGRAM OUTPUT

$$H(z) = \frac{2 + 0 * z^{-1} + 2 * z^{-2}}{3 + 0 * z^{-1} + 1 * z^{-2}}$$

binomdst

#include "mathlib.h"
double binomdst(int *n*, int *k0*, double *p*, Real *pb*[ ]);

DESCRIPTION

Binomial distributions are useful in a variety of statistical applications. They are frequently used to predict the outcomes from a repetitive sampling of a population. The underlying assumption behind the binomial distribution is that after each sample is taken, it is returned to the population before the next sample is taken. Hence the phrase "sampling with replacement" is often used in describing binomial distributed events. A binomial distribution represents a discrete process, since only integer valued outcomes are possible.

The *binomdst* function computes the cumulative probability that $k0$ or fewer of a particular outcome will occur after n trials of a binomial distributed experiment. In a binomial experiment, there are two possible mutually exclusive outcomes for each random sampling (e. g., heads or tails for a coin flip).

The p argument specifies the probability of a particular outcome. For example, p might represent the probability that heads will be the outcome of a coin flip. The probability of the opposite outcome (in this case, tails) is $1 - p$. These probabilities remain the same for all the n repetitions of the experiment. For a coin flip, both p and $1 - p$ would be equal to 0.5.

If the cumulative probability of a specific outcome is represented by a random variable X, then:

$$P(X \leq k0) = \sum_{i=0}^{k0} \binom{n}{i} p^i (1 - p)^{n-i}$$

The *pb*[ ] argument is used to return the probability that exactly i outcomes will occur (where $i = 0, 1, 2, \ldots, k0$). Therefore, the *pb*[ ]) argument should be declared as **Real *pb*[*k0* + 1]**; in order to be able to store all of the discrete probabilities. If the discrete probabilities are not needed, then the *pb*[ ] argument should be specified as NULL.

RETURNS

A double precision whole number representing the cumulative probability that after n trials, a binomial random variable X is less than or equal to the integer $k0$, where $0 \leq k0 \leq n$). The individual probabilities of exactly $0, 1, \ldots,$ and $k0$ outcomes are also returned in the array *pb*[ ].

SEE ALSO

hyperdst(), and poissdst().

REFERENCE

L. L. Lapin (1983, pp. 94–104).

```
#include "mathlib.h"
#include <stdio.h>
main()
{
int k0,i,n;
Real pb[5],p;
double ans;
/* out of 10 coin tosses, compute the odds that 4, or less, are heads:    */
     n = 10;  /* this is the total number of coin tosses                   */
     k0 = 4;  /* this is the maximum number of heads desired               */
     p = 0.5; /* p = probability of a head on one toss, if game is fair */
     ans = binomdst(n, k0, p, pb);/*compute Binomial distribution          */
     for (i=0; i <= k0; i++) {
          printf("the odds that exactly %d heads occur is %f\n", i,pb[i]);
     }
     printf("the odds that %d or fewer heads occurring is %f\n", k0, ans);
     /* out of 10 coin tosses, the odds that 4, or less,
        are heads is 0.376953 */
     return 0;
}
```

Cadd

#include "mathlib.h"
Complex Cadd(Complex *x*, Complex *y*);

DESCRIPTION

The *Cadd* function adds the two complex numbers x and y. Both x and y must be declared as **Complex** data types. The resulting sum is also a **Complex** data type.

RETURNS

z, where $z = x + y$ and x, y, and z are all complex numbers.

SEE ALSO

Csub(), Cdiv(), Cmul().

```
#include "mathlib.h"
#include <stdio.h>
main()
{
Complex x,y,z;

        x.r = 1;
        x.i = 1;
        y.r = 2;
        y.i = 2;
        z = Cadd(x, y); /* add x and y */
        printf("the real part of z = %f\n", z.r);
        printf("the imaginary part of z = %f\n", z.i);
        /* the answer should be:  z.r = 3;  z.i = 3;  */
        return 0;
}
```

Cdiv

#include "mathlib.h"
Complex Cdiv(Complex *x*, Complex *y*);

DESCRIPTION

The *Cdiv* function divides the complex number x by y. Both x and y must be declared as **Complex** data types. The resulting division is also a **Complex** data type.

RETURNS

z, where $z = x/y$ and x, y, and z are all complex numbers.

SEE ALSO

Cadd(), Csub(), Cmul().

```
#include "mathlib.h"
#include <stdio.h>
main()
{
Complex x,y,z;

        x.r = 1;
        x.i = 1;
        y.r = 2;
        y.i = 2;
        z = Cdiv(x, y); /* divide x by y  */
        printf("the real part of z = %f\n", z.r);
        printf("the imaginary part of z = %f\n", z.i);
        /* the answer should be:  z.r = 0.5;  z.i = 0;      */
        return 0;
}
```

chebser

#include "mathlib.h"
double chebser(double *x*, double *coef*[ ], unsigned *n*, int *even*);

DESCRIPTION

Chebyshev polynomials are often useful in approximating trigonometric and transcendental functions (see the following example for e^x). Chebyshev series are optimum in the sense that they have the minimum approximation error when compared to other polynomials with the same number of terms. The procedure for obtaining the Chebyshev coefficients, *coef*[ ], is beyond the scope of this description. However, it is recommended that the interested reader consult the excellent references given below for more details.

With the *chebser* function, it is possible to evaluate two types of Chebyshev series. If *even* is 0, the standard Chebyshev series is computed:

$$0.5 * \text{coef}[0] + T_1(x) * \text{coef}[1] + T_2(x) * \text{coef}[2] + \ldots + T_{n-1}(x) * \text{coef}[n-1]$$

where $T_i(x)$ is the ith Chebyshev polynomial function of x, and n is the number of terms in the Chebyshev series. If *even* is nonzero, then the following even Chebyshev series is computed:

$$0.5 * \text{coef}[0] + T_2(x) * \text{coef}[1] + T_4(x) * \text{coef}[2] + \ldots + T_{2*n-2}(x) * \text{coef}[n-1]$$

RETURNS

A double precision value representing $f(x)$, where $f(x)$ is a function approximated by the n coefficients *coef*[ ] of a Chebyshev series.

SEE ALSO

pseries().

REFERENCES

C. W. Clenshaw (1962), and Cecil Hastings (1955).

```
#include "mathlib.h"
#include <stdio.h>
#include <math.h>
double coef[7]= { 2.53213, 1.13032, 0.27149, 0.04434,
                  0.00547, 0.00054, 0.00004 };
main()
{
int i,even,n,m=20;
double dx,x,approx;
/* evaluate the 7-term Chebyshev series approximation to exp(x), -1 < x < 1*/
    dx = 2.0/(double) m;
    x=-1;      /* start the approximation at x = -1                   */
    n = 7;     /* this is the number of terms in the Chebyshev series */
    even = 0; /* do a standard Chebyshev polynomial series            */
    printf("   x      exp(x)  approximation\n");
    for(i=0; i<=m; i++) {
            approx = chebser(x, coef, n, even);/* evaluate polynomial  */
            printf("%f %f %f\n", x, exp(x), approx);
            x += dx;
    }
    return 0;
}
```

Clineqn

#include "mathlib.h"
Complex__Vector Clineqn(Complex__Matrix *m*, Complex__Vector *v*, unsigned *n*);

DESCRIPTION
Occasionally engineers and physicists are required to find solutions to systems of simultaneous complex linear equations. These equations frequently arise in the theory of "Fields and Waves" and Acoustics. Just as for linear systems with real coefficients, a unique solution to a complex system exists only if the determinant of the system matrix is nonzero. More precisely, since the determinant of a complex matrix is a complex number, its magnitude must be nonzero if a solution exists. Thus, the case of no solution occurs whenever the magnitude of the system determinant is zero (i. e., the system matrix is singular, or noninvertible).

The *Clineqn* function solves the following set of complex simultaneous linear equations:

$$m[n][n] * x[n] = v[n]$$

where m is an n by n complex system matrix, v is a complex constraining vector of length n, and x is the solution vector (also complex and of length n).

Note that v and m must be allocated by the *Cvalloc* and *Cmxalloc* functions prior to calling the *Clineqn* function. The solution vector x is automatically allocated by the *Clineqn* function.

The user may select the option of whether or not to write over the input constraining vector $v[\ ]$. If memory is a limitation, and $v[\ ]$ is no longer needed, this may prove useful.   This option can be selected by setting the global variable **useinput_ = 1** prior to calling *Clineqn*. The default option is to not destroy the input arrays (i. e., **useinput_ = 0**). If **useinput_ = 1**, then the solution is returned in the input vector v and freeing v is the same as freeing the solution vector.

RETURNS
A complex vector x of length n, where x is the solution vector for a set of simultaneous linear equations m, constrained by v:

$$m[n][n] * x[n] = v[n]$$

or NULL if no solution exists (i. e., m is singular). If the determinant of m is 0, then a "Singular matrix" error message (math_errno = E__MSING) is generated.

SEE ALSO
lineqn().

REFERENCE
J. H. Wilkinson (1965).

```c
#include "mathlib.h"
#include <stdio.h>

main()
{
Complex_Vector v,x;
Complex_Matrix m;

/* solve the following complex set of 2 linear equations &  2 unknowns:  */
/*                                                                        */
/*          m[2][2]*x[2] = v[2]                                           */
/*                                                                        */
/* where m[][] and v[] are complex and are given below:                   */

        m = Cmxalloc(NULL, 2, 2);   /* allocate 2 by 2 system matrix     */
        v = Cvalloc(NULL,2);        /* allocate constraining vector      */
        m[0][0].r = 2.321;  m[0][0].i = -1.27; /* define row 0 of matrix m */
        m[0][1].r =-3.14;   m[0][1].i = 1.21;
        m[1][0].r = 1.414;  m[1][0].i = 0.866; /* define row 1 of matrix m */
        m[1][1].r = -0.932; m[1][1].i = 0.111;

        v[0].r = 1;  v[0].i = 1;    /* define constraining vector v[]     */
        v[1].r = 8;  v[1].i = 8;

        x = Clineqn( m, v, 2);  /* now solve for x[] */

        /* the answer should be :  x[0].r = 8.483701, x[0].i = -1.778161
                                   x[1].r = 6.255770, x[1].i = -2.653481   */
        printf("x[%d].r = %f, x[%d].i = %f\n", 0, x[0].r, 0, x[0].i);
        printf("x[%d].r = %f, x[%d].i = %f\n", 1, x[1].r, 1, x[1].i);

        Cmxfree(m);    /* free matrix m[][] */
        Cvfree(v);     /* free vector v[]    */
        Cvfree(x);     /* free vector x[]    */
        return 0;
}
```

Cmag

#include "mathlib.h"
Real Cmag(Complex x);

DESCRIPTION
The *Cmag* function computes the magnitude of the complex number x. The x argument must be declared as a **Complex** data type. The resulting magnitude is of type **Real**.

RETURNS
y, where $y = |x| = $ sqrt $(x.r * x.r + x.i * x.i)$, y is a real number, and x is a complex number.

```
#include "mathlib.h"
#include <stdio.h>
main()
{
Complex x;
Real y;

        x.r = 1;
        x.i = 1;
        y = Cmag(x); /* compute the magnitude of x        */
        printf("the magnitude of x = %f\n", y);
        /* the answer should be:  y = sqrt(2.) = 1.414214;      */
        return 0;
}
```

Cmul

#include "mathlib.h"
Complex Cmul(Complex *x*, Complex *y*);

DESCRIPTION

The *Cmul* function multiplies the two complex numbers x and y. Both x and y must be declared as **Complex** data types. The resulting product is also a **Complex** number.

RETURNS

z, where $z = x * y$ and x, y, and z are all complex numbers.

SEE ALSO

Cadd(), Csub(), Cdiv().

```
#include "mathlib.h"
#include <stdio.h>
main()
{
Complex x,y,z;

        x.r = 1;
        x.i = 1;
        y.r = 2;
        y.i = 2;
        z = Cmul(x, y); /* multiply x by y            */
        printf("the real part of z = %f\n", z.r);
        printf("the imaginary part of z = %f\n", z.i);
        /* the answer should be:  z.r = 0;  z.i = 4;        */
        return 0;
}
```

Cmxadd

#include "mathlib.h"

Complex__Matrix Cmxadd(Complex__Matrix *m*1, Complex__Matrix *m*2, unsigned *rows*, unsigned *cols*);

DESCRIPTION

The *Cmxadd* function adds the two complex matrices $m1$ and $m2$. The *rows* and *cols* arguments specify the number of rows and columns in each matrix.

Both $m1$ and $m2$ must be allocated by the *Cmxalloc* function prior to calling the *Cmxadd* function. The resulting complex matrix, representing the sum of $m1$ and $m2$, is automatically allocated by the *Cmxadd* function.

The user may select the option of whether or not to write over the input matrix $m1$ with "in-place" computations. If memory is a limitation, and $m1$ is no longer needed, this may prove useful. This option can be selected by setting the global variable **useinput_ = 1** prior to calling *Cmxadd*. If this option is selected, the resulting matrix overwrites the input matrix $m1$. The default option is to not destroy the input matrix $m1$ (i. e., **useinput_ = 0**).

RETURNS

A complex matrix m of size *rows* by *cols*, where $m = m1 + m2$, or NULL if an error occurs.

SEE ALSO

Cmxsub(), mxadd(), mxsub(), Cvadd(), Cvsub(), vadd(), vsub().

```c
#include "mathlib.h"
#include <stdio.h>
main()
{
int i,j;
Complex_Matrix m,m1,m2;
/* add the following two matrices:                                 */
/*                                                                 */
/*          m[][] = m1[][] + m2[][]                                */
/*                                                                 */
/* where m[][] and v[] are complex and are given below:            */
        m1 = Cmxalloc(NULL, 2, 2);  /* allocate a 2 by 2 matrix        */
        m2 = Cmxalloc(NULL, 2, 2); /* allocate another 2 by 2 matrix    */
        m1[0][0].r = 2;   m1[0][0].i = 2;   m1[0][1].r = 2;   m1[0][1].i = 2;
        m1[1][0].r = 2;   m1[1][0].i = 2;   m1[1][1].r = 2;   m1[1][1].i = 2;
        m2[0][0].r = 3;   m2[0][0].i = 3;   m2[0][1].r = 3;   m2[0][1].i = 3;
        m2[1][0].r = 3;   m2[1][0].i = 3;   m2[1][1].r = 3;   m2[1][1].i = 3;

        m = Cmxadd(m1, m2, 2, 2); /* add m1 and m2 */

        /* the answer should be : m[0][0].r = 5;   m[0][0].i = 5;
                                  m[0][1].r = 5;   m[0][1].i = 5;
                                  m[1][0].r = 5;   m[1][0].i = 5;
                                  m[1][1].r = 5;   m[1][1].i = 5;      */
        for (i=0; i<2; i++) {
            for (j=0; j<2; j++) {
                printf("m[%d][%d].r = %f, m[%d][%d].i = %f\n",
                        i, j, m[i][j].r, i, j, m[i][j].i);
            }
        }
        Cmxfree(m);     /* free matrix m  */
        Cmxfree(m1);    /* free matrix m1 */
        Cmxfree(m2);    /* free matrix m2 */
        return 0;
}
```

Cmxalloc

#include "mathlib.h"

Complex__Matrix Cmxalloc(Complex__Ptr *address*, **unsigned** *rows*, **unsigned** *cols*);

DESCRIPTION

The *Cmxalloc* function allocates memory for a complex matrix. The *rows* and *cols* arguments specify the number of rows and columns in the matrix.

The *address* argument is a pointer to a **Complex** value. If the *address* argument is NULL, the *Cmxalloc* function allocates space for the matrix from the available dynamic memory. If the *address* argument is not NULL, the *Cmxalloc* function assumes that it is the address of a two-dimensional array of **Complex** values. In this case, only an array of pointers is allocated and initialized to point to the beginning of each row of the two-dimensional array. This provides a mechanism for using preinitialized arrays.

RETURNS

A *rows* by *cols* complex matrix if successful, or NULL if an error occurs. If the matrix is too large, an "Array too large" error message is generated (math_errno = E_MSIZE). If there is not enough dynamic memory, an "Insufficient dynamic memory available" error message is generated (math_errno = E_MALLOC).

SEE ALSO

mxalloc(), valloc(), Cvalloc().

```
#include "mathlib.h"
#include <stdio.h>
Complex MAT1[2][2] = { 1, 2, 3, 4, 5, 6, 7, 8 } ;
main()
{
Complex_Matrix  mat1,mat2;
unsigned i,j,rows=2,cols=2;

        mat2 = Cmxalloc(NULL, rows, cols); /* allocate complex matrix mat2   */
        mat1 = Cmxalloc(MAT1, rows, cols); /* allocate and initialize mat1   */

                /* now initialize complex matrix mat2 */
        mat2[0][0].r = 1; mat2[0][0].i = 2; mat2[0][1].r = 3; mat2[0][1].i = 4;
        mat2[1][0].r = 5; mat2[1][0].i = 6; mat2[1][1].r = 7; mat2[1][1].i = 8;

/* the matrices mat1[][] and mat2[][] should be equal   */

        for (i=0; i<rows; i++) {
            for (j=0; j<cols; j++) {
                printf("mat1[%d][%d].r = %3.1f, mat1[%d][%d].i = %3.1f   ",
                        i, j, mat1[i][j].r, i, j, mat1[i][j].i);
                printf("mat2[%d][%d].r = %3.1f, mat2[%d][%d].i = %3.1f\n",
                        i, j, mat2[i][j].r, i, j, mat2[i][j].i);
            }
        }
        Cmxfree(mat1);  /* free matrix mat1[][] */
        Cmxfree(mat2);  /* free matrix mat2[][] */
        return 0;
}
```

Cmxconjg

#include "mathlib.h"
Complex__Matrix Cmxconjg(Complex__Matrix *m*, unsigned *rows*, unsigned *cols*);

DESCRIPTION

The *Cmxconjg* function computes the conjugate of the complex matrix m. The *rows* and *cols* arguments specify the number of rows and columns in the matrix.

The complex matrix m must be allocated by the *Cmxalloc* function prior to calling the *Cmxconjg* function. The resulting conjugate matrix is automatically allocated by the *Cmxconjg* function.

The user may select the option of whether or not to write the resulting matrix over the input matrix m. If memory is a limitation, and m is no longer needed, this may prove useful. This option can be selected by setting the global variable **useinput__ = 1** prior to calling *Cmxconjg*. The default option is to not destroy the input matrix m (i. e., **useinput__ = 0**).

RETURNS

A *rows* by *cols* complex matrix representing the conjugate of m, or NULL if an error occurs.

SEE ALSO

conjg(), Cvconjg().

```c
#include "mathlib.h"
#include <stdio.h>
Complex M[2][2] = { 1,-1, 2,-2, 3,-3, 4,-4 };
main()
{
Complex_Matrix m,m1;
unsigned rows=2,cols=2;
int i,j;

        m = Cmxalloc(M, rows, cols); /* allocate and initialize m[][] */

        m1 = Cmxconjg(m, rows,cols); /* compute the conjugate of m[][]*/

        /*   m[][].r should be equal to m1[][].r and
             m[][].i should be equal to -m1[][].i                         */

        for (i=0; i<rows; i++) {
            for (j=0; j<cols; j++) {
                printf("m[%d][%d].r = %3.1f, m[%d][%d].i = %3.1f   ",
                        i, j, m[i][j].r, i, j, m[i][j].i);
                printf("m1[%d][%d].r = %3.1f, m1[%d][%d].i = %3.1f\n",
                        i, j, m1[i][j].r, i, j, m1[i][j].i);
            }
        }

        Cmxfree(m);      /* free matrix m[][]  */
        Cmxfree(m1);     /* free matrix m1[][] */
        return 0;
}
```

Cmxcopy

```
#include "mathlib.h"
void Cmxcopy(Complex_Matrix dest, Complex_Matrix src, unsigned rows, unsigned cols);
```

DESCRIPTION

The *Cmxcopy* function copies the complex matrix *src* to the complex matrix *dest*. The *rows* and *cols* arguments specify the number of rows and columns in each matrix.

The *dest* and *src* matrices must be allocated by the *Cmxalloc* function prior to calling the *Cmxcopy* function.

RETURNS

None.

SEE ALSO

mxcopy(), vcopy(), Cvcopy(), Cmxdup(), mxdup(), Cvdup(), vdup().

```c
#include "mathlib.h"
#include <stdio.h>
Complex SRC[2][2] = { 1, 2, 3, 4, 5, 6, 7, 8 } ;
main()
{
Complex_Matrix  src,dest;
unsigned i,j,rows=2,cols=2;
        dest = Cmxalloc(NULL, rows, cols);/*allocate complex matrix dest[][]*/
        src = Cmxalloc(SRC, rows, cols);  /*allocate and initialize src[][] */
        Cmxcopy(dest, src, rows, cols);  /* copy the complex matrix src[][] */
/* the matrices src[][] and dest[][] should be equal  */
        for (i=0; i<rows; i++) {
            for (j=0; j<cols; j++) {
                printf("src[%d][%d].r = %3.1f, src[%d][%d].i = %3.1f   ",
                        i, j, src[i][j].r, i, j, src[i][j].i);
                printf("dest[%d][%d].r = %3.1f, dest[%d][%d].i = %3.1f\n",
                        i, j, dest[i][j].r, i, j, dest[i][j].i);
            }
        }
        Cmxfree(src);    /* free matrix src[][]  */
        Cmxfree(dest);   /* free matrix dest[][] */
        return 0;
}
```

Cmxdeter

#include "mathlib.h"

Complex Cmxdeter(Complex__Matrix *m*, unsigned *n*);

DESCRIPTION

The *Cmxdeter* function computes the determinant of the complex matrix *m*, where *m* contains *n* rows and *n* columns. The complex matrix *m* must be allocated by *Cmxalloc* prior to calling the *Cmxdeter* function.

The user may select the option of whether or not to write over the input matrix *m* with "in-place" computations. If memory is a limitation, and *m* is no longer needed, this may prove useful. This option can be selected by setting the global variable **useinput__ = 1** prior to calling *Cmxdeter*. The default option is to not destroy *m* (i. e., **useinput__ = 0**).

The *Cmxdeter* function can be used to determine if a system of simultaneous complex linear equations has a solution. Just as for linear systems with real coefficients, a unique solution to a complex system exists only if the determinant of the system matrix is nonzero. More precisely, since the determinant of a complex matrix is a complex number, its magnitude must be nonzero if a solution exists. Thus, the case of no solution occurs whenever the magnitude of the system determinant is zero (i. e., the system matrix is singular, or noninvertible).

RETURNS

A complex value representing the determinant of the *n* by *n* complex matrix *m* if successful (also stored in global variable **Cdeterm__**), or 0 (i. e., both the real and imaginary parts of the complex value are 0) if an error occurs. If the determinant is 0, then a "Singular matrix" error is generated (math__errno = E__MSING).

SEE ALSO

mxdeterm().

REFERENCE

J. H. Wilkinson (1965).

```
#include "mathlib.h"
#include <stdio.h>
#include <math.h>
Complex M[2][2] = { 1,2, 3,4, 5,6, 7,8 } ;

main()
{
Complex det;
Complex_Matrix m;

/* compute the determinant of the complex 2 by 2 matrix m[][]            */

     m = Cmxalloc(M, 2, 2);   /* allocate & initialize system matrix   */

     det = Cmxdeter(m, 2);     /* now find the determinant of m[][]     */

     /* the answer should be : det.r = 0      det.i = -16                */

     printf("det.r = %f, det.i = %f\n", det.r, det.i);

     Cmxfree(m);     /* free matrix m[][] */
     return 0;
}
```

Cmxdup

#include "mathlib.h"
Complex__Matrix Cmxdup(Complex__Matrix *m*, unsigned *rows*, unsigned *cols*);

DESCRIPTION

The *Cmxdup* function makes a duplicate copy of the complex matrix m. The *rows* and *cols* arguments specify the number of rows and columns in the matrix.

The complex matrix m must be allocated by *Cmxalloc* prior to calling the *Cmxdup* function. The *Cmxdup* function automatically allocates memory for the duplicate copy.

RETURNS

A *rows* by *cols* complex matrix that is a duplicate copy of m if successful, or NULL if an error occurs. If there is not enough dynamic memory, an "Insufficient dynamic memory available" error message is generated (math_errno = E_MALLOC).

SEE ALSO

mxdup(), vdup(), Cvdup(), Cmxcopy(), mxcopy(), vcopy(), Cvcopy().

```
#include "mathlib.h"
#include <stdio.h>
Complex M1[2][2] = { 1, 2, 3, 4, 5, 6, 7, 8 } ;


main()
{
Complex_Matrix  m1,m2;
unsigned i,j,rows=2,cols=2;

        m1 = Cmxalloc(M1, rows, cols);  /* allocate and initialize m1[][]    */

        m2 = Cmxdup(m1, rows, cols);    /* now duplicate the matrix m1[][]    */

/* the matrices m1[][] and m2[][] should be equal  */

        for (i=0; i<rows; i++) {
            for (j=0; j<cols; j++) {
                printf("m1[%d][%d].r = %3.1f, m1[%d][%d].i = %3.1f   ",
                        i, j, m1[i][j].r, i, j, m1[i][j].i);
                printf("m2[%d][%d].r = %3.1f, m2[%d][%d].i = %3.1f\n",
                        i, j, m2[i][j].r, i, j, m2[i][j].i);
            }
        }

        Cmxfree(m1);    /* free matrix m1[][] */
        Cmxfree(m2);    /* free matrix m2[][] */
        return 0;
}
```

Cmxeigen

#include "mathlib.h"
Real_Vector Cmxeigen(Complex_Matrix *m*, unsigned *n*, unsigned *vflag*);

DESCRIPTION
Eigensystems play a crucial role in the theory of electrical and mechanical resonance, and also in the theory of statics. Eigenvalues and eigenvectors can be used to determine the stability of feedback systems and to control the convergence of associated tracking algorithms. The concepts of similarity transforms and matrix diagonalization are of fundamental importance to eigenanalysis because the eigenvalues are invariant to such transforms.

A nonsingular $n \times n$ complex Hermitian matrix has n real eigenvalues and n distinct eigenvectors. A Hermitian matrix exhibits complex conjugate symmetry around its main diagonal, which also means that the elements along the main diagonal must be real. Thus, a real symmetric matrix is also Hermitian, since the imaginary parts of its elements are all zero. In fact, the eigenanalysis of any Hermitian matrix can be transformed into an equivalent problem that involves a $2n \times 2n$ real symmetric matrix. The real symmetric eigensystem can then be solved with Jacobi transformations.

The *Cmxeigen* function computes the eigenvalues, and optionally the eigenvectors of a complex Hermitian matrix.

The n argument specifies the number of rows and columns in the square Hermitian matrix m.

The *vflag* argument specifies whether or not to compute the eigenvectors in addition to the eigenvalues. If *vflag* is 0, the eigenvectors are not computed. If *vflag* is nonzero, the eigenvectors are computed and returned in matrix m. In this case, each column of matrix m will contain one of the eigenvectors. The eigenvalues and associated eigenvectors are sorted in descending order.

The *Cmxeigen* function calls the *mxeigen* function.

RETURNS
A real vector of length n representing the ordered eigenvalues of the complex Hermitian matrix m if successful, or NULL if an error occurs. If matrix m does not have the proper symmetry of a Hermitian matrix, the error message "Input matrix is non-Hermitian" is generated (math_errno = E_Hermitian).

SEE ALSO
mxeigen().

REFERENCES
J. H. Wilkinson (1965); W. H. Press, et al. (1986).

```
#include <math.h>
#include <stdio.h>
#include "mathlib.h"

Complex M[3][3] = {  7,0,    -2,2,   1,1,
                    -2,-2,   10,0,  -2,-3,
                     1,-1,   -2,3,   7,0  };
main()
{
Complex_Matrix eigenvectors;
Real_Vector eigenvalues;
unsigned i, vflag=1, n=3;

/* Find eigenvalues and eigenvectors of A[][] */

        eigenvectors = Cmxalloc(M, n, n);
        eigenvalues = Cmxeigen(eigenvectors, n, vflag);

        printf("eigenvalues              complex eigenvectors \n");
        for (i=0;i<n; i++){
           printf("%9.6f  <(%9.6f,%9.6f), (%9.6f,%9.6f), (%9.6f,%9.6f)>\n",
               eigenvalues[i],
               eigenvectors[i][0].r,eigenvectors[i][0].i,
               eigenvectors[i][1].r,eigenvectors[i][1].i,
               eigenvectors[i][2].r,eigenvectors[i][2].i);
        }
        vfree(eigenvalues);
        Cmxfree(eigenvectors);
        return 0;
}
```

PROGRAM OUTPUT

```
eigenvalues              complex eigenvectors
13.782416  <(-0.409665,-0.049625), ( 0.767786,-0.202369), ( 0.283851,-0.344517)>
 6.069389  <( 0.543908, 0.544336), ( 0.239306,-0.138772), ( 0.493517, 0.296274)>
 4.148195  <(-0.468876, 0.133152), (-0.540636,-0.027193), ( 0.679051,-0.091043)>
```

Cmxfree

#include "mathlib.h"
void Cmxfree(Complex__Matrix *m*);

DESCRIPTION

The *Cmxfree* function frees the memory allocated to the complex matrix m. The memory for the complex matrix m must have been allocated by the *Cmxalloc* function prior to calling the *Cmxfree* function.

RETURNS

None.

SEE ALSO

mxfree(), vfree(), Cvfree(), mathfree().

```
#include "mathlib.h"
#include <stdio.h>
Complex M1[2][2] = { 1,2, 3,4, 5,6, 7,8};
main()
{
Complex_Matrix m,m1;
int i,j;

        m = Cmxalloc(NULL, 2, 2); /* allocate dynamic memory for m[][]    */
        m1 = Cmxalloc(M1, 2, 2);  /* initialize static array m1[][] and
                                     allocate dynamic memory for the 1-D
                                     array of pointers to its rows        */

    for (i=0; i<2; i++) {
            for (j=0; j<2; j++) {
                    m[i][j] = m1[i][j]; /* set m[][] equal to m1[][] */
                    printf("m[%d][%d].r = %f, m[%d][%d].i = %f\n",
                            i, j, m[i][j].r,  i, j, m[i][j].i);
            }
    }
      /* now free all the dynamic memory occupied by m[][] and m1[][] */
    Cmxfree(m);  /*  free the dynamic memory occupied by m[][]        */
    Cmxfree(m1); /*  free the dynamic memory used by the 1-D array of */
                 /*    pointers to the rows of m1[][]                 */
    return 0;
}
```

Cmxinit

```
#include "mathlib.h"
void Cmxinit(Complex_Matrix m, unsigned rows, unsigned cols, Complex value);
```

DESCRIPTION

The *Cmxinit* function initializes each element of the complex matrix m with the complex scalar specified by *value*. The *rows* and *cols* arguments specify the number of rows and columns in the matrix. The complex matrix m must be allocated by *Cmxalloc* prior to calling the *Cmxinit* function.

RETURNS

None.

SEE ALSO

mxinit(), Cvinit(), vinit().

```
#include "mathlib.h"
#include <stdio.h>
main()
{
Complex_Matrix m;
Complex zero;
int i,j;

        m = Cmxalloc(NULL, 2, 2);       /* allocate matrix m[][]             */

        zero.r = zero.i = 0;

        Cmxinit(m, 2, 2, zero);         /* set matrix m[][] to zero          */
        m[0][0].r = m[0][0].i = 1;      /* set diagonal terms to 1 and       */
        m[1][1].r = m[1][1].i = 1;      /* create the 2 by 2 identity matrix */

        printf("the following is the complex 2 by 2 identity matrix:\n");
        for (i=0; i<2; i++) {
                for (j=0; j<2; j++) {
                        printf(" %f %f    ", m[i][j].r, m[i][j].i);
                }
                printf("\n");
        }
        return 0;
}
```

Cmxinv

#include "mathlib.h"
Complex_Matrix Cmxinv(Complex_Matrix *m*, unsigned *n*);

DESCRIPTION
The *Cmxinv* function computes the inverse of the complex matrix *m*. The *n* argument specifies the number of rows and columns in the square matrix *m*.

The complex matrix *m* must be allocated by the *Cmxalloc* function prior to calling the *Cmxinv* function. The resulting complex matrix representing the inverse of *m* is automatically allocated by the *Cmxinv* function.

The user may select the option of whether or not to write over the input matrix *m* with "in-place" computations. If memory is a limitation, and *m* is no longer needed, this may prove useful. This option can be selected by setting the global variable **useinput_ = 1** prior to calling *Cmxinv*. The default option is to not destroy the input matrix *m* (i. e., **useinput_ = 0**).

Matrix inversion can be used to solve sets of linear equations. Given *n* equations with *n* unknowns, the linear set of equations can be written as:

$$M * v = b$$

If M^{-1} exists, then the solution vector *v* can be found by premultiplying both sides of the above equation by M^{-1}. Since $M^{-1} * M$ is equal to the identity matrix, the solution is

$$v = M^{-1} * b$$

Although matrix inversion is a convenient method for solving systems of linear equations, it is not the most efficient, nor the most accurate. Direct matrix reduction techniques such as LU decomposition (see the *Clineqn* and *lineqn* functions) and Gaussian elimination often provide superior speed and accuracy. This is especially true for large matrices that are nearly singular. However, for small matrices, matrix inversion is often adequate.

RETURNS
An *n* by *n* complex matrix representing the inverse of the complex matrix *m* if successful, or NULL if an error occurs. If the inverse of *m* does not exist, a "Singular matrix" error is generated (math_errno = E_MSING).

SEE ALSO
Cmxinv22(), Cmxinv33(), mxinv(), mxinv22(), mxinv33(), Clineqn(), lineqn().

```
#include "mathlib.h"
#include <stdio.h>
Complex M[2][2] = { 1,2, 3,4, 5,6, 7,8 };
main()
{
int i,j;
Complex_Matrix m,minv;

        m = Cmxalloc(M, 2, 2); /* allocate & initialize 2 by 2 system matrix */

        minv = Cmxinv(m, 2);                    /* now find the inverse of m[][]    */

        /* the answer should be :  */

        /* minv[0][0].r = -0.500000, minv[0][0].i = 0.437500                */
        /* minv[0][1].r = 0.250000, minv[0][1].i = -0.187500               */
        /* minv[1][0].r = 0.375000, minv[1][0].i = -0.312500               */
        /* minv[1][1].r = -0.125000, minv[1][1].i = 0.062500               */

        for (i=0; i<2; i++) {
                for (j=0; j<2; j++) {
                        printf("minv[%d][%d].r = %f, minv[%d][%d].i = %f\n",
                                i, j, minv[i][j].r,  i, j, minv[i][j].i);
                }
        }
        Cmxfree(m);        /* free matrix m[][]    */
        Cmxfree(minv);     /* free matrix minv[][] */
        return 0;
}
```

Cmxinv22

> #include "mathlib.h"
> Complex_Matrix Cmxinv22(Complex_Matrix *m*);

DESCRIPTION

The *Cmxinv22* function computes the inverse of the two by two square complex matrix m. The complex matrix m must be allocated by the *Cmxalloc* function prior to calling the *Cmxinv22* function. The resulting complex matrix representing the inverse of m is automatically allocated by the *Cmxinv22* function.

The user may select the option of whether or not to write over the input matrix m with "in-place" computations. If memory is a limitation, and m is no longer needed, this may prove useful. This option can be selected by setting the global variable **useinput_ = 1** prior to calling *Cmxinv22*. The default option is to not destroy the input matrix m (i. e., **useinput_ = 0**).

The *Cmxinv22* function is specific to two by two matrices and has some size and speed advantages over the more general *Cmxinv* function.

RETURNS

A two by two complex matrix representing the inverse of the complex matrix m if successful, or NULL if an error occurs. If the inverse of m exists, the global variable **Cdeterm_** is set equal to the determinant of the resulting matrix. Otherwise, a "Singular matrix" error is generated (math_errno = E_MSING).

SEE ALSO

Cmxinv(), Cmxinv33(), mxinv(), mxinv22(), mxinv33(), Clineqn(), lineqn().

```
#include "mathlib.h"
#include <stdio.h>
Complex M[2][2] = { 1,2, 3,4, 5,6, 7,8 };
main()
{
int i,j;
Complex_Matrix m,minv;

        m = Cmxalloc(M, 2, 2); /* allocate & initialize 2 by 2 system matrix */

        minv = Cmxinv22(m);                 /* now find the inverse of m[][]    */

        /* the answer should be :  */

        /* minv[0][0].r = -0.500000, minv[0][0].i = 0.437500            */
        /* minv[0][1].r = 0.250000, minv[0][1].i = -0.187500           */
        /* minv[1][0].r = 0.375000, minv[1][0].i = -0.312500           */
        /* minv[1][1].r = -0.125000, minv[1][1].i = 0.062500           */

        for (i=0; i<2; i++) {
                for (j=0; j<2; j++) {
                        printf("minv[%d][%d].r = %f, minv[%d][%d].i = %f\n",
                                i, j, minv[i][j].r,  i, j, minv[i][j].i);
                }
        }
        Cmxfree(m);        /* free matrix m[][]    */
        Cmxfree(minv);     /* free matrix minv[][] */
        return 0;
}
```

Cmxinv33

#include "mathlib.h"
Complex__Matrix Cmxinv33(Complex__Matrix *m*);

DESCRIPTION

The *Cmxinv33* function computes the inverse of the three by three square complex matrix *m*. The complex matrix *m* must be allocated by the *Cmxalloc* function prior to calling the *Cmxinv33* function. The resulting complex matrix representing the inverse of *m* is automatically allocated by the *Cmxinv33* function.

The user may select the option of whether or not to write over the input matrix *m* with "in-place" computations. If memory is a limitation, and *m* is no longer needed, this may prove useful. This option can be selected by setting the global variable **useinput_ = 1** prior to calling *Cmxinv33*. The default option is to not destroy the input matrix *m* (i. e., **useinput_ = 0**).

The *Cmxinv33* function is specific to three by three matrices and has some size and speed advantages over the more general *Cmxinv* function.

RETURNS

A three by three complex matrix representing the inverse of the complex matrix *m* if successful, or NULL if an error occurs. If the inverse of *m* exists, the global variable **Cdeterm_** is set equal to the determinant of the resulting matrix. Otherwise, a "Singular matrix" error is generated (math_errno = E_MSING).

SEE ALSO

Cmxinv(), Cmxinv22(), mxinv(), mxinv22(), mxinv33(), Clineqn(), lineqn().

```
#include "mathlib.h"
#include <stdio.h>
Complex M[3][3] = { 1,2, 0,0, 0,0, 0,0, 9,10, 0,0, 0,0, 0,0, 1,1 };
main()
{
int i,j;
Complex_Matrix m,minv;

        m = Cmxalloc(M, 3, 3); /* allocate & initialize 3 by 3 system matrix */

        minv = Cmxinv33(m);                 /* now find the inverse of m[][]    */

        /* the answer should be :  */

        /*    minv[0][0].r = 0.200000, minv[0][0].i = -0.400000  */
        /*    minv[0][1].r = 0.000000, minv[0][1].i = -0.000000  */
        /*    minv[0][2].r = 0.000000, minv[0][2].i = -0.000000  */
        /*    minv[1][0].r = 0.000000, minv[1][0].i = -0.000000  */
        /*    minv[1][1].r = 0.049724, minv[1][1].i = -0.055249  */
        /*    minv[1][2].r = 0.000000, minv[1][2].i = -0.000000  */
        /*    minv[2][0].r = 0.000000, minv[2][0].i = -0.000000  */
        /*    minv[2][1].r = 0.000000, minv[2][1].i = -0.000000  */
        /*    minv[2][2].r = 0.500000, minv[2][2].i = -0.500000  */

        for (i=0; i<3; i++) {
                for (j=0; j<3; j++) {
                        printf("minv[%d][%d].r = %f, minv[%d][%d].i = %f\n",
                                i, j, minv[i][j].r,  i, j, minv[i][j].i);
                }
        }
        Cmxfree(m);        /* free matrix m[][]    */
        Cmxfree(minv);     /* free matrix minv[][] */
        return 0;
}
```

Cmxmag

> #include "mathlib.h"
> Real__Matrix Cmxmag(Complex__Matrix *m*, unsigned *rows*, unsigned *cols*);

DESCRIPTION
The *Cmxmag* function computes the magnitude of the complex matrix m. For each complex value in matrix m, a real value is calculated as the square root of the sum of the squares of the real and imaginary parts.

The *rows* and *cols* arguments specify the number of rows and columns in the matrix. The complex matrix m must be allocated by *Cmxalloc* prior to calling the *Cmxmag* function. The resulting real matrix representing the magnitude of m is automatically allocated by the *Cmxmag* function.

RETURNS
A *rows* by *cols* real matrix that represents the magnitude of each complex value in matrix m if successful, or NULL if an error occurs.

SEE ALSO
Cmag(), Cvmag(), vmag().

```
#include "mathlib.h"
#include <stdio.h>
Complex M[2][2] = { 1,1, 1,1, 1,1, 1,1 };
main()
{
int i,j;
Complex_Matrix m;
Real_Matrix m1;
        m = Cmxalloc(M, 2, 2);/* initialize m[][] to static complex array M[]*/
        m1 = Cmxmag(m, 2, 2); /* compute the magnitude of m[][]             */
        /* the answer should be :  */
        /*  m1[0][0] = 1.414214  m1[0][1] = 1.414214  */
        /*  m1[1][0] = 1.414214  m1[1][1] = 1.414214  */
        for (i=0; i<2; i++) {
                for (j=0; j<2; j++) {
                        printf("m1[%d][%d] = %f\n", i, j, m1[i][j]);
                }
        }
        Cmxfree(m);       /* free complex matrix m[][]    */
        mxfree(m1);       /* free real matrix m1[][]    */
        return 0;
}
```

Cmxmaxvl

```
#include "mathlib.h"
Real Cmxmaxvl(Complex__Matrix m, unsigned rows, unsigned cols, unsigned *imx,
              unsigned *jmx);
```

DESCRIPTION

The *Cmxmaxvl* function finds the element in the complex matrix m that has the largest magnitude. The magnitude of each complex value is calculated as the square root of the sum of the squares of the real and imaginary parts.

The *rows* and *cols* arguments specify the number of rows and columns in the matrix. The *imx* argument is the address of an unsigned integer where *Cmxmaxvl* stores the row number of the largest value. The *jmx* argument is the address of an unsigned integer where *Cmxmaxvl* stores the column number of the largest value.

RETURNS

A real value representing the magnitude of the largest value in the complex matrix m. The row and column of the value with the largest magnitude is returned in *imx* and *jmx*, respectively.

SEE ALSO

Cmxminvl(), Cvmaxval(), Cvminval(), mxmaxval(), mxminval(), vmaxval(), vminval().

```
#include "mathlib.h"
#include <stdio.h>
Complex M[2][2] = { 1,2, 3,4, 10,10, 5,6 };
main()
{
Complex_Matrix m;
unsigned imax,jmax;
Real biggest;

    m = Cmxalloc(M, 2, 2);     /* allocate and initialize matrix m[][]   */

    biggest = Cmxmaxvl(m, 2, 2, &imax, &jmax); /* find largest magnitude */

    printf("the maximum magnitude is = %f\n", biggest);
    printf("it is in row %d and column %d\n", imax, jmax);
    printf("the real component is %f and the imaginary component is %f\n",
            m[imax][jmax].r, m[imax][jmax].i);
    Cmxfree(m); /* free matrix m[][] */
    return 0;
}
```

Cmxminvl

```
#include "mathlib.h"
Real Cmxminvl(Complex_Matrix m, unsigned rows, unsigned cols unsigned *imn,
              unsigned *jmn);
```

DESCRIPTION

The *Cmxminvl* function finds the element in the complex matrix m that has the smallest magnitude. The magnitude of each complex value is calculated as the square root of the sum of the squares of the real and imaginary parts.

The *rows* and *cols* arguments specify the number of rows and columns in the matrix. The *imn* argument is the address of an unsigned integer where *Cmxminvl* stores the row number of the smallest value. The *jmn* argument is the address of an unsigned integer where *Cmxminvl* stores the column number of the smallest value.

RETURNS

A real value representing the magnitude of the smallest value in the complex matrix m. The row and column of the value with the smallest magnitude is returned in *imn* and *jmn*, respectively.

SEE ALSO

Cmxmaxvl(), Cvmaxval(), Cvminval(), mxmaxval(), mxminval(), vmaxval(), vminval().

```
#include "mathlib.h"
#include <stdio.h>
Complex M[2][2] = { 1,2, 3,4, 0,1, 5,6 };
main()
{
Complex_Matrix m;
unsigned imin,jmin;
Real smallest;

        m = Cmxalloc(M, 2, 2);      /* allocate and initialize matrix m[][]   */

        smallest = Cmxminvl(m, 2, 2, &imin, &jmin);/*find smallest magnitude */

        printf("the minimum magnitude is = %f\n", smallest);
        printf("it is in row %d and column %d\n", imin, jmin);
        printf("the real component is %f and the imaginary component is %f\n",
               m[imin][jmin].r, m[imin][jmin].i);
        Cmxfree(m); /* free matrix m[][] */
        return 0;
}
```

Cmxmul

```
#include "mathlib.h"
Complex_Matrix Cmxmul(Complex_Matrix m1, Complex_Matrix m2,
                      unsigned n1, unsigned n2, unsigned n3);
```

DESCRIPTION

The *Cmxmul* function multiplies the two complex matrices $m1$ and $m2$.

The number of rows and columns in $m1$ is specified by $n1$ and $n2$ respectively (i. e., $m1[n1][n2]$). The number of rows and columns in $m2$ is specified by $n2$ and $n3$ respectively (i. e., $m2[n2][n3]$). The number of columns in matrix $m1$ must be equal to the number of rows in matrix $m2$.

The complex matrices $m1$ and $m2$ must be allocated by *Cmxalloc* prior to calling the *Cmxmul* function. The resulting complex matrix representing the product of $m1$ and $m2$ is automatically allocated by the *Cmxmul* function. The resulting matrix contains $n1$ rows and $n3$ columns.

RETURNS

An $n1$ by $n3$ complex matrix representing the product of $m1$ and $m2$ if successful, or NULL if an error occurs.

SEE ALSO

Cmxmul1(), Cmxmul2(), mxmul(), mxmul1(), mxmul2().

```c
#include "mathlib.h"
#include <stdio.h>
Complex M1[2][2] = { 2,2, 2,2, 2,2, 2,2 };
Complex M2[2][2] = { 3,3, 3,3, 3,3, 3,3 };
main()
{
int i,j;
Complex_Matrix m,m1,m2;
/* multiply the following two matrices:                                 */
/*        m[][] = m1[][] * m2[][]                                        */
/* where m1[][] and m2[][] are complex and are given below:             */
      m1 = Cmxalloc(M1, 2, 2);  /* allocate a 2 by 2 matrix             */
      m2 = Cmxalloc(M2, 2, 2); /* allocate another 2 by 2 matrix        */
      m  = Cmxmul(m1, m2, 2, 2, 2); /* multiply m1[][] by m2[][]        */
      /* the answer should be : m[i][j].r = 0;  m[i][j].i = 24;         */
      /*                                      for all i and j           */
      for (i=0; i<2; i++) {
          for (j=0; j<2; j++) {
              printf("m[%d][%d].r = %f, m[%d][%d].i = %f\n",
                      i, j, m[i][j].r, i, j, m[i][j].i);
          }
      }
      Cmxfree(m);     /* free matrix m[][]  */
      Cmxfree(m1);    /* free matrix m1[][] */
      Cmxfree(m2);    /* free matrix m2[][] */
      return 0;
}
```

Cmxmul1

```
#include "mathlib.h"
Complex_Matrix Cmxmul1(Complex_Matrix m1, Complex_Matrix m2,
                       unsigned n1, unsigned n2, unsigned n3);
```

DESCRIPTION

The *Cmxmul1* function multiplies the transpose of the complex matrix $m1$ by the complex matrix $m2$ (i. e., $m1^T * m2$).

The number of rows and columns in $m1$ is specified by $n2$ and $n1$ respectively (i. e., $m1[n2][n1]$). The number of rows and columns in $m2$ is specified by $n2$ and $n3$ respectively (i. e., $m2[n2][n3]$). The number of rows in matrix $m1$ must be equal to the number of rows in matrix $m2$.

The complex matrices $m1$ and $m2$ must be allocated by *Cmxalloc* prior to calling the *Cmxmul1* function. The resulting complex matrix representing the product of $m1^T$ and $m2$ is automatically allocated by the *Cmxmul1* function. The resulting matrix contains $n1$ rows and $n3$ columns.

RETURNS

An $n1$ by $n3$ complex matrix representing the product of $m1^T$ and $m2$ if successful, or NULL if an error occurs.

SEE ALSO

Cmxmul(), Cmxmul2(), mxmul(), mxmul1(), mxmul2().

```
#include "mathlib.h"
#include <stdio.h>
Complex M1[2][3] = { 1,1, 1,1, 1,1,   2,2, 2,2, 2,2 };
Complex M2[2][2] = { 1,1, 1,1,  2,2, 2,2 };
main()
{
int i,j;
Complex_Matrix m,m1,m2;
/* multiply the following two matrices:                                   */
/*        m[][] = T(m1[][]) * m2[][]                                      */
/* where m1[][] and m2[][] are complex and are given below:               */
      m1 = Cmxalloc(M1, 2, 3);   /* allocate a 2 by 3 matrix              */
      m2 = Cmxalloc(M2, 2, 2);  /* allocate a 2 by 2 matrix               */
      m  = Cmxmul1(m1, m2, 3, 2, 2);/*multiply TRANSPOSE{m1[][]} by m2[][] */

/* the answer is :  m[i][j].r = 0, and m[i][j].i = 10, for all i and j     */

      for (i=0; i<3; i++) {
          for (j=0; j<2; j++) {
              printf("m[%d][%d].r = %f, m[%d][%d].i = %f\n",
                      i, j, m[i][j].r, i, j, m[i][j].i);
          }
      }
      Cmxfree(m);      /* free matrix m[][]  */
      Cmxfree(m1);     /* free matrix m1[][] */
      Cmxfree(m2);     /* free matrix m2[][] */
      return 0;
}
```

Cmxmul2

#include "mathlib.h"
Complex__Matrix Cmxmul2(Complex__Matrix $m1$, Complex__Matrix $m2$,
                  unsigned $n1$, unsigned $n2$, unsigned $n3$);

DESCRIPTION

The *Cmxmul2* function multiplies the complex matrix $m1$ by the transpose of the complex matrix $m2$ (i. e., $m1 * m2^T$).

The number of rows and columns in $m1$ is specified by $n1$ and $n2$ respectively (i. e., $m1[n1][n2]$). The number of rows and columns in $m2$ is specified by $n3$ and $n2$ respectively (i. e., $m2[n3][n2]$). The number of columns in matrix $m1$ must be equal to the number of columns in matrix $m2$.

The complex matrices $m1$ and $m2$ must be allocated by *Cmxalloc* prior to calling the *Cmxmul2* function. The resulting complex matrix representing the product of $m1$ and $m2^T$ is automatically allocated by the *Cmxmul2* function. The resulting matrix contains $n1$ rows and $n3$ columns.

RETURNS

An $n1$ by $n3$ complex matrix representing the product of $m1$ and $m2^T$ if successful, or NULL if an error occurs.

SEE ALSO

Cmxmul(), Cmxmul1(), mxmul(), mxmul1(), mxmul2().

```
#include "mathlib.h"
#include <stdio.h>
Complex M1[2][2] = { 1,1, 1,1,  2,2, 2,2 };
Complex M2[3][2] = { 1,1, 1,1,  2,2, 2,2,  3,3, 3,3 };
main()
{
int i,j;
Complex_Matrix m,m1,m2;
/* multiply the following two matrices:                              */
/*       m[][] = m1[][] * T(m2[][])                                  */
/* where m1[][] and m2[][] are complex and are given below:          */
        m1 = Cmxalloc(M1, 2, 2);  /* allocate a 2 by 2 matrix        */
        m2 = Cmxalloc(M2, 3, 2);  /* allocate a 3 by 2 matrix        */
        m  = Cmxmul2(m1, m2, 2, 2, 3);/*multiply m1[][] by TRANSPOSE{ m2[][]} */

        /* the answer is :  m[i][j].r = 0,  for all i and j; and     */
        /*  m[0][0].i = 4,  m[0][1].i = 8,  m[0][2].i = 12           */
        /*  m[1][0].i = 8,  m[1][1].i = 16, m[1][2].i = 24           */
        for (i=0; i<2; i++) {
            for (j=0; j<3; j++) {
                printf("m[%d][%d].r = %f, m[%d][%d].i = %f\n",
                        i, j, m[i][j].r, i, j, m[i][j].i);
            }
        }
        Cmxfree(m);     /* free matrix m[][]  */
        Cmxfree(m1);    /* free matrix m1[][] */
        Cmxfree(m2);    /* free matrix m2[][] */
        return 0;
}
```

CmxscalC

#include "mathlib.h"
Complex_Matrix CmxscalC(Complex_Matrix *m*, unsigned *rows*, unsigned *cols*,
Complex *value*);

DESCRIPTION

The *CmxscalC* function multiplies the complex matrix *m* by the complex scalar *value*. The *rows* and *cols* arguments specify the number of rows and columns in the matrix.

The complex matrix *m* must be allocated by *Cmxalloc* prior to calling the *Cmxscal* function. The resulting complex matrix representing the product of *m* and *value* is automatically allocated by the *CmxscalC* function.

The user may select the option of whether or not to write over the input matrix *m* with "in-place" computations. If memory is a limitation, and *m* is no longer needed, this may prove useful. This option can be selected by setting the global variable **useinput_ = 1** prior to calling *CmxscalC*. The default option is to not destroy the input matrix *m* (i. e., **useinput_ = 0**).

RETURNS

A *rows* by *cols* complex matrix representing the product of the complex matrix *m* with the complex *value* if successful, or NULL if an error occurs.

SEE ALSO

CmxscalR(), mxscale(), CvscalC(), CvscalR(), vscale().

```
#include "mathlib.h"
#include <stdio.h>
Complex M[2][2] = { 1,1, 1,1, 1,1, 1,1 };
main()
{
Complex_Matrix m,m1;
unsigned rows=2,cols=2;
int i,j;
Complex value;

        value.r = 2; value.i = 2;  /* define a complex scalar          */
        m=Cmxalloc(M, rows, cols); /* allocate and initialize m[][]     */
        m1 = CmxscalC(m, rows, cols, value);/* now scale complex matrix m[][]*/

        /* the answer should be : m1[0][0].r = 0;  m1[0][0].i = 4;
                                  m1[0][1].r = 0;  m1[0][1].i = 4;
                                  m1[1][0].r = 0;  m1[1][0].i = 4;
                                  m1[1][1].r = 0;  m1[1][1].i = 4;         */
        for (i=0; i<2; i++) {
            for (j=0; j<2; j++) {
                printf("m1[%d][%d].r = %f, m1[%d][%d].i = %f\n",
                        i, j, m1[i][j].r, i, j, m1[i][j].i);
            }
        }
        Cmxfree(m);    /* free matrix m[][]  */
        Cmxfree(m1);   /* free matrix m1[][] */
        return 0;
}
```

CmxscalR

#include "mathlib.h"
Complex__Matrix CmxscalR(Complex__Matrix *m*, unsigned *rows*, unsigned *cols*, Real *value*);

DESCRIPTION

The *CmxscalR* function multiplies the complex matrix m by the real scalar *value*. The *rows* and *cols* arguments specify the number of rows and columns in the matrix.

The complex matrix m must be allocated by *Cmxalloc* prior to calling the *CmxscalR* function. The resulting complex matrix representing the product of m and *value* is automatically allocated by the *CmxscalR* function.

The user may select the option of whether or not to write over the input matrix m with "in-place" computations. If memory is a limitation, and m is no longer needed, this may prove useful. This option can be selected by setting the global variable **useinput_ = 1** prior to calling *CmxscalR*. The default option is to not destroy the input matrix m (i. e., **useinput_ = 0**).

RETURNS

A *rows* by *cols* complex matrix created by multiplying the complex matrix m by the real *value* if successful, or NULL if an error occurs.

SEE ALSO

CmxscalC(), mxscale(), CvscalC(), CvscalR(), vscale().

```
#include "mathlib.h"
#include <stdio.h>
Complex M[2][2] = { 1,1,  1,1,  1,1,  1,1 };
main()
{
Complex_Matrix m,m1;
unsigned rows=2,cols=2;
int i,j;
Real value;

        value = 2;   /* define a real scalar                        */
        m=Cmxalloc(M, rows, cols); /* allocate and initialize m[][]  */
        m1 = CmxscalR(m, rows, cols, value);/* now scale complex matrix m[][]*/

        /* the answer should be : m1[0][0].r = 2;   m1[0][0].i = 2;
                                  m1[0][1].r = 2;   m1[0][1].i = 2;
                                  m1[1][0].r = 2;   m1[1][0].i = 2;
                                  m1[1][1].r = 2;   m1[1][1].i = 2;        */
        for (i=0; i<2; i++) {
            for (j=0; j<2; j++) {
                printf("m1[%d][%d].r = %f, m1[%d][%d].i = %f\n",
                       i, j, m1[i][j].r, i, j, m1[i][j].i);
            }
        }
        Cmxfree(m);    /* free matrix m[][]  */
        Cmxfree(m1);   /* free matrix m1[][] */
        return 0;
}
```

Cmxsub

#include "mathlib.h"
Complex_Matrix Cmxsub(Complex_Matrix *m*1, Complex_Matrix *m*2, unsigned *rows*, unsigned *cols*);

DESCRIPTION

The *Cmxsub* function subtracts the complex matrix $m2$ from the complex matrix $m1$. The *rows* and *cols* arguments specify the number of rows and columns in each matrix.

The complex matrices $m1$ and $m2$ must be allocated by *Cmxalloc* prior to calling the *Cmxsub* function. The resulting complex matrix representing the difference between $m1$ and $m2$ is automatically allocated by the *Cmxsub* function.

The user may select the option of whether or not to write over the input matrix $m1$ with "in-place" computations. If memory is a limitation, and $m1$ is no longer needed, this may prove useful. This option can be selected by setting the global variable **useinput_ = 1** prior to calling *Cmxsub*. If this option is selected, the resulting matrix overwrites the input matrix $m1$. The default option is to not destroy the input matrix $m1$ (i. e., **useinput_ = 0**).

RETURNS

A *rows* by *cols* complex matrix representing the difference between $m1$ and $m2$ if successful, or NULL if an error occurs.

SEE ALSO

Cmxadd(), mxadd(), mxsub(), Cvadd(), Cvsub(), vadd(), vsub().

```c
#include "mathlib.h"
#include <stdio.h>
main()
{
int i,j;
Complex_Matrix m,m1,m2;
/* subtract the following two matrices:                         */
/*                                                              */
/*         m[][] = m1[][] - m2[][]                              */
/*                                                              */
/* where m1[][] and m2[][] are complex and are given below:     */
        m1 = Cmxalloc(NULL, 2, 2);  /* allocate a 2 by 2 matrix        */
        m2 = Cmxalloc(NULL, 2, 2); /* allocate another 2 by 2 matrix   */
        m1[0][0].r = 2;   m1[0][0].i = 2;   m1[0][1].r = 2;    m1[0][1].i = 2;
        m1[1][0].r = 2;   m1[1][0].i = 2;   m1[1][1].r = 2;    m1[1][1].i = 2;
        m2[0][0].r = 3;   m2[0][0].i = 3;   m2[0][1].r = 3;    m2[0][1].i = 3;
        m2[1][0].r = 3;   m2[1][0].i = 3;   m2[1][1].r = 3;    m2[1][1].i = 3;

        m = Cmxsub(m1, m2, 2, 2);            /* subtract m2[][] from m1[][] */

        /* the answer should be : m[0][0].r = -1;  m[0][0].i = -1;
                                  m[0][1].r = -1;  m[0][1].i = -1;
                                  m[1][0].r = -1;  m[1][0].i = -1;
                                  m[1][1].r = -1;  m[1][1].i = -1;        */
        for (i=0; i<2; i++) {
            for (j=0; j<2; j++) {
                printf("m[%d][%d].r = %f, m[%d][%d].i = %f\n",
                    i, j, m[i][j].r, i, j, m[i][j].i);
            }
        }
        Cmxfree(m);     /* free matrix m[][]  */
        Cmxfree(m1);    /* free matrix m1[][] */
        Cmxfree(m2);    /* free matrix m2[][] */
        return 0;
}
```

Cmxtrace

#include "mathlib.h"

Complex Cmxtrace(Complex__Matrix *m*, unsigned *n*);

DESCRIPTION

The *Cmxtrace* function computes the trace of the complex matrix *m*. The trace of a matrix is the sum of its diagonal terms:

$$m[0][0] + m[1][1] + \ldots + m[n][n]$$

The *n* argument specifies the number of rows and columns in the square matrix *m*. The complex matrix *m* must be allocated by the *Cmxalloc* function prior to calling the *Cmxtrace* function.

The trace of a matrix can be used to normalize the matrix coefficients. If the matrix is nonsingular (i. e., the determinant is nonzero), then it can be diagonalized. It is interesting to note that the trace of the diagonalized matrix is always equal to the trace of the original matrix. That is, the trace of a matrix is invariant to matrix diagonalization.

RETURNS

A complex value representing the sum of the diagonal elements of the complex matrix *m*.

SEE ALSO

mxtrace().

```
#include "mathlib.h"
#include <stdio.h>
Complex M[2][2] = { 1,1, 0,0, 0,0, 1,1 };
main()
{
Complex trace;
Complex_Matrix m;
unsigned n=2;
        m = Cmxalloc(M, n, n);    /* allocate and initialize m[][]      */
        trace = Cmxtrace(m, n);   /* now compute the trace of m[][]     */
        /* the answer should be:  trace.r = 2;   trace.i = 2;           */
        printf("the real part of the trace = %f\n", trace.r);
        printf("the imaginary part of the trace = %f\n", trace.i);
        Cmxfree(m); /*  free m[][]  */
        return 0;
}
```

Cmxtrans

#include "mathlib.h"
Complex__Matrix Cmxtrans(Complex__Matrix *m*, unsigned *rows*, unsigned *cols*);

DESCRIPTION

The *Cmxtrans* function forms the transpose of the complex matrix m. The *rows* and *cols* arguments specify the number of rows and columns in the matrix.

The transpose is obtained by interchanging the rows and columns of matrix m. Therefore, the number of rows in the tranposed matrix is equal to *cols*, and the number columns is equal to *rows*.

The complex matrix m must be allocated by the *Cmxalloc* function prior to calling the *Cmxtrans* function. The resulting complex matrix representing the transpose of m is automatically allocated by the *Cmxtrans* function.

If m is square, the user may select the option of whether or not to write the resulting output over the input matrix m. If memory is a limitation, and m is no longer needed, this may prove useful. This option can be selected by setting the global variable **useinput__ = 1** prior to calling *Cmxtrans*. The default option is to not destroy the input matrix m (i. e., **useinput__ = 0**).

RETURNS

A *cols* by *rows* complex matrix representing the transpose of matrix m (i. e. m^T if successful, or NULL if an error occurs).

SEE ALSO

mxtransp().

```
#include "mathlib.h"
#include <stdio.h>
Complex M[2][3] = { 0,0,   1,1,   2,2,
                    3,3,   4,4,   5,5 };
main()
{
Complex_Matrix m,mt;
unsigned i, j, rows=2, cols=3;

    m=Cmxalloc(M, rows, cols);  /* allocate and initialize matrix m[][] */
    mt = Cmxtrans(m, rows, cols); /* form the matrix transpose of m[][] */

    /* the answer should be: mt[i][j] = m[j][i];
       i=0,1,... cols; j=0,1,...,rows                                    */
    for(i=0; i<cols; i++) {
        for(j=0; j<rows; j++) {
            printf("mt[%d][%d].r = %5.2f, mt[%d][%d].i = %5.2f,"
                   " m[%d][%d].r = %5.2f, m[%d][%d].i = %5.2f\n",
                    i, j, mt[i][j].r, i, j, mt[i][j].i,
                    j, i, m[j][i].r, j, i, m[j][i].i);
        }
    }
    Cmxfree(m);   /* free matrix m[][]  */
    Cmxfree(mt);  /* free matrix mt[][]  */
    return 0;
}
```

combin

#include "mathlib.h"
double combin(int *n*, int *r*);

DESCRIPTION

The *combin* function computes the number of different ways of choosing r objects from a total of n, where $r \leq n$, without regard to order. The number of such combinations, $_nC_r$, is given by

$$_nC_r = n!/(r! * (n-r)!)$$

There are several useful identities involving combinations. Perhaps the most important of these is

$$_nC_r = {_nC_{n-r}}$$

This formula is intuitively clear, since when we explicitly decide to choose r objects at a time, we are implicitly deciding to exclude $n - r$ objects at a time.

RETURNS

A double precision whole number representing the number of combinations of n things taken r at a time (i. e., $_nC_r$), where $r \leq n$.

SEE ALSO

fact(), permut().

```
#include "mathlib.h"
#include <stdio.h>
main()
{
double nCr;
int n=4,r=2;
        /* find the number of ways of choosing 2 things from 4 without
            regard to order                                             */
        nCr = combin(n, r);
        /* the answer should be nCr= 4!/(2!*(4-2)!) = 6                 */
        printf("# of combinations of %d things taken %d at a time = %f\n",
                n, r, nCr);
        return 0;
}
```

Complx

#include "mathlib.h"
Complex Complx(Real _x_, Real _y_);

DESCRIPTION

The _Complx_ function forms a complex number from a pair of real numbers, x and y. The x argument corresponds to the real part of the complex number. The y argument corresponds to the imaginary part of the complex number.

RETURNS

A complex number z, where $z.r = x$ and $z.i = y$.

```
#include "mathlib.h"
#include <stdio.h>
main()
{
Real x,y;
Complex z;

        x = 1;
        y = 2;
        z = Complx(x, y); /* form the complex number z  */
        printf("the real part of z = %f\n", z.r);
        printf("the imaginary part of z = %f\n", z.i);
        /* the answer should be:  z.r = 1;  z.i = 2;    */
        return 0;
}
```

Conjg

#include "mathlib.h"
Complex Conjg(Complex x);

DESCRIPTION
The *Conjg* function computes the conjugate of the complex number x.

RETURNS
A complex number y, where $y = x.r - i * x.i$

SEE ALSO
Cmxconjg(), Cvconjg().

```
#include "mathlib.h"
#include <stdio.h>
main()
{
Complex x,y;

        x.r = 1;
        x.i = 1;
        y = Conjg(x); /* find the conjugate of x   */
        printf("the real part of the conjugate of x = %f\n", y.r);
        printf("the imaginary part of the conjugate of x = %f\n", y.i);
        /* the answer should be:  y.r = 1; and y.i = -1      */
        return 0;
}
```

conjgrad

#include "mathlib.h"
Real conjgrad(Real (*_funct_)( ), unsigned _n_, Real__Vector _v_, Real _est_, Real _eps_, unsigned _limit_);

DESCRIPTION

One of the most difficult problems in mathematical analysis is the minimization of multidimensional nonlinear equations. For example, consider the following equation:

$$z = 302 + 6x^2 - 48x + 8^2 - 80y$$

Factoring this expression shows that this equation represents the surface of an *elliptic paraboloid* with a minimum at $x = 4$, $y = 5$, $z = 6$.

$$z = 6 + 6(x - 4)^2 + 8(y - 5)^2$$

The shape of an elliptic paraboloid resembles that of an automobile headlight, as shown in Figure 9.4. Although, for this case, the minimum can be readily determined, most multidimensional minimization problems are not this simple.

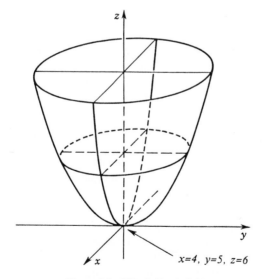

Figure 9.4 Elliptic Paraboloid

Gradient minimization approaches are ideal for problems such as these. The gradient at a point on a surface is the vector that is perpendicular to the tangent plane (i. e., the plane that is tangent to the surface at that point). The negated gradient vector points in the direction of the greatest decrease in the function value from that point.

If the gradient exists, and if unimodality can be assumed, the minimum of the function can be found by successively following the surface in the direction of its steepest descent. This approach is sometimes called the **method of steepest descent**.

For multimodal surfaces, multiple starting points are often required. Gradient approaches work best when the scales of the independent variables are chosen so that the components of the typical gradient vector are of comparable magnitude. These search procedures can be very inefficient if a few components of the gradient are consistently much smaller than the rest.

The **conjugate gradient** technique is a more efficient search strategy than the method of steepest descent. This n-dimensional approach searches along "conjugate directions," which are weighted averages of the current gradient with the gradients of the last n steps. This averaging tends to smooth the search path, resulting in a much more rapid convergence than the method of steepest descent. For a more detailed description of the conjugate gradient approach, the reader should refer to the original work of Fletcher and Reeves (1964).

One very nice feature of the conjugate gradient technique is that for simple surfaces, the number of required iterations is usually insensitive to the initial guess. This is because the step size is weighted by the gradient of the function. For unimodal surfaces, the worse the initial guess, the larger the gradient.

The *conjgrad* function minimizes the user-defined function *funct*, consisting of n independent variables. The prototype for the user-defined function is

Real *funct*(**unsigned** *n*, **Real__Vector** *v*, **Real__Vector** *grad*);

where n specifies the number of independent variables, v specifies the value of each independent variable at a particular point, and *grad* is used to return the gradient of the function at that particular point. The *grad* vector should return the first derivative of the function with respect to each of the independent variables (i.e., *grad*[0] is the first derivative with respect to the first variable, *grad*[1] is the first derivative with respect to the second variable, etc.). The return value of *funct* is the value of the function at the specified point.

The v argument specifies the initial value of each independent variable where the iteration starts. For example, $v[0]$ specifies the value of the first variable, $v[1]$ specifies the value of the second variable, and so on. On return, this vector will contain the values for the independent variables where the function minimum occurs.

The *est* argument specifies an estimate of the minimum function value.

The *eps* argument specifies the expected absolute error. If the type **Real** has six significant digits, then a reasonable value for *eps* would be $1.0e-6$.

The *limit* argument specifies the maximum number of iterations to perform. If the function minimum is not found within the specified number of iterations, then the *conjgrad* function returns the best estimate of the function minimum at the point the *limit* is reached.

RETURNS

A real value representing the minimum value of the user-defined function *funct*. The values of the independent variables at the minimum are returned in the v vector. If the number of iterations exceeds *limit*, then the error message "No convergence in LIMIT iterations" is generated (math_errno = E_DIVERGE), and a best estimate for the minimum value is returned. The error messages "Encountered errors in calculating the gradient" (math_errno = E_GRAD_ERR), or "Minimum probably does not exist" (math_errno = E_NO_MIN), may also be generated. The E_NO_MIN error condition occurs if when stepping in one of the computed directions, the function will never increase by a tolerable amount. This error may also occur if the interval where the function increases is small, and the initial guess is relatively far away from the minimum such that the minimum was skipped. This is due to the linear search technique, which doubles the step size until a point is found where the function increases.

REFERENCE

R. Fletcher and C. M. Reeves (1964).

```
#include "mathlib.h"
#include <stdio.h>
#include <math.h>
#include <stdlib.h>

Real funct(unsigned n, Real_Vector x, Real_Vector grad)
{ /* function with two independent variables: x and y
      f(x, y) = 6 + 6(x-4)^2 + 8(y-5)^2
      fx(x, y) = 12(x-4)          (first derivative with respect to x)
      fy(x, y) = 16(y-5)          (first derivative with respect to y)
                                                                    */
unsigned n1 = n-1;
        grad[0] = 12*(x[0]-4);    /* first derivative with respect to x */
        grad[n1] = 16*(x[n1]-5);  /* first derivative with respect to y */
        return (Real) 6 + 6*(x[0]-4)*(x[0]-4) + 8*(x[n1]-5)*(x[n1]-5);
}


/* FIND THE MINIMUM OF THE ABOVE ELLIPTIC PARABOLOID VIA CONJUGATE GRADIENTS */
main()
{
unsigned n,limit;
Real_Vector v;
Real f, est, eps;

        n=2;                        /* number of independent variables    */
        v = valloc(NULL, n);        /* allocate variable vector           */
        v[0] = 100.0;               /* initial guess for x                */
        v[1] = 2000.0;              /* initial guess for y                */
        est= -3000.0;               /* estimated function value at minimum */
        eps=1.e-7;                  /* absolute error tolerance           */
        limit=100;                  /* maximum number of iterations       */
        f = conjgrad(funct, n, v, est, eps, limit);
        printf("v[0] = %10.7f, v[1] = %10.7f, and f(x, y) = %10.7f\n",
                v[0], v[1], f);

        /* answer is: v[0] =  4.0000000, v[1] =  5.0000000,
                  and f(x, y) =  6.0000000 */
        return 0;
}
```

conv2dft

```
#include "mathlib.h"
Real__Matrix conv2dft(Real__Matrix a, Real__Matrix b, unsigned arows,
                      unsigned acols, unsigned brows, unsigned bcols,
                      unsigned * rows, unsigned * cols, int last);
```

DESCRIPTION

A 2-D convolution is a mathematical technique used to implement 2-D FIR digital filters. A 2-D FFT can be used to efficiently convolve large sets of 2-D data (i. e., rows > 64) and columns > 64).

The *conv2dft* function uses a 2-D FFT approach to compute the linear 2-D convolution of real matrix *a* with real matrix *b*.

The *arows* and *acols* arguments specify the number of rows and columns in matrix *a*. The *brows* and *bcols* arguments specify the number of rows and columns in matrix *b*.

The *last* argument specifies whether or not this function will be called more than once with the same reference matrix *a* (e. g., same 2-D FIR filter). If so, then one 2-D FFT computation can be eliminated. If *last* = 0), then the 2-D FFT of the reference matrix *a* is saved internally for the next call. If *last* = 1, then the 2-D FFT of the reference matrix *a* is freed before exiting the function. The final call to the *conv2dft* function should specify *last* as 1 in order to free the allocated memory.

The convolution output matrix is the return value of this function and is automatically allocated by the *conv2dft* function. The numbers of rows and columns in the output matrix are returned by the *rows* and *cols* arguments. This resulting output convolution matrix should be freed when it is no longer needed.

The *conv2dft* function calls the *fft2d* function.

RETURNS

A *rows* by *cols* real matrix representing the 2-D convolution of the real matrix *a* with the real matrix *b* if successful, or NULL if an error occurs.

SEE ALSO

convolve(), convofft().

REFERENCE

E. O. Brigham (1974).

```c
#include <stdio.h>
#include <stdlib.h>
#include <math.h>
#include "mathlib.h"
main()
{
int last;
unsigned i, j, row, col, row0=2, col0=2, row1=4, col1=4;
Real_Matrix m0, m1, m2, convolution;

    m0 = mxalloc(NULL, row0, col0);
    m1 = mxalloc(NULL, row1, col1);
    m2 = mxalloc(NULL, row1, col1);

    for(i=0; i<row0; i++) {
        for(j=0; j<col0; j++) m0[i][j] = 1.0;
    }
    for(i=0; i<row1; i++) {
        for(j=0; j<col1; j++) m1[i][j] = m2[i][j] = 1.0;
    }

    last = 0;          /* last = 0 ==> m0[][] will be used again */
    convolution = conv2dft(m0, m1, row0, col0, row1, col1, &row, &col, last);

    mxfree(convolution);
    printf("first convolution is finished\n");

    last = 1;          /* last = 1 ==> m0[][] will not be used again */
    convolution = conv2dft(m0, m2, row0, col0, row1, col1, &row, &col, last);

    printf("second convolution is finished:\n");
    for(i=0; i<row; i++) {
       for(j=0; j<col; j++) {
          printf("%4.0f",convolution[i][j]);
       }
       printf("\n");
    }
    return 0;
}
```

PROGRAM OUTPUT

```
first convolution is finished
second convolution is finished:
    1   2   2   2   1
    2   4   4   4   2
    2   4   4   4   2
    2   4   4   4   2
    1   2   2   2   1
```

convofft

```
#include "mathlib.h"
Real_Vector convofft(Real_Vector a, unsigned na, Real_Vector b, unsigned nb, int last);
```

DESCRIPTION

Convolution is the mathematical technique required to implement an FIR digital filter. The FFT can be used to efficiently convolve digital sequences.

The *convofft* function uses an FFT to compute the linear convolution of vector a with vector b. The na argument specifies the length of vector a and the nb argument specifies the length of vector b.

The *last* argument specifies whether or not this function will be called more than once with the same reference vector a (e.g., same FIR filter). If so, then one FFT computation can be eliminated. If $last = 0$, then the FFT of the reference vector a is saved internally for the next call. If $last = 1$, then the FFT of the reference vector a is freed before exiting the function. The final call to the *convofft* function should specify *last* as 1 in order to free the allocated memory.

The convolution output vector is the return value of this function and is automatically allocated by the *convofft* function. The length of the output vector is $na + nb - 1$. This resulting output convolution vector should be freed when it is no longer needed.

The *convofft* function calls the *fftr_inv* and the *fftreal* functions.

RETURNS

A real vector of length $na + nb - 1$ representing the convolution of the real vector a if successful, or NULL if an error occurs.

SEE ALSO

convolve(), conv2dft().

REFERENCE

E. O. Brigham (1974).

```c
#include <stdio.h>
#include <stdlib.h>
#include <math.h>
#include "mathlib.h"
main()
{
Real_Vector signal1, signal2, signal3, convolution;
int i,last;
unsigned n=8,n3=3;

/* main program to convolve two vectors of data  */

        signal1 = valloc(NULL, n);
        signal2 = valloc(NULL, n);
        signal3 = valloc(NULL, n3);
        for (i=0; i<n; i++) signal1[i] = signal2[i] = 0.0;
        for (i=n/4; i<3*n/4; i++) {
           signal1[i] = 1.0;
           signal2[i] = 1.0;
        }
        for (i=0; i<n3; i++) signal3[i] = 1.0;

/*convolve signal1 with signal2 (last = 0 ==> signal1 will be used again*/
        last = 0;
        convolution = convofft(signal1, n, signal2, n, last);
        printf(" signal1 convolved with signal2\n");
        for (i=0; i<n+n-1; i++) {
            printf("%4d      %6.2f\n", i, convolution[i]);
        }
        vfree(convolution); /*free convolution vector for next subroutine call */

/*convolve signal1 with signal3 (last = 1 ==> signal1 won't be used again*/
        last = 1;
        convolution = convofft(signal1, n, signal3, n3, last);
        printf("\n");
        printf(" signal1 convolved with signal3:\n");
        for (i=0; i<n+n3-1; i++) {
            printf("%4d      %6.2f\n", i, convolution[i]);
        }
        mathfree();
        return 0;
}
```

PROGRAM OUTPUT

```
signal1 convolved with signal2
     0          0.00
     1          0.00
     2          0.00
     3          0.00
     4          1.00
     5          2.00
     6          3.00
     7          4.00
     8          3.00
     9          2.00
    10          1.00
    11          0.00
    12          0.00
    13          0.00
    14          0.00


signal1 convolved with signal3
     0          0.00
     1          0.00
     2          1.00
     3          2.00
     4          3.00
     5          3.00
     6          2.00
     7          1.00
     8          0.00
     9          0.00
```

convolve

```
#include "mathlib.h"
Real__Vector convolve(Real__Vector data, int ndata, Real__Vector weights, int nweights,
                      int ndec, int itype, int isym, int * length);
```

DESCRIPTION

Convolution is the mathematical technique that is used to implement filtering processes. The purpose of most filtering applications is either to enhance desirable signal frequencies, or to reject undesirable frequencies (e. g., noise). The tone control of high-fidelity audio systems is perhaps the most common example of such a filter.

It is beyond the scope of this manual to provide all the mathematical development necessary to understand digital filter theory. For more information about digital filtering, the interested reader is advised to consult the excellent reference given below. Instead, an example of a very simple digital lowpass filter, the "n-point" averager, is described (see the programmed example below, where $n = nweights = 4$).

The "n-point" averager provides a good intuitive understanding of the digital lowpass filtering process. Suppose that some measurement data is degraded by noise. For example, the data might represent a sinusoid, even though the noise makes it difficult to determine its amplitude, frequency, and phase. First, it is important that enough samples are taken to properly represent the signal (i.e., the sinusoid should be oversampled). One approach to reduce the noise is to replace each data point with the average of itself and several of its neighbors. If the average is taken over n neighbors, the noise will usually be reduced by slightly more than a factor of $1/n$. In digital filter terminology, the "n-point" averager is an example of a "finite impulse response" (FIR) digital filter.

All the weights of an "n-point" averager are equal. The "n-point" averager is a symmetric (even) function. Although the weights of more sophisticated FIR lowpass filters are not all equal, they are often symmetric. Similarly, high-pass and differentiating FIR filters are frequently antisymmetric (odd) functions. These symmetric properties can be exploited to reduce the number of multiplications (and array storage locations) required to implement the filter.

FIR lowpass filters can also be used to lower the sampling rate (decimate) of a digital signal by an integral amount. These are often referred to as decimating FIR filters (see the *downsamp* function).

The *convolve* function performs the discrete convolution necessary to implement any FIR digital filter. The *data* argument is the input vector of length *ndata*. The *weights* argument is the vector of filter weights of length *nweights*. The *ndec* argument is the integer factor by which the sampling rate can be decreased (decimated) at the filter output. Normally, *ndec* = 1, since the desired sampling rate of the output is usually the same as that of the input. If *ndec* = 2, then the output will be sampled at half the input rate (i.e., the returned vector will be about half as long as the input vector *data*).

The *itype* argument indicates the type of symmetry (if any) in the FIR filter. If the filter is an even function (many lowpass filters are even), then *itype* should be 1. If the filter is an odd function, then *itype* should be −1. Hilbert transform filters and some differentiating filters are odd functions.

If the FIR filter is not an even or an odd function, then the *isym* argument should be 0. In this case, the *itype* argument is not used. Otherwise, if the filter has symmetry (or antisymmetry)), then *isym* should be 1. This allows the *convolve* function to exploit the filter's symmetry (or antisymmetry), reducing the number of multiplies by about a factor of two.

The actual number of points in the output vector is returned via the *length* argument. The value of *length* is determined according to the following formula:

$$length = (ndata \text{ - } nweights + 1)/ndec$$

RETURNS

A real vector of the calculated *length*, which is the filtered version of the input vector *data*, or NULL if an error occurs.

SEE ALSO

lowpass(), highpass(), bandpass(), downsamp(), deriv(), interp().

REFERENCE

A. V. Oppenheim and R. W. Schafer (1975).

```
#include "mathlib.h"
#include <stdio.h>
#include <stdlib.h>
#include <math.h>

#define NWEIGHTS 4
#define NDATA    100

Real weights[NWEIGHTS] = { 0.25, 0.25, 0.25, 0.25 };

main()
{
Real_Vector data, y;
int i, ny, ndec=1, itype=1, isym=1;
double truth, noise, off, init_error=0.0, filter_error=0.0;
    data=valloc(NULL, NDATA);/* allocate a vector for the data */
    for(i=0; i<NDATA; i++) {/* generate data, one cycle of a sinusoid+noise */
        truth = sin(twopi_*i/(double) NDATA);
        noise = 2*((double) rand()/RAND_MAX - 0.5); /* uniform from +1 to -1*/
        data[i] = truth + noise;
        init_error += noise*noise/(double) NDATA; /* noise = truth - data   */
    } /* now filter the data:                                               */
    y = convolve(data, NDATA, (Real_Vector) weights, NWEIGHTS,
                ndec, itype, isym, &ny);
    off= (double) (NWEIGHTS-1.)/2.; /* align truth with filtered data       */
    for(i=0; i<ny; i++) { /* compute the error after filtering             */
        truth = sin(twopi_*(i+off)/(double) NDATA);
        filter_error += (truth - y[i])*(truth - y[i])/(double) ny;
    }
    /* the 4-point averager should remove a little more than 1/4th of the
       noise:                                                               */
    printf("the error before filtering is %f\n", init_error); /* = 0.345443 */
    printf("the error after filtering is %f\n", filter_error);/* = 0.073728 */
    return 0;
}
```

Cpowspec

#include "mathlib.h"
Real__Vector Cpowspec(Complex__Vector *v*, unsigned *nv*, unsigned *npw*, Real__Vector *w*);

DESCRIPTION

The *Cpowspec* function computes the power spectrum of the complex vector v of length nv. The method used is Welch's periodogram. Most modern spectrum analyzers use this approach. The signal length nv is divided into several nonoverlapping segments, each of length npw, where npw is a "power-of-two." Each segment is Fourier transformed, and all the transforms are averaged, yielding the desired power spectrum.

If windowing of the data segments is desired, the user must supply the window vector w of length npw. In this case, the method is called a *modified periodogram*. The *tdwindow* function provides several different windows for this purpose. If no window is desired (i.e., a rectangular window is chosen), then NULL should be specified for w.

If the complex sinusoid $A * [\cos(2 * \pi * k * i/n) + j * \sin(2 * \pi * k * i/n)]$ is input to this function, the corresponding output will consist of one spectral component at the $(k + 1)$th element of the power spectrum, where $k = 0, 1, 2, \ldots, npw/2$, $i = 0, 1, 2, \ldots, n - 1$, and $\pi = 3.141592654 \ldots$. With no windowing, the magnitude of the component is A.

If the complex sinusoid $A * [\cos(2 * \pi * k * i/n) - j * \sin(2 * \pi * k * i/n)]$ is input to this function, the corresponding output will consist of one spectral component at the $n - k + 1)$th element of the power spectrum. With no windowing, the magnitude of the component is also A.

The *Cpowspec* function calls the *fftrad2* FFT function.

RETURNS

A real vector of length nv representing the power spectrum of the complex vector v if successful, or NULL if an error occurs. If $nv < npw$, the following error message "Length of data must be greater than or equal to FFT size" is generated (math_errno = E__FFTSIZE).

SEE ALSO

powspec(), fftrad2(), fft42(), tdwindow().

REFERENCES

P. D. Welch (1970); A. V. Oppenheim and R. W. Schafer (1975).

```c
#include "mathlib.h"
#include <stdio.h>
#include <stdlib.h>
#include <math.h>

main()
{
Complex_Vector v;
unsigned npw=16, i, nv=128;
Complex noise;
Real_Vector w,spectrum1,spectrum2;

    v=Cvalloc(NULL, nv);  /* allocate an array for data to be analyzed      */
    w = tdwindow(npw, 7); /* 7 is a Hamming window, see the tdwindow() routine */
    for(i=0; i<nv; i++) { /* generate nv/npw cycles of a sinusoid + noise    */
        noise.r = 2*((double) rand()/RAND_MAX - 0.5);/* uniform from +1 to -1  */
        noise.i = 2*((double) rand()/RAND_MAX - 0.5);/* uniform from +1 to -1  */
        v[i].r = cos(twopi_*4*i/(double) npw) + noise.r; /*signal=4th harmonic */
        v[i].i = sin(twopi_*4*i/(double) npw) + noise.i;
    }
    spectrum1=Cpowspec(v, npw, npw, NULL);/*compute spectrum of 1st segment */
    spectrum2=Cpowspec(v, nv, npw, NULL); /* compute spectrum of entire data*/
    /* 4th harmonic of 2nd spectrum should be higher and
       noise floor should be smoother                                       */
    printf("            spectrum of    spectrum of\n");
    printf("harmonic #  1st segment    all segments\n\n");

    for(i=0; i<npw; i++) {
        printf("   %2d        %9.6f         %9.6f\n",
                                 i, spectrum1[i],spectrum2[i]);
    }
    Cvfree(v);          /* free v[]         */
    vfree(spectrum1);   /* free spectrum1[] */
    vfree(spectrum2);   /* free spectrum2[] */
    vfree(w);           /* free w[]         */
    return 0;
}
```

PROGRAM OUTPUT

harmonic #	spectrum of 1st segment	spectrum of all segments
0	0.001394	0.019613
1	0.024732	0.042699
2	0.059887	0.035658
3	0.000588	0.016009
4	0.721915	1.074491
5	0.004981	0.041846
6	0.081498	0.046221
7	0.042043	0.067726
8	0.050548	0.049407
9	0.062981	0.042475
10	0.020899	0.039128
11	0.185170	0.037621
12	0.022579	0.033128
13	0.049184	0.049911
14	0.066707	0.052321
15	0.011472	0.015523

cros2dft

```
#include "mathlib.h"
Real__Matrix cros2dft(Real__Matrix a, Real__Matrix b, unsigned * rows, unsigned * cols,
                int last);
```

DESCRIPTION

Crosscorrelation (sometimes called covariance) is a useful tool in random signal and power spectral analysis. The 2-D FFT can be used to efficiently correlate moderately large sets of 2-D data (i. e., rows > 64) and columns > 64).

The *cros2dft* function uses a 2-D FFT to compute the crosscorrelation of real matrix *a* with real matrix *b* (which leads and lags the reference matrix *a*).

The *rows* and *cols* arguments specify the number of rows and columns in both of the input matrices, *a* and *b*. Notice that these arguments are passed as pointers, since they are also used to return the number of rows and columns in the resulting crosscorrelation matrix.

The zero lag value is always the centermost element of the resulting crosscorrelation matrix x (e. g., $x[rows/2][cols/2]$). For best results, the correlation lags should be within 10 percent of the central lag point. This will usually ensure that the estimate of each of the lags is accurate. For example, if matrix *a* has 100 rows and columns, the crosscorrelation matrix x will have 99 rows and columns and the central lag will be at $x[49][49]$. The most accurate lags are in the rectangle bounded by the two points $x[44][44]$ and $x[54][54]$.

The *last* argument specifies whether or not this function will be called more than once with the same reference matrix *a* (e. g., same 2-D FIR filter). If so, then one 2-D FFT computation can be eliminated. If *last* = 0, then the 2-D FFT of the reference matrix *a* is saved internally for the next call. If *last* = 1, then the 2-D FFT of the reference matrix *a* is freed before exiting the function. The final call to the *cros2dft* function should specify *last* as 1 in order to free the allocated memory.

The crosscorrelation output matrix is the return value of this function and is automatically allocated by the *cros2dft* function. The numbers of rows and columns in the output matrix are returned by the *rows* and *cols* arguments. This resulting output crosscorrelation matrix should be freed when it is no longer needed.

The *cros2dft* function calls the *fft2d* function.

RETURNS

A *rows* by *cols* real matrix representing the 2-D crosscorrelation of the real matrix *a* with the real matrix *b* if successful, or NULL if an error occurs.

SEE ALSO

autofft(), autocor(), crosscor(), crossfft(), auto2dft().

REFERENCES

R. B. Blackman and J. W. Tukey (1959); E. O. Brigham (1974); A. V. Oppenheim and R. W. Schafer (p. 539, 1975).

```c
#include <stdio.h>
#include <stdlib.h>
#include <math.h>
#include "mathlib.h"
main()
{
int last;
unsigned i, j, cros_rows, cros_cols, rows=9, cols=9;
double yval0, xval0, yval1, xval1, sigma=0.5, r0=4.0, c0=4.0, r1=4.0, c1=4.0;
Real_Matrix m0, m1, m2, crosscorrelation;

/* Find the 2-D crosscorrelation of a 2-D Gaussian pulse centered at (r0,c0)  */
/* with another 2-D Gaussian pulse centered at (r1,c1)  */

    m0 = mxalloc(NULL, rows, cols);
    m1 = mxalloc(NULL, rows, cols);
    m2 = mxalloc(NULL, rows, cols);

    for(i=0; i<rows; i++) {      /* create two 2-dimensional Gaussian pulses */
        xval0 = (i-r0)*(i-r0);   /* with centers at (r0,c0) and (r1,c1) with */
        xval1 = (i-r1)*(i-r1);   /* variances of sigma                       */
        for(j=0; j<cols; j++) {
            yval0 = (j-c0)*(j-c0);
            yval1 = (j-c1)*(j-c1);
            m0[i][j] = exp(-(xval0+yval0)/(2*sigma));
            m1[i][j] = m2[i][j] = exp(-(xval1+yval1)/(2*sigma));
        }
    }
    cros_rows = rows;
    cros_cols = cols;
    last = 0;           /* last = 0 ==> m0[][] will be used again */
    crosscorrelation = cros2dft(m0, m1, &cros_rows, &cros_cols, last);
    mxfree(crosscorrelation);
    printf("first crosscorrelation is finished\n");

    cros_rows = rows;
    cros_cols = cols;
    last = 1;       /* last = 1 ==> m0[][] will not be used again */
    crosscorrelation = cros2dft(m0, m2, &cros_rows, &cros_cols, last);

    printf("second crosscorrelation is finished:\n");
    for(i=0; i<cros_rows; i++) {
        for(j=0; j<cros_cols; j++) {
            printf("%4.0f",fabs(90*crosscorrelation[i][j]));
        }
        printf("\n");
    }
    mathfree();
    return 0;
}
```

PROGRAM OUTPUT

```
first crosscorrelation is finished
second crosscorrelation is finished:
     0   0   0   0   0   0   0   0   0
     0   0   0   0   0   0   0   0   0
     0   0   0   0   0   0   0   0   0
     0   0   0   0   1   0   0   0   0
     0   0   0   1   2   1   0   0   0
     0   0   0   0   1   1   0   0   0
     0   0   0   0   0   0   0   0   0
     0   0   0   0   0   0   0   0   0
     0   0   0   0   0   0   0   0   0
```

crosscor

```
#include "mathlib.h"
int crosscor(Real__Vector a, Real__Vector b, unsigned n, unsigned nlags, Real__Vector lag,
        Real__Vector lead);
```

DESCRIPTION

Crosscorrelation (sometimes called covariance) is a useful tool in random signal and power spectral analysis. Unlike autocorrelation, the crosscorrelation function is not generally symmetric.

The *crosscor* function computes the crosscorrelation of vector a with vector b (which leads and lags the reference vector a). The n argument specifies the length of the input vectors a and b.

The *nlags* argument specifies the length of the crosscorrelation output vectors *lag* and *lead*. The first element in each output vector is the central lag of the crosscorrelation function. For best results, the number of desired correlation lags should be less than 10 percent of the number of input data samples (i. e., $nlags < .10 * n$). This will usually ensure that the estimate of each of the lags is accurate.

The input and output vectors must be allocated prior to calling the *crosscor* function and should be freed when no longer needed.

RETURNS

0 if successful or -1 if an error occurs. If successful, vectors *lag* and *lead* of length *nlags* represent the crosscorrelation of the real vector a with real vector b, where b leads and lags a. If the number of data samples n is less than or equal to the number of lags *nlags*, the error message "Not enough input data" is generated (math_errno = E_NOTENOUGH).

SEE ALSO

autocor(), autofft(), auto2dft(), crossfft(), cros2dft().

REFERENCES

R. B. Blackman and J. W. Tukey (1959); A. V. Oppenheim and R. W. Schafer (p. 539, 1975).

```c
#include <stdio.h>
#include <stdlib.h>
#include <math.h>
#include "mathlib.h"
main()
{
Real_Vector signal1, signal2, lag, lead;
int i;
unsigned n=128,nlags=13;

/* main program to crosscorrelate two vectors of data  */

        signal1 = valloc(NULL, n);
        signal2 = valloc(NULL, n);
        lag = valloc(NULL, nlags);
        lead = valloc(NULL, nlags);
        for (i=0; i<n; i++) signal1[i] = signal2[i] = 0.0;
        for (i=n/4; i<3*n/4; i++) {
           signal1[i] = 1.0;
           signal2[i] = (Real) i;
        }
        crosscor(signal1, signal2, n, nlags, lag, lead);
        printf("      lags     leads\n");
        for (i=0; i<nlags; i++) {
            printf("%2d %f %f\n", i, lag[i], lead[i]);
        }
        mathfree();
        return 0;
}
```

PROGRAM OUTPUT

```
     lags      leads
 0 15.875000 15.875000
 1 15.623031 15.126969
 2 15.359127 14.375000
 3 15.083000 13.619000
 4 14.794355 12.858871
 5 14.492886 12.094512
 6 14.178279 11.325820
 7 13.850207 10.552686
 8 13.508333 9.775000
 9 13.152311 8.992647
10 12.781780 8.205508
11 12.396368 7.413462
12 11.995690 6.616379
```

crossfft

```
#include "mathlib.h"
int crossfft(Real__Vector a, Real__Vector b, unsigned n, unsigned nlags, Real__Vector lag,
            Real__Vector lead, int last);
```

DESCRIPTION

Crosscorrelation (sometimes called covariance) is a useful tool in random signal and power spectral analysis. Unlike autocorrelation, the crosscorrelation function is not generally symmetric.

Like the *crosscor* function, the *crossfft* function computes the crosscorrelation of vector a with vector b (which leads and lags the reference vector a). However, unlike the *crosscor* function, the *crossfft* function is designed to efficiently correlate moderately large data sets ($n > 512$) or $nlags > 256$).

The n argument specifies the length of the input vectors a and b. The *nlags* argument specifies the length of the crosscorrelation output vectors *lag* and *lead*. The first element in each output vector is the central lag of the crosscorrelation function. For best results, the number of desired correlation lags should be less than 10 percent of the number of input data samples (i. e., $nlags < .10 * n$). This will usually ensure that the estimate of each of the lags is accurate.

The *last* argument specifies whether or not this function will be called more than once with the same reference vector a. If so, then one FFT computation can be eliminated. If *last* = 0, then the FFT of the reference vector a is saved internally for the next call. If *last* = 1, then the FFT of the reference vector a is freed before exiting the function. The final call to the *crossfft* function should specify *last* as 1 in order to free the allocated memory.

The input and output vectors must be allocated prior to calling the *crossfft* function and should be freed when no longer needed.

The *crossfft* function calls the *fftr__inv* and the *fftreal* functions.

RETURNS

0 if successful or -1 if an error occurs. If successful, vectors *lag* and *lead* of length *nlags* represent the crosscorrelation of the real vector a with real vector b, where b leads and lags a. If the number of data samples n is less than or equal to the number of lags *nlags*, the error message "Not enough input data" is generated (math_errno = E_NOTENOUGH).

SEE ALSO

autocor(), autofft(), auto2dft(), crosscor(), cros2dft().

REFERENCES

R. B. Blackman and J. W. Tukey (1959); E. O. Brigham (1974); A. V. Oppenheim and R. W. Schafer (p. 539, 1975).

```c
#include <stdio.h>
#include <stdlib.h>
#include <math.h>
#include "mathlib.h"
main()
{
Real_Vector signal1, signal2, signal3, lag, lead;
int i,last;
unsigned n=128,nlags=8;

/* main program to crosscorrelate two vectors of data  */

        signal1 = valloc(NULL, n);
        signal2 = valloc(NULL, n);
        signal3 = valloc(NULL, n);
        lag = valloc(NULL, nlags);
        lead = valloc(NULL, nlags);
        for (i=0; i<n; i++) signal1[i] = signal2[i] = signal3[i] = 0.0;
        for (i=n/4; i<3*n/4; i++) {
            signal1[i] = 1.0;
            signal2[i] = (Real) i;
            signal3[i] = (Real) -2*i+13;
        }
/*crosscorrelate signal1 with signal2 (last = 0 ==> signal1 will be used again*/
        last = 0;
        crossfft(signal1, signal2, n, nlags, lag, lead, last);
        printf(" signal1 with signal2:\n");
        printf("       lags      leads\n");
       for (i=0; i<nlags; i++) {
            printf("%2d %f %f\n", i, lag[i], lead[i]);
        }
/*crosscorrelate signal1 with signal3 (last = 1 ==>signal1 won't be used again*/
        last = 1;
        crossfft(signal1, signal3, n, nlags, lag, lead, 1);
        printf("\n");
        printf(" signal1 with signal3:\n");
        printf("       lags      leads\n");
        for (i=0; i<nlags; i++) {
            printf("%2d %f %f\n", i, lag[i], lead[i]);
        }
        mathfree();
        return 0;
}
```

PROGRAM OUTPUT

```
signal1 with signal2:
      lags     leads
0 15.875000 15.875000
1 15.623031 15.126969
2 15.359127 14.375000
3 15.083000 13.619000
4 14.794355 12.858871
5 14.492886 12.094512
6 14.178279 11.325820
7 13.850207 10.552686

signal1 with signal3:
      lags     leads
0 -28.500000 -28.500000
1 -28.098425 -27.106299
2 -27.674603 -25.706349
3 -27.228000 -24.300000
4 -26.758065 -22.887097
5 -26.264228 -21.467480
6 -25.745902 -20.040984
7 -25.202479 -18.607438
```

CscalR

$$\text{#include "mathlib.h"}$$
$$\textbf{Complex CscalR(Complex } x, \textbf{ Real } y);$$

DESCRIPTION

The *CscalR* function multiplies the complex value x by the real value y. The resulting product is a complex value.

RETURNS

A complex value z, where $z = y * (x.r + i * x.i)$.

```
#include "mathlib.h"
#include <stdio.h>
main()
{
Complex x,z;
Real y;

        x.r = 1;
        x.i = 1;
        y = 2;
        z = CscalR(x, y);
        printf("the real part of z = %f\n", z.r);
        printf("the imaginary part of z = %f\n", z.i);
        /* the answer should be:  z.r = 2;   z.i = 2;      */
        return 0;
}
```

Csqrt

#include "mathlib.h"
Complex Csqrt(Complex x);

DESCRIPTION

The *Csqrt* function computes the principal square root of the complex value *x*. The resulting complex square root is also a complex value.

All complex numbers can be represented in polar or rectangular coordinates. However, the meaning of the square root of a complex number is best understood in polar coordinates. (See Figure 9.5). Let r be the magnitude of the complex number x, and let θ be the angle formed by the vector that points to x in the complex plane. The principal square root of x is the complex number that results from rotating this vector counterclockwise to half the angle, $\theta/2$. Like the inverse tangent, the square root function is periodic and is ambiguous for angles greater than 360 degrees. The term "principal square root" is used when the angles are restricted to less than 360 degrees. For this range of angles, the complex square root function is unique.

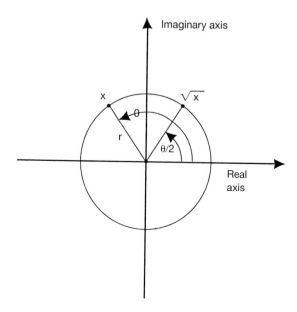

Figure 9.5   Square Root of Complex Number x.

RETURNS

A complex value y representing the principal square root of the complex value x, where

$$y = \text{sqrt}(r) * \{\cos(\theta/2) + i * \sin(\theta/2)\}$$
$$r = \text{sqrt}(x.r * x.r + x.i * x.i)$$
$$\theta = \text{atan2}(x.i, x.r)$$

```
#include "mathlib.h"
#include <stdio.h>
main()
{
Complex x,y;

        x.r = 0;
        x.i = 1;
        y = Csqrt(x); /* compute the principal square root of x  */
        printf("the real part of the square root of x = %f\n", y.r);
        printf("the imaginary part of the square root of x = %f\n", y.i);
        /* the answer should be:  y.r = y.i = sqrt(2)/2 = 0.707107; */
        return 0;

}
```

Csub

#include "mathlib.h"
Complex Csub(Complex *x*, Complex *y*);

DESCRIPTION
The *Csub* function subtracts the complex value y from the complex value x. The resulting difference is also a complex value.

RETURNS
A complex value z, where $z = x - y$.

SEE ALSO
Cadd(), Cdiv(), Cmul().

```
#include "mathlib.h"
#include <stdio.h>
main()
{
Complex x,y,z;

        x.r = 1;
        x.i = 1;
        y.r = 2;
        y.i = 2;
        z = Csub(x, y); /* subtract y from x  */
        printf("the real part of z = %f\n", z.r);
        printf("the imaginary part of z = %f\n", z.i);
        /* the answer should be:  z.r = -1;  z.i = -1;      */
        return 0;
}
```

curvreg

```
#include "mathlib.h"
Real_Vector curvreg(Real_Vector vx, Real_Vector vy, unsigned n, unsigned terms,
                    Real * mse);
```

DESCRIPTION

The *curvreg* function uses the least-squares method to determine the coefficients of a polynomial equation that best fits the x-y data points specified by the real vectors vx and vy.

The n argument specifies the number of data points in vectors vx and vy. The *terms* argument specifies the number of terms in the polynomial equation. The number of terms is always one greater than the degree of the polynomial.

The mean-square error between the data points and the resulting polynomial equation is returned via the *mse* argument. The *mse* argument must be the address of a variable of type **Real**. A small mean- square error ($<<0.1$) means that the polynomial equation closely approximates the data, whereas a large mean-square error ($>>0.1$) means that the fit is not very good.

The polynomial equation may be expressed as follows:

$$f(x) = \text{coef}[0] + x * \text{coef}[1] + x^2 * \text{coef}[2] + \ldots + x^{\text{terms}-1} * \text{coef}[\text{terms} - 1]$$

where coef[0] ... coef[terms − 1] are the calculated coefficients, and terms − 1 is the degree of the polynomial.

The mean-square error between the polynomial equation and the data points may be expressed as follows:

$$\text{mse} = (f(vx[0]) - vy[0])^2 + (f(vx[1]) - vy[1])^2 + \ldots$$

For certain sets of data, the least-squares polynomial approximation may be ill-conditioned. This is more likely to occur when the degree of the polynomial is large (greater than seven). The least-squares polynomial solution is obtained by solving a linear set of equations (the normal equations). The determinant of the normal equations can become very small for polynomials of large order. If the determinant becomes zero, then this least-squares method will fail.

The *curvreg* function calls the *least_sq* function.

RETURNS

A real vector of length *terms* representing the coefficients of a *terms* − 1 degree polynomial if successful, or NULL if an error occurs. If the determinant of the normal equations is zero, then the error message "Singular Matrix" is generated (math_errno = E_MSING).

SEE ALSO
least_sq(), linreg(), pseudinv(), pseries().

REFERENCE
R. W. Hamming (1973).

```c
#include "mathlib.h"
#include <stdio.h>
Real VX[7] = { -3, -2, -1, 0, 1, 2, 3 };
Real VY[7] = { 4, 2, 3, 0, -1, -2, -5 };
main()
{
Real_Vector vx,vy,coeff;
Real mse;
int n=7,terms=3;

/* fit the above "x" and "y" points with the best quadratic polynomial       */

        vx = valloc(VX, n); /* allocate and initialize the "x" values       */
        vy = valloc(VY, n); /* allocate and initialize the "y" values       */

        coeff = curvreg(vx, vy, n, terms, &mse);/*fit the data with a quadratic*/

        printf("The best fit quadratic equation is: ");
        printf("y = %f  %f*x  %f*x*x\n", coeff[0], coeff[1], coeff[2]);
        /* The answer is: y = 0.666667  -1.392857*x  -0.130952*x*x */
        return 0;
}
```

Cvadd

```
#include "mathlib.h"
Complex_Vector Cvadd(Complex_Vector v1, Complex_Vector v2, unsigned n);
```

DESCRIPTION

The *Cvadd* function adds the two complex vectors $v1$ and $v2$. The resulting sum is also a complex vector.

The n argument specifies the number of elements in vectors $v1$ and $v2$. The complex vectors $v1$ and $v2$ must be allocated by the *Cvalloc* function prior to calling the *Cvadd* function. The resulting complex vector representing the sum of $v1$ and $v2$ is automatically allocated by the *Cvadd* function.

The user may select the option of whether or not to write over the input vector $v1$ with "in-place" computations. If memory is a limitation, and $v1$ is no longer needed, this may prove useful. This option can be selected by setting the global variable **useinput_** $= 1$ prior to calling *Cvadd*. If this option is selected, the resulting sum overwrites the input vector $v1$. The default option is to not destroy the input vector $v1$ (i. e., **useinput_** $= 0$).

RETURNS

A complex vector v of length n, where $v = v1 + v2$.

SEE ALSO

Cmxadd(), Cmxsub(), mxadd(), mxsub(), Cvsub(), vadd(), vsub().

```
#include "mathlib.h"
#include <stdio.h>
main()
{
int i;
Complex_Vector v,v1,v2;
unsigned n=2;
/* add the following two vectors:                                */
/*                                                               */
/*          v[] = v1[] + v2[]                                    */
/*                                                               */
/* where v[] and v1[] are complex and are given below:          */
        v1 = Cvalloc(NULL, n);  /* allocate vector of length 2   */
        v2 = Cvalloc(NULL, n);  /* allocate another vector of length 2   */
        v1[0].r = 2;   v1[0].i = 2;   v1[1].r = 2;   v1[1].i = 2;
        v2[0].r = 3;   v2[0].i = 3;   v2[1].r = 3;   v2[1].i = 3;

        v = Cvadd(v1, v2, n); /* add v1[] and v2[]               */

        /* the answer should be : v[0].r = 5;   v[0].i = 5;
                                  v[1].r = 5;   v[1].i = 5;       */
        for (i=0; i<n; i++) {
            printf("v[%d].r = %f, v[%d].i = %f\n", i, v[i].r, i, v[i].i);
        }
        Cvfree(v);    /* free vector v[]  */
        Cvfree(v1);   /* free vector v1[] */
        Cvfree(v2);   /* free vector v2[] */
        return 0;
}
```

Cvalloc

```
#include "mathlib.h"
Complex_Vector Cvalloc(Complex_Ptr address, unsigned n);
```

DESCRIPTION

The *Cvalloc* function allocates memory for a complex vector of length n.

The *address* argument is a pointer to a **Complex** value. If the *address* argument is NULL, the *Cvalloc* function allocates space for the vector from the available dynamic memory. If the *address* argument is not NULL, the *Cvalloc* function assumes that it is the address of a single dimensioned array of **Complex** values. In this case, no memory is allocated; the *Cvalloc* function simply returns a pointer to the beginning of the single dimensioned array. This provides a mechanism for using preinitialized arrays.

RETURNS

A complex vector of length n if successful, or NULL if an error occurs. If the vector is too large, an "Array too large" error message is generated (math_errno = E_MSIZE). If there is not enough dynamic memory, an "Insufficient dynamic memory available" error message is generated (math_errno = E_MALLOC).

SEE ALSO

Cmxalloc(), mxalloc(), valloc().

```
#include "mathlib.h"
#include <stdio.h>
Complex VECT1[4] = { 1, 2, 3, 4, 5, 6, 7, 8 } ;
main()
{
Complex_Vector  vect1,vect2;
unsigned n=4;
int i;

        vect2 = Cvalloc(NULL, n);    /* allocate complex vector vect2    */
        vect1 = Cvalloc(VECT1, n);   /* allocate and initialize vect1    */

        /* now initialize complex vector vect2 */
        vect2[0].r = 1; vect2[0].i = 2; vect2[1].r = 3; vect2[1].i = 4;
        vect2[2].r = 5; vect2[2].i = 6; vect2[3].r = 7; vect2[3].i = 8;

/* the vectors vect1[] and vect2[] should be equal  */

        for (i=0; i<n; i++) {
                printf("vect1[%d].r = %3.1f, vect1[%d].i = %3.1f   ",
                        i, vect1[i].r, i, vect1[i].i);
                printf("vect2[%d].r = %3.1f, vect2[%d].i = %3.1f\n",
                        i, vect2[i].r, i, vect2[i].i);
        }
        Cvfree(vect2);    /*  free vector vect2[] */
        return 0;
}
```

Cvconjg

#include "mathlib.h"
Complex_Vector Cvconjg(Complex_Vector *v*, unsigned *n*);

DESCRIPTION

The *Cvconjg* function computes the conjugate of the complex vector v of length n.

The complex vector v must be allocated by the *Cvalloc* function prior to calling the *Cvconjg* function. The resulting conjugate vector is automatically allocated by the *Cvconjg* function.

The user may select the option of whether or not to write the resulting vector over the input vector v. If memory is a limitation, and v is no longer needed, this may prove useful. This option can be selected by setting the global variable **useinput_** = **1** prior to calling *Cvconjg*. The default option is to not destroy the input vector v (i. e., **useinput_** = **0**).

RETURNS

A complex vector of length n representing the conjugate of v, or NULL if an error occurs.

SEE ALSO

conjg(), Cmxconjg().

```
#include "mathlib.h"
#include <stdio.h>
Complex V[4] = { 1,-1, 2,-2, 3,-3, 4,-4 };
main()
{
Complex_Vector v,v1;
unsigned n=4;
int i;

        v = Cvalloc(V, n); /* allocate and initialize v[]      */

        v1 = Cvconjg(v, n); /* compute the conjugate of v[]    */

/*   v[].r should be equal to v1[].r and
     v[].i should be equal to -v1[].i         */

        for (i=0; i<n; i++) {
                printf("v[%d].r = %3.1f, v[%d].i = %3.1f    ",
                        i, v[i].r, i, v[i].i);
                printf("v1[%d].r = %3.1f, v1[%d].i = %3.1f\n",
                        i, v1[i].r, i, v1[i].i);
        }
        Cvfree(v1);    /* free vector v1[]  */
        return 0;
}
```

Cvcopy

```
#include "mathlib.h"
void Cvcopy(Complex__Vector dest, Complex__Vector src, unsigned n);
```

DESCRIPTION

The *Cvcopy* function copies the complex vector *src* to the complex vector *dest*.

The *n* argument specifies the number of elements in each vector. The *dest* and *src* vectors must be allocated by the *Cvalloc* function prior to calling the *Cvcopy* function.

RETURNS

None.

SEE ALSO

Cmxcopy(), mxcopy(), vcopy(), Cmxdup(), mxdup(), Cvdup(), vdup().

```
#include "mathlib.h"
#include <stdio.h>
Complex SRC[4] = { 1, 2, 3, 4, 5, 6, 7, 8 } ;
main()
{
Complex_Vector  src,dest;
unsigned n=4;
int i;

        dest = Cvalloc(NULL, n);     /* allocate complex vector dest[]    */
        src = Cvalloc(SRC, n);       /* allocate and initialize src[]     */

        Cvcopy(dest, src, n); /* now copy the complex vector src[]        */

/* the vectors src[] and dest[] should be equal  */

        for (i=0; i<n; i++) {
                printf("src[%d].r = %3.1f, src[%d].i = %3.1f    ",
                        i, src[i].r, i, src[i].i);
                printf("dest[%d].r = %3.1f, dest[%d].i = %3.1f\n",
                        i, dest[i].r, i, dest[i].i);
        }
        Cvfree(dest);    /* free vector  dest[]  */
        return 0;
}
```

Cvdot

```
#include "mathlib.h"
Complex Cvdot(Complex_Vector v1, Complex_Vector v2, unsigned n);
```

DESCRIPTION

The *Cvdot* function computes the dot product of the two complex vectors $v1$ and $v2$ of length n.

The dot product (sometimes called the inner product) of two vectors is given by

$$v1 * v2 = v1[0] * v2[0] + v1[1] * v2[1] + \ldots + v1[n-1] * v2[n-1]$$

where n is the length of both vectors and * denotes complex multiplication.

RETURNS

A complex value representing the dot product of the two complex vectors $v1$ and $v2$ of length n.

SEE ALSO

vdot(), vcross().

```
#include <stdio.h>
#include <stdlib.h>
#include <math.h>
#include "mathlib.h"

Complex V1[3] = { 0,1, 2,3, 4,5 };
Complex V2[3] = { 6,7, 8,9, 10,11 };

main()
{
unsigned n=3;
Complex_Vector v1, v2;
Complex dot;

    v1 = Cvalloc(V1, n);
    v2 = Cvalloc(V2, n);
    dot = Cvdot(v1, v2, n);
    printf("dot.r = %6.2f dot.i = %6.2f\n", dot.r, dot.i);
    return 0;
}
```

PROGRAM OUTPUT

```
dot.r = -33.00 dot.i = 142.00
```

Cvdup

#include "mathhlib.h"
Complex__Vector Cvdup(Complex__Vector *v*, unsigned *n*);

DESCRIPTION

The *Cvdup* function makes a duplicate copy of the complex vector *v*.

The *n* argument specifies the number of elements in the vector. The complex vector *v* must be allocated by *Cvalloc* prior to calling the *Cvdup* function. The *Cvdup* function automatically allocates memory for the duplicate copy.

RETURNS

A complex vector of length *n* that is a duplicate copy of *v* if successful, or NULL if an error occurs. If there is not enough dynamic memory, an "Insufficient dynamic memory available" error message is generated (math_errno = E_MALLOC).

SEE ALSO

Cmxdup(), mxdup(), vdup(), Cmxcopy(), mxcopy(), vcopy(), Cvcopy().

```
#include "mathlib.h"
#include <stdio.h>
Complex V1[2] = { 1,2, 3,4 } ;
main()
{
Complex_Vector  v1,v2;
unsigned i;
unsigned n=2;
        v1 = Cvalloc(V1, n);          /* allocate and initialize v1[]    */
        v2 = Cvdup(v1, n);            /* now duplicate the vector v1[]    */
/* the vectors v1[] and v2[] should be equal  */
        for (i=0; i<n; i++) {
                printf("v1[%d].r = %3.1f, v1[%d].i = %3.1f   ",
                        i, v1[i].r, i, v1[i].i);
                printf("v2[%d].r = %3.1f, v2[%d].i = %3.1f\n",
                        i, v2[i].r, i, v2[i].i);
        }
        Cvfree(v2);   /* free vector v2[]  */
        return 0;
}
```

Cvectomx

#include "mathlib.h"
Complex__Matrix Cvectomx(Complex__Vector *v1*, Complex__Vector *v2*, unsigned *n*);

DESCRIPTION
The *Cvectomx* function multiplies the transpose of the complex vector $v1$ by the complex vector $v2$.

The n argument specifies the number of elements in vectors $v1$ and $v2$. The resulting product is a complex matrix of n rows by n columns. Using row-vector notation, we have

$$v1^T[n] * v2[n] = m[n][n]$$

The complex vectors $v1$ and $v2$ must be allocated by *Cvalloc* prior to calling the *Cvectomx* function. If desired, $v1$ and $v2$ can be the same vector. The resulting n by n complex matrix is automatically allocated by the *Cvectomx* function.

RETURNS
An n by n complex matrix m, where $m[n][n] = v1^T[n] * v2[n]$ if successful, or NULL if an error occurs.

SEE ALSO
Cvmxmul(), Cvmxmul1(), vectomat().

```
#include "mathlib.h"
#include <stdio.h>
#include <math.h>
#include <stdlib.h>
Complex V1[5] = { 2,2, 2,2, 2,2, 2,2, 2,2 };
Complex V2[5] = { 3,3, 3,3, 3,3, 3,3, 3,3 };
main()
{
int i,j,n=5;
Complex_Vector v1,v2;
Complex_Matrix m;

        v1 = Cvalloc(V1, n);
        v2 = Cvalloc(V2, n);
        m = Cvectomx(v1, v2, n); /* multiply the two vectors into a matrix     */
        for (i=0; i<n; i++) {
              for (j=0; j<n; j++) {
                    printf("%d %d %f %f\n", i, j, m[i][j].r, m[i][j].i);
              }
        }
  /* the answer is m[i][j].r = 0 and m[i][j].i = 12; for all i and j= 0,1,..4 */
        Cmxfree(m) ; /* free matrix m[][] */
        return 0;
}
```

Cvfree

#include "mathlib.h"
void Cvfree(Complex__Vector v);

DESCRIPTION

The *Cvfree* function frees the memory allocated to the complex vector *v*. The memory for the complex vector *v* must have been allocated by the *Cvalloc* function prior to calling the *Cvfree* function.

RETURNS

None.

SEE ALSO

Cmxfree(), mxfree(), vfree(), mathfree().

```
#include "mathlib.h"
#include <stdio.h>
Complex V1[2] = { 1,2, 3,4 };
main()
{
Complex_Vector v,v1;
int i;

        v = Cvalloc(NULL, 2); /* allocate dynamic memory for v[]        */
        v1 = Cvalloc(V1, 2);   /* initialize static array v1[]; no
                                  dynamic memory is allocated for v1[]   */
        for (i=0; i<2; i++) {
                v[i] = v1[i];           /* set v[] equal to v1[]         */
                printf("v[%d].r = %f, v[%d].i = %f\n",i,v[i].r,i,v[i].i);
        }
        Cvfree(v);  /*  free the dynamic memory occupied by v[]        */
        return 0;

}
```

Cvinit

```
#include "mathlib.h"
void Cvinit(Complex_Vector v, unsigned n, Complex value);
```

DESCRIPTION

The *Cvinit* function initializes each of the n elements of the complex vector v with the complex scalar *value*.

The complex vector v must be allocated by *Cvalloc* prior to calling the *Cvinit* function.

RETURNS

None.

SEE ALSO

Cmxinit(), mxinit(), vinit().

```
#include "mathlib.h"
#include <stdio.h>
main()
{
Complex_Vector v;
Complex zero;
int i;
unsigned npts=2;

        v = Cvalloc(NULL, npts);        /* allocate vector m[][]          */

        zero.r = zero.i = 0;

        Cvinit(v, npts, zero);          /* set vector v[] to zero         */

        printf("the following complex vector of length 2 should be zero:\n");
        for (i=0; i<npts; i++) {
                printf(" %f %f\n", v[i].r, v[i].i);
        }
        return 0;
}
```

Cvmag

#include "mathlib.h"
Real__Vector Cvmag(Complex__Vector *v*, unsigned *n*);

DESCRIPTION

The *Cvmag* function computes the magnitude of the complex vector v. For each complex value in vector v, a real value is calculated as the square root of the sum of the squares of the real and imaginary parts.

The n argument specifies the number of elements in the vector. The complex vector v must be allocated by *Cvalloc* prior to calling the *Cvmag* function. The resulting real vector representing the magnitude of v is automatically allocated by the *Cvmag* function.

RETURNS

A real vector of length n that represents the magnitude of each complex value in vector v if successful, or NULL if an error occurs.

SEE ALSO

Cmag(), Cmxmag(), vmag().

```
#include "mathlib.h"
#include <stdio.h>
Complex V[4] = { 1,1, 1,1, 1,1, 1,1 };
main()
{
Complex_Vector v;
Real_Vector v1;
int i;

        v = Cvalloc(V, 4);  /* initialize v[] to static complex array V[]    */

        v1 = Cvmag(v, 4); /* compute the magnitude of v[]                      */

        /* the answer should be : v1[0] = v1[1] =  v1[2] = v1[3] = 1.414214  */

        for (i=0; i<4; i++) {
                printf("v1[%d] = %f\n", i, v1[i]);
        }
        vfree(v1);        /* free vector v1[]    */
        return 0;
}
```

Cvmaxval

#include "mathlib.h"
Real Cvmaxval(Complex__Vector *v*, unsigned *n*, unsigned **imx*);

DESCRIPTION

The *Cvmaxvl* function finds the element in the complex vector v that has the largest magnitude. The magnitude of each complex value is calculated as the square root of the sum of the squares of the real and imaginary parts.

    The n argument specifies the number of elements in the vector. The **imx* argument is the address of an unsigned integer where *Cvmaxval* stores the element (i. e., index) number of the largest value.

RETURNS

A real value representing the magnitude of the largest value in the complex vector v. The index of the element corresponding to the largest value is returned via the *imx* argument.

SEE ALSO

Cmxmaxvl(), Cmxminvl(), Cvminval(), mxmaxval(), mxminval(), vmaxval(), vminval().

```
#include "mathlib.h"
#include <stdio.h>
Complex V[4] = { 1,2, 3,4, 0,1, 5,6 };
main()
{
Complex_Vector v;
unsigned imax;
Real biggest;

    v = Cvalloc(V, 4);              /* allocate and initialize vector v[]   */

    biggest = Cvmaxval(v, 4, &imax);/*  find largest magnitude in v[]   */

    printf("the maximum magnitude is = %f\n", biggest);
    printf("it occurs at element # %d of vector v[]\n", imax);
    printf("the real component is %f and the imaginary component is %f\n",
            v[imax].r, v[imax].i);
    return 0;
}
```

Cvminval

```
#include "mathlib.h"
Real Cvminval(Complex__Vector v, unsigned n, unsigned *imn);
```

DESCRIPTION

The *Cvminval* function finds the element in the complex vector *v* that has the smallest magnitude. The magnitude of each complex value is calculated as the square root of the sum of the squares of the real and imaginary parts.

The *n* argument specifies the number of elements in the vector. The **imn* argument is the address of an unsigned integer where *Cvminval* stores the element (i. e., index) number of the smallest value.

RETURNS

A real value representing the magnitude of the smallest value in the complex vector *v*. The index of the value with the smallest magnitude is returned in *imn*.

SEE ALSO

Cmxmaxvl(), Cmxminvl(), Cvmaxval(), mxmaxval(), mxminval(), vmaxval(), vminval().

```
#include "mathlib.h"
#include <stdio.h>
Complex V[4] = { 1,2, 3,4, 0,1, 5,6 };
main()
{
Complex_Vector v;
unsigned imin;
Real smallest;

        v = Cvalloc(V, 4);               /* allocate and initialize vector v[]    */

        smallest = Cvminval(v, 4, &imin);/*  find smallest magnitude in v[]  */

        printf("the minimum magnitude is = %f\n", smallest);
        printf("it occurs at element # %d of vector v[]\n", imin);
        printf("the real component is %f and the imaginary component is %f\n",
                v[imin].r, v[imin].i);
        return 0;
}
```

Cvmxmul

#include "mathlib.h"
Complex__Vector Cvmxmul(Complex__Matrix *m*, Complex__Vector *v*, unsigned *n1*,
unsigned *n2*);

DESCRIPTION

The *Cvmxmul* function multiplies the complex matrix *m* by the complex vector *v*.

The *n*1 argument specifies the number of rows in matrix *m*. The *n*2 argument specifies the number of columns in matrix *m*, as well as the number of elements in vector *v*.

The complex matrix *m* and the complex vector *v* must be allocated by *Cmxalloc* and *Cvalloc* respectively prior to calling the *Cvmxmul* function. The resulting complex vector of length *n*1, representing the product of *m* and *v*, is automatically allocated by the *Cvmxmul* function.

RETURNS

A complex vector *v*1 of length *n*1, where $v1[n1] = m[n1][n2] * v[n2]$ if successful, or NULL if an error occurs.

SEE ALSO

vmxmul(), Cvmxmul1(), Cvectomx(), vectomat().

```
#include "mathlib.h"
#include <stdio.h>
#include <math.h>
#include <stdlib.h>
Complex V[3] = { 4,4, 4,4, 4,4 };
Complex M[2][3] = { 1,2, 3,4, 5,6, 7,8, 9,0, 1,2 };
main()
{
int i,n1=2,n2=3;
Complex_Vector v,v1;
Complex_Matrix m;

        v = Cvalloc(V, n2);     /* allocate and initialize v[]        */
        m = Cmxalloc(M, n1, n2);/* allocate and initialize m[][]      */
        v1 = Cvmxmul(m, v, n1, n2);  /* compute the product v1[] = m[][]*v[]  */
        for (i=0; i<n1; i++) {
                printf("%d %f %f\n", i, v1[i].r, v1[i].i);
        }
  /* the answer is v1[0].r = -12, v1[0].i = 84; v1[1].r = 28, v1[1].i = 108  */
        Cmxfree(m) ; /* free matrix m[][] */
        return 0;
}
```

Cvmxmul1

#include "mathlib.h"
Complex_Vector Cvmxmul1(Complex_Matrix *m*, Complex_Vector *v*, unsigned *n1*,
                                                    unsigned *n2*);

DESCRIPTION
The *Cvmxmul1* function multiplies the transpose of the complex matrix m by the complex vector v.

The $n1$ argument specifies the number of columns in matrix m. The $n2$ argument specifies the number of rows in matrix m, as well as the number of elements in vector v.

The complex matrix m and the complex vector v must be allocated by *Cmxalloc* and *Cvalloc* respectively prior to calling the *Cvmxmul1* function. The resulting complex vector of length $n1$, representing the product of m^T, and v is automatically allocated by the *Cvmxmul1* function.

RETURNS
A complex vector $v1$ of length $n1$, where $v1[n1] = m^T[n2][n1] * v[n2]$ if successful, or NULL if an error occurs.

SEE ALSO
vmxmul(), Cvmxmul(), Cvectomx(), vectomat().

```
#include "mathlib.h"
#include <stdio.h>
#include <math.h>
#include <stdlib.h>
Complex V[3] = { 4,4, 4,4, 4,4 };
Complex M[3][2] = { 1,2, 7,8, 3,4, 9,0, 5,6, 1,2 };
main()
{
int i,n1=2,n2=3;
Complex_Vector v,v1;
Complex_Matrix m;

        v = Cvalloc(V, n2);      /* allocate and initialize v[]          */
        m = Cmxalloc(M, n2, n1);/* allocate and initialize m[][]         */
        v1 = Cvmxmul1(m, v, n1, n2);/* compute the product v1[] = m'[][]*v[] */
        for (i=0; i<n1; i++) {
                printf("%d %f %f\n", i, v1[i].r, v1[i].i);
        }
  /* the answer is v1[0].r = -12, v1[0].i = 84; v1[1].r = 28, v1[1].i = 108 */
        Cmxfree(m) ; /* free matrix m[][] */
        return 0;
}
```

Cvread

#include "mathlib.h"
Complex__Vector Cvread(FILE *fp, unsigned n);

DESCRIPTION

The *Cvread* function reads the next n complex values from the file pointed to by fp. The file must have been opened as a binary file (e. g., fp = fopen("filename", "rb")), since the *Cvread* function assumes that the complex values are stored in binary format. The number of bytes read from the file is equal to $n*$ **sizeof(Complex)**. The file pointer should be advanced this number of bytes toward the end-of-file if the n complex values are read successfully.

The *Cvread* function automatically allocates a complex vector of length n to store the complex values that are read from the file.

The *Cvwrite* function may be used to create a file of binary complex values.

RETURNS:

A complex vector of length n if successful, or NULL if an error occurs. If a read error occurs, an "Error reading from file" error message is generated (math_errno = E_READ).

SEE ALSO

Cvwrite(), vwrite(), vread(), xywrite(), xyread().

```
#include "mathlib.h"
#include <stdio.h>
#include <stdlib.h>
Complex V1[3] = { 1,1, 2,2, 3,3 };
Complex V2[3] = { 4,4, 5,5, 6,6 };
Complex V3[3] = { 7,7, 8,8, 9,9 };
main()
{
FILE *fp;
Complex_Vector v1,v2,v3,x1,x2,x3;
int i;

    v1 = Cvalloc(V1, 3);
    v2 = Cvalloc(V2, 3);
    v3 = Cvalloc(V3, 3);

    fp = fopen("binary.dat", "w+b");/*open file "binary.dat" for binary write */
    if (fp == NULL) {                    /*see if any errors were generated     */
            printf("can't open binary.dat\n");
            exit(1);
    }

    Cvwrite(fp, v1, 3);     /* write vector v1[] to "binary.dat"   */
    Cvwrite(fp, v2, 3);     /* write vector v2[] to "binary.dat"   */
    Cvwrite(fp, v3, 3);     /* write vector v3[] to "binary.dat"   */

    fseek(fp, 2*3*sizeof(Complex), SEEK_SET);/* move pointer to 3rd vector    */
    x3 = Cvread(fp, 3);                  /* read vector x3[] from binary.dat    */

    fseek(fp, 3*sizeof(Complex), SEEK_SET);/* move pointer to 2nd vector      */
    x2 = Cvread(fp, 3);                  /* read vector x2[] from binary.dat    */

    fseek(fp, 0, SEEK_SET);/* move pointer to start of 1st vector              */
    x1 = Cvread(fp, 3);                  /* read vector x1[] from binary.dat    */

    for( i=0; i<3; i++){
        printf("x3[%d].r = %3.1f, x2[%d].r = %3.1f, x1[%d].r = %3.1f\n",
                i,x3[i].r, i,x2[i].r, i,x1[i].r);
        printf("x3[%d].i = %3.1f, x2[%d].i = %3.1f, x1[%d].i = %3.1f\n",
                i,x3[i].i, i,x2[i].i, i,x1[i].i);
        printf("\n");
    }
    fclose(fp);
    return 0;
}
```

PROGRAM OUTPUT:

```
x3[0].r = 7.0, x2[0].r = 4.0, x1[0].r = 1.0
x3[0].i = 7.0, x2[0].i = 4.0, x1[0].i = 1.0

x3[1].r = 8.0, x2[1].r = 5.0, x1[1].r = 2.0
x3[1].i = 8.0, x2[1].i = 5.0, x1[1].i = 2.0

x3[2].r = 9.0, x2[2].r = 6.0, x1[2].r = 3.0
x3[2].i = 9.0, x2[2].i = 6.0, x1[2].i = 3.0
```

CvscalC

```
#include "mathlib.h"
Complex_Vector CvscalC(Complex_Vector v, unsigned n, Complex value);
```

DESCRIPTION

The *CvscalC* function multiplies the complex vector v by the complex scalar *value*.

The n argument specifies the number of elements in the vector. The complex vector v must be allocated by *Cvalloc* prior to calling the *CvscalC* function. The resulting complex vector representing the product of v and *value* is automatically allocated by the *CvscalC* function.

The user may select the option of whether or not to write over the input vector v with "in-place" computations. If memory is a limitation, and v is no longer needed, this may prove useful. This option can be selected by setting the global variable **useinput_** = **1** prior to calling *CvscalC*. The default option is to not destroy the input vector v (i. e., **useinput_** = **0**).

RETURNS

An n element complex vector representing the product of the complex vector v with the complex *value* if successful, or NULL if an error occurs.

SEE ALSO

CmxscalC(), CmxscalR(), mxscale(), CvscalR(), vscale().

```
#include "mathlib.h"
#include <stdio.h>
Complex V[2] = { 1,1, 1,1 };
main()
{
Complex_Vector v,v1;
unsigned n=2;
int i;
Complex value;

        value.r = 2; value.i = 2;  /* define a complex scalar            */
        v=Cvalloc(V, n);           /* allocate and initialize v[]        */
        v1 = CvscalC(v, n, value); /* now scale complex vector v[] by "value"*/
        /* the answer should be : v1[0].r = 0;   v1[0].i = 4;
                                  v1[0].r = 0;   v1[0].i = 4;             */
        for (i=0; i<2; i++) {
                printf("v1[%d].r = %f, v1[%d].i = %f\n",
                                                i, v1[i].r, i, v1[i].i);
        }
        Cvfree(v1);    /* free vector v1[] */
        return 0;
}
```

CvscalR

```
#include "mathlib.h"
Complex__Vector CvscalR(Complex__Vector v, unsigned n, Real value);
```

DESCRIPTION

The *CvscalR* function multiplies the complex vector v by the real scalar *value*.

The n argument specifies the number of elements in the vector. The complex vector v must be allocated by *Cvalloc* prior to calling the *CvscalR* function. The resulting complex vector representing the product of v and *value* is automatically allocated by the *CvscalR* function.

The user may select the option of whether or not to write over the input vector v with "in-place" computations. If memory is a limitation, and v is no longer needed, this may prove useful. This option can be selected by setting the global variable **useinput_ = 1** prior to calling *CvscalR*. The default option is to not destroy the input vector v (i. e., **useinput_ = 0**).

RETURNS

An n element complex vector representing the product of the complex vector v with the real *value* if successful, or NULL if an error occurs.

SEE ALSO

CmxscalC(), CmxscalR(), mxscale(), CvscalC(), vscale().

```c
#include "mathlib.h"
#include <stdio.h>
Complex V[2] = { 1,1, 1,1 };
main()
{
Complex_Vector v,v1;
unsigned n=2;
int i;
Real value;

        value = 2;                  /* define a real scalar             */
        v=Cvalloc(V, n);            /* allocate and initialize v[]      */
        v1 = CvscalR(v, n, value); /* now scale complex vector v[] by "value"*/

        /* the answer should be : v1[0].r = 2;  v1[0].i = 2;
                                  v1[0].r = 2;  v1[0].i = 2;            */
        for (i=0; i<2; i++) {
                printf("v1[%d].r = %f, v1[%d].i = %f\n",
                                        i, v1[i].r, i, v1[i].i);
        }
        Cvfree(v1);   /* free vector v1[] */
        return 0;
}
```

Cvsub

#include "mathlib.h"

Complex_Vector Cvsub(Complex_Vector *v1*, Complex_Vector *v2*, unsigned *n*);

DESCRIPTION

The *Cvsub* function subtracts the complex vector $v2$ from the complex vector $v1$.

The n argument specifies the number of elements in each vector. The complex vectors $v1$ and $v2$ must be allocated by *Cvalloc* prior to calling the *Cvsub* function. The resulting complex vector representing the difference between $v1$ and $v2$ is automatically allocated by the *Cvsub* function.

The user may select the option of whether or not to write over the input vector $v1$ with in-place computations. If memory is a limitation, and $v1$ is no longer needed, this may prove useful. This option can be selected by setting the global variable **useinput_ = 1** prior to calling *Cvsub*. If this option is selected, the resulting vector overwrites the input vector $v1$. The default option is to not destroy the input vector $v1$ (i. e., **useinput_ = 0**).

RETURNS

A complex vector of length n representing the difference between $v1$ and $v2$ if successful, or NULL if an error occurs.

SEE ALSO

Cmxadd(), Cmxsub(), mxadd(), mxsub(), Cvadd(), vadd(), vsub().

```
#include "mathlib.h"
#include <stdio.h>
main()
{
int i;
Complex_Vector v,v1,v2;
unsigned n=2;
/* subtract the following two vectors:                          */
/*                                                              */
/*          v[] = v1[] - v2[]                                   */
/*                                                              */
/* where v1[] and v2[] are complex and are given below:        */
        v1 = Cvalloc(NULL, n);  /* allocate vector of length 2   */
        v2 = Cvalloc(NULL, n);  /* allocate another vector of length 2   */
        v1[0].r = 2;   v1[0].i = 2;  v1[1].r = 2;    v1[1].i = 2;
        v2[0].r = 3;   v2[0].i = 3;  v2[1].r = 3;    v2[1].i = 3;

        v = Cvsub(v1, v2, n); /* subtract v2[] from v1[]          */

        /* the answer should be : v[0].r = -1;   v[0].i = -1;
                                  v[1].r = -1;   v[1].i = -1;      */
        for (i=0; i<n; i++) {
            printf("v[%d].r = %f, v[%d].i = %f\n", i, v[i].r, i, v[i].i);
        }
        Cvfree(v);     /* free vector v[]  */
        Cvfree(v1);    /* free vector v1[] */
        Cvfree(v2);    /* free vector v2[] */
        return 0;
}
```

Cvwrite

#include "mathlib.h"
int Cvwrite(FILE *fp, Complex_Vector v, unsigned n);

DESCRIPTION

The *Cvwrite* function writes the n complex values in the complex vector v to the file pointed to by fp. The file must have been opened as a binary file (e. g., fp = fopen("filename", "wb")), since the *Cvwrite* function outputs the complex values in binary format. The number of bytes written to the file is equal to $n*$ **sizeof(Complex)**. The file pointer should be advanced this number of bytes toward the end-of-file if the n complex values are written successfully.

The complex vector v must be allocated by *Cvalloc* prior to calling the *Cvwrite* function.

The *Cvread* function may be used to read complex values from a binary file.

RETURNS

0 if the complex vector v is successfully written to the file, or -1 if an error occurs. If a write error occurs, an "Error writing to file" error message is generated (math_errno = E_WRITE).

SEE ALSO

Cvread(), vread(), xywrite(), xyread().

```
#include <stdio.h>
#include <stdlib.h>
#include "mathlib.h"
Complex V1[3] = { 1,1, 2,2, 3,3 };
Complex V2[3] = { 4,4, 5,5, 6,6 };
Complex V3[3] = { 7,7, 8,8, 9,9 };
main()
{
FILE *fp;
Complex_Vector v1,v2,v3,x1,x2,x3;
int i;

    v1 = Cvalloc(V1, 3);
    v2 = Cvalloc(V2, 3);
    v3 = Cvalloc(V3, 3);

    fp = fopen("binary.dat", "w+b");/*open file "binary.dat" for binary write */
    if (fp == NULL) {                     /*see if any errors were generated      */
            printf("can't open binary.dat\n");
            exit(1);
    }

    Cvwrite(fp, v1, 3);                  /* write vector v1[] to "binary.dat"   */
    Cvwrite(fp, v2, 3);                  /* write vector v2[] to "binary.dat"   */
    Cvwrite(fp, v3, 3);                  /* write vector v3[] to "binary.dat"   */
    fseek(fp, 0, SEEK_SET);      /* position file indicator to top of file    */
    x1 = Cvread(fp, 3);                  /* read vector x1[] from "binary.dat"  */
    x2 = Cvread(fp, 3);                  /* read vector x2[] from "binary.dat"  */
    x3 = Cvread(fp, 3);                  /* read vector x3[] from "binary.dat"  */

    for( i=0; i<3; i++) {
            printf("x1[%d].r = %3.1f, x2[%d].r = %3.1f, x3[%d].r = %3.1f\n",
                    i,x1[i].r, i,x2[i].r, i,x3[i].r);
            printf("x1[%d].i = %3.1f, x2[%d].i = %3.1f, x3[%d].i = %3.1f\n",
                    i,x1[i].i, i,x2[i].i, i,x3[i].i);
            printf("\n");
    }
    fclose(fp);
    return 0;
}
```

PROGRAM OUTPUT

```
x1[0].r = 1.0, x2[0].r = 4.0, x3[0].r = 7.0
x1[0].i = 1.0, x2[0].i = 4.0, x3[0].i = 7.0

x1[1].r = 2.0, x2[1].r = 5.0, x3[1].r = 8.0
x1[1].i = 2.0, x2[1].i = 5.0, x3[1].i = 8.0

x1[2].r = 3.0, x2[2].r = 6.0, x3[2].r = 9.0
x1[2].i = 3.0, x2[2].i = 6.0, x3[2].i = 9.0
```

dct

#include "mathlib.h"
Real__Vector dct(Real__Vector x, unsigned n int isign);

DESCRIPTION

The *dct* function computes the discrete cosine transform (**DCT**) of the real vector x of length n.

The *isign* argument determines the direction of the transform. If *isign* $= -1$, then a forward DCT is performed. If *isign* $= +1$, then an inverse DCT is performed.

If the real sinusoid $A * \cos(\omega * k * (2 * i + 1))$ is used to generate the x vector, the transformed vector will consist of one frequency component at the kth harmonic, where $k = 0, 1, 2, \ldots, n - 1$, $i = 0, 1, 2, \ldots, n - 1$, $\omega = \pi/(2 * n)$, and $\pi = 3.141592654\ldots$. If k is greater than zero, the magnitude of the component is $A(n/2)^{1/2}$.

RETURNS:

A real vector of length n representing the discrete cosine transform of the input vector x if successful, or NULL if an error occurs.

SEE ALSO

dct2d(), fftreal(), fftr__inv(), fftrad2(), fft42().

REFERENCE

N. S. Jayant and P. Noll (1984).

```
#include "mathlib.h"
#include <stdio.h>
#include <stdlib.h>
#include <math.h>
main()
{
Real_Vector data, t1, t2;
unsigned n=8, i, k = 7;

/* Compute the forward & inverse Discrete Cosine Transform of a cosine signal */

    data=valloc(NULL, n);   /* allocate an array for test data          */
    for(i=0; i<n; i++) {    /*  generate kth harmonic of                */
        data[i] = cos(pi_*(2*i+1)*k/(double)(2*n));
    }
                                    /* make the input data purely real  */
    t1 = dct(data, n, -1);          /* do a forward cosine transform     */
    t2 = dct(t1, n , 1);            /* do an inverse cosine transform    */
    printf("   original data     data passed through dct twice\n");
    for(i=0; i<n; i++) {
        printf("%d   %9.6f                     %9.6f\n", i, data[i], t2[i]);
    }
    vfree(data);
    vfree(t1);
    vfree(t2);
    return 0;
}
```

PROGRAM OUTPUT

	original data	data passed through dct twice
0	0.195090	0.195090
1	-0.555570	-0.555570
2	0.831470	0.831470
3	-0.980785	-0.980785
4	0.980785	0.980785
5	-0.831470	-0.831470
6	0.555570	0.555570
7	-0.195090	-0.195090

dct2d

#include "mathlib.h"
int dct2d(Real__Matrix *m*, unsigned *n*, int *isign*, int *last*);

DESCRIPTION

The discrete cosine transform (**DCT**) can be used to efficiently compress images. The JPEG committee has adopted the 8 by 8 2-D DCT as part of its standard for compressing still images.

The *dct2d* function computes the 2-D DCT of the square matrix *m*. A fast block matrix approach is used to compute the 2-D DCT. The *n* argument specifies the number of rows and columns in the matrix. The *isign* argument specifies the direction of the transform. If *isign* = −1, then a forward 2-D DCT is performed. If *isign* = +1, then an inverse 2-D DCT is performed.

The *last* argument specifies whether or not this function will be called more than once with the same size *n* by *n* matrix. This can speed things up, since calculation of the cosine matrix may be eliminated for all but the first call. If *last* = 0, then the *n* by *n* cosine matrix is calculated and saved internally for the next call. If *last* = 1, then the cosine matrix is freed before exiting the function. The final call to the *dct2d* function should specify *last* as 1 in order to free the allocated memory.

The discrete cosine transformed matrix is stored in matrix *m* (i. e., the input matrix is overwritten).

If the real 2-D sinusoid $A * \cos(\omega * k * (2 * i + 1)) * \cos(\omega * l * (2 * j + 1))$ is used to generate the matrix *m*, the transformed matrix will consist of one frequency component at the *k*th row and the *l*th column, where $k = 0, 1, 2, \ldots, n - 1$, $l = 0, 1, 2, \ldots, n - 1$, $i = 0, 1, 2, \ldots, n - 1$, $j = 0, 1, 2, \ldots, n - 1$, $\omega = \pi/(2 * n)$, and $\pi = 3.141592654\ldots$. If *k* and *l* are greater than zero, the magnitude of the component is $A * n/2$.

RETURNS

0 if successful, or −1 if an error occurs. If successful, matrix *m* will be an *n* by *n* discrete cosine transformed matrix. If there is not enough memory to allocate space for the cosine matrix, the error message "Insufficient dynamic memory available" is generated (math__errno = E__MALLOC).

SEE ALSO

dct(), fft2d(), fft2d__r().

REFERENCE

N. S. Jayant and P. Noll (1984).

```
#include <stdio.h>
#include <stdlib.h>
#include <math.h>
#include "mathlib.h"
main()
{
int i,j;
unsigned n=8;
double arg_row, arg_col, i0=2, j0=3;
Real_Matrix m;

/* Compute the forward & inverse 2-D Discrete Cosine Transform of a 2-D
   cosine signal */

        m = mxalloc(NULL, n, n);
        printf("input matrix:\n");
        for(i=0; i<n; i++) {  /* create 2-D real cosine data */
                arg_row = pi_*(2*i+1)/2.0;
                for(j=0; j<n; j++) {
                        arg_col = pi_*(2*j+1)/2.0;
                        m[i][j] = cos(arg_row*i0/n) * cos(arg_col*j0/n);
                        printf("%8.4f", m[i][j]);
                }
                printf("\n");
        }    /* call forward 2-D cosine transform */

        printf("matrix passed twice through 2-D DCT:\n");
        dct2d(m, n, -1, 0);
        dct2d(m, n,  1, 1);
        for(i=0; i<n; i++) {
                for(j=0; j<n; j++) {
                        printf("%8.4f", m[i][j]);
                }
                printf("\n");
        }
        mxfree(m);
        return 0;
}
```

dct2d

PROGRAM OUTPUT

```
input matrix:
  0.7682 -0.1802 -0.9061 -0.5133  0.5133  0.9061  0.1802 -0.7682
  0.3182 -0.0747 -0.3753 -0.2126  0.2126  0.3753  0.0747 -0.3182
 -0.3182  0.0747  0.3753  0.2126 -0.2126 -0.3753 -0.0747  0.3182
 -0.7682  0.1802  0.9061  0.5133 -0.5133 -0.9061 -0.1802  0.7682
 -0.7682  0.1802  0.9061  0.5133 -0.5133 -0.9061 -0.1802  0.7682
 -0.3182  0.0747  0.3753  0.2126 -0.2126 -0.3753 -0.0747  0.3182
  0.3182 -0.0747 -0.3753 -0.2126  0.2126  0.3753  0.0747 -0.3182
  0.7682 -0.1802 -0.9061 -0.5133  0.5133  0.9061  0.1802 -0.7682
matrix passed twice through 2-D DCT:
  0.7682 -0.1802 -0.9061 -0.5133  0.5133  0.9061  0.1802 -0.7682
  0.3182 -0.0747 -0.3753 -0.2126  0.2126  0.3753  0.0747 -0.3182
 -0.3182  0.0747  0.3753  0.2126 -0.2126 -0.3753 -0.0747  0.3182
 -0.7682 ·0.1802  0.9061  0.5133 -0.5133 -0.9061 -0.1802  0.7682
 -0.7682  0.1802  0.9061  0.5133 -0.5133 -0.9061 -0.1802  0.7682
 -0.3182  0.0747  0.3753  0.2126 -0.2126 -0.3753 -0.0747  0.3182
  0.3182 -0.0747 -0.3753 -0.2126  0.2126  0.3753  0.0747 -0.3182
  0.7682 -0.1802 -0.9061 -0.5133  0.5133  0.9061  0.1802 -0.7682
```

deriv

#include "mathlib.h"

Real_Vector deriv(Real_Vector *fx*, Real *dx*, int *n*);

DESCRIPTION

The *deriv* function numerically differentiates the equally spaced data represented by the real vector fx of length n. The method of weighted central differences is used to calculate the derivatives.

The dx argument is the fixed increment between samples. For example, if fx represents time samples of a waveform, dx is the sampling interval:

$$dx = T/(n-1)$$

where T is the duration of the waveform and n is the number of data points (i. e., the length of the fx vector).

This approach approximates the derivative, $f'(x)$, at the point x_j with

$$f'(x_j) = [h_{-2}f(x_{j-2}) + h_{-1}f(x_{j-1}) + h_1 f(x_{j+1}) + h_2 f(x_{j+2})]/dx$$

where the h_i's are the differentiating filter weights. Usually, h_i is an antisymmetric (odd) function:

$$h_i = -h_{-i}$$

The *deriv* function differentiates the central part of the real vector fx with the following five filter weights:

$$h_{-2} = 1/12, \qquad h_{-1} = -2/3, \qquad h_0 = 0, \qquad h_1 = 2/3, \qquad h_2 = -1/12$$

The error in estimating the derivative, $R(x)$, over the central region of the fx vector is approximately

$$R(x) = 0.1 * dx * |f'(x)|$$

For a more exact error expression, see Abramowitz and Stegun (1964, p. 914).

The lower and upper edges of the fx vector (i. e., $i = 0, 1$ and $i = n-1, n-2$) are differentiated with 2-point (see the *highpass* function) and 3-point difference filters. There must be at least five data points or nothing is done.

The *deriv* function calls the *convolve* function.

RETURNS

A real vector of length n representing the numerical derivative of the real vector fx if successful, or NULL if an error occurs. If $n < 5$, the error message "Not enough input data" is generated (math_errno = E_NOTENOUGH).

SEE ALSO

deriv1(), integrat(), romberg(), highpass().

REFERENCE

Abramowitz and Stegun (1964, p. 883, eq. 25.3.6).

```
#include "mathlib.h"
#include <stdio.h>
#include <math.h>
#include <stdlib.h>
main()
{
int i,n=20;
Real dx;
Real_Vector y, fx;

    fx = valloc(NULL, n);
    dx = twopi_/(double)(n-1);  /* generate one cycle of a sinusoid */
    for (i=0; i < n; i++) fx[i]=sin(dx*i);

    y = deriv(fx, dx, n);   /* numerically differentiate sin(x)     */

    /* if y = sin(x) the derivative should be: y' = cos(x)          */
    printf("                              numerical   true derivative:\n");
    printf(" i       x          sin(x)     derivative      cos(x)\n\n");
    for (i=0; i < n ; i++) {
        printf("%2d    %9.6f    %9.6f    %9.6f    %9.6f\n",
                        i,dx*i,fx[i], y[i], cos(dx*i));

    }
    return 0;
}
```

PROGRAM OUTPUT

i	x	sin(x)	numerical derivative	true derivative: cos(x)
0	0.000000	0.000000	0.981873	1.000000
1	0.330694	0.324699	0.928672	0.945817
2	0.661388	0.614213	0.788830	0.789141
3	0.992082	0.837166	0.546733	0.546948
4	1.322776	0.969400	0.245389	0.245485
5	1.653470	0.996584	-0.082547	-0.082579
6	1.984164	0.915773	-0.401537	-0.401695
7	2.314858	0.735724	-0.677015	-0.677282
8	2.645552	0.475947	-0.879128	-0.879474
9	2.976246	0.164595	-0.985973	-0.986361
10	3.306940	-0.164595	-0.985973	-0.986361
11	3.637634	-0.475947	-0.879128	-0.879474
12	3.968328	-0.735724	-0.677015	-0.677282
13	4.299022	-0.915773	-0.401537	-0.401695
14	4.629715	-0.996584	-0.082547	-0.082579
15	4.960409	-0.969400	0.245389	0.245485
16	5.291103	-0.837166	0.546733	0.546948
17	5.621797	-0.614213	0.788830	0.789141
18	5.952491	-0.324699	0.928672	0.945817
19	6.283185	-0.000000	0.981873	1.000000

deriv1

#include "mathlib.h"
double deriv1(double (*fx)(), double x, double dx);

DESCRIPTION
The *deriv1* function numerically differentiates the user-defined function fx at the point x, using the method of divided central differences with Romberg extrapolation.

The dx argument is the initial step size. This value should be small enough so that the function fx is defined and reasonably smooth over the interval $[x - dx, x + dx]$, yet large enough to avoid roundoff problems. The user-defined function must be differentiable over this interval.

The approach begins by approximating the derivative $f'(x)$ at the point x with the first central difference:

$$f_0'(x) = [f(x + dx) - f(x - dx)]/(2 * dx)$$

The step size dx is then reduced and Romberg extrapolation is used to refine the estimated derivative. The truncation error due to this extrapolation is fourth order. The algorithm then repeats, extrapolating each new estimate of the derivative with higher order differences. Upon completion, an accuracy of more than 11 significant digits is usually achieved.

The prototype for the user-defined function fx is

```
double fx(double x);
```

where, given x, the user-defined function must return $f(x)$.

RETURNS
A double precision value representing the numerical derivative of the user-defined function fx at the point x. If $dx = 0$, the error message "Step size must be nonzero to estimate the derivative" is generated (math_errno = E_STEPSIZE), and HUGE_VAL is returned.

SEE ALSO
deriv(), integrat(), romberg(), highpass().

REFERENCE
S. Fillipi and H. Engles (1966).

```
#include "mathlib.h"
#include <stdio.h>
#include <math.h>
#include <stdlib.h>

/* example subroutine to evaluate the value of f(x0) at x0  */
double fx(double x0)
{
        return sin(x0);
}
main()
{
double yprime, x=0, dx=0.1;

/* numerically differentiate sin(x) at x = 0                            */

    yprime=deriv1(fx, x, dx);   /* estimate derivative with divided     */
                                /* central differences approach         */

/* if y = sin(x) the derivative should be: cos(x)                       */

    printf("Divided Central Difference estimate of derivative of sin(x)\n");
    printf("                    at x = 0\n\n");

    printf("    true derivative = %f\n", cos(x));
    printf("numerical derivative = %f\n", yprime);
    printf("   the difference is = %f\n", yprime-cos(x));

    return 0;
}
```

PROGRAM OUTPUT

```
Divided Central Difference estimate of derivative of sin(x)
                    at x = 0

    true derivative = 1.000000
numerical derivative = 1.000000
   the difference is = -0.000000
```

downsamp

#include "mathlib.h"
Real__Vector downsamp(Real__Vector *data*, int *ndata*, int *ndec*);

DESCRIPTION

The *downsamp* function uses a Kaiser lowpass filter (see the *lowpass* function) to lower the sampling rate of the real vector *data* of length *ndata*. This "down-sampling" is sometimes called integer decimation since the data length *ndata* is reduced by an integer amount.

The *ndec* argument is the integer factor by which the sampling rate (and data length) is reduced. For example, if *ndata* = 100 and *ndec* = 5, then the result is a real vector of length 20. These 20 points represent all of the low-frequency information of the original 100 points, and are not simply a subset formed by deleting the other 90 points. The *ndec* argument must be greater than or equal to 2 and less than or equal to 10.

The real vector *data* must be allocated by *valloc* prior to calling the *downsamp* function. The resulting reduced length (i. e., decimated) real vector is automatically allocated by the *downsamp* function. The length of this vector is

$$(\text{int})((ndata - 1)/ndec) + 1$$

When one is decimating by large factors, some care must be taken that the signals of interest are sufficiently oversampled, or they will be filtered out with the noise. It is important to note that even if the input data is appropriately oversampled, all resampling approaches introduce some errors. These errors can be controlled by adjusting the length, *nweights*, of the Kaiser filter and its stopband attenuation, d_s (see the *lowpass* function).

The *downsamp* function sets the resampling error to $d_s = 0.001$, which corresponds to a filter stopband attenuation of -60 dB. The lowpass filter length is automatically chosen by the following empirical formula:

$$nweights = [1.08 * |\log_{10}(d_s)| + 1.174)] * ndec + 0.5$$

or

$$nweights = (\text{int})(4.414 * \text{ ndec } + 0.5)$$

This sets the corner frequencies of the lowpass decimating filter to the following arbitrary but reasonable values:

$$F_1 = 0.1235/(F_s/ndec)$$
$$F_2 = 1.253/(F_s/ndec)$$

where F_s is the original sampling rate in Hertz.

Due to edge effects, this function filters only the central n points of the *data* vector, where

$$n = (int)((ndata - nweights)/ndec) + 1)$$

The points at the edges are not filtered. The number of points at the edges is defined as

$$nedge = nweights/(2 * ndec)$$

Therefore, the points corresponding to the lower and upper edges (i. e., $i < nedge$ and $i > n + nedge - 1$) are simply copied from the original input vector *data*.

There must be at least $ndec + nweights - 1$ data points, or nothing is done. The *downsamp* function calls the *convolve* and *lowpass* functions.

RETURNS

A real vector of length (int) $((ndata - 1)/ndec) + 1$, representing the "down-sampled" version of the input vector *data* of length *ndata* if successful, or NULL if an error occurs. If $ndata < ndec + nweights - 1$, the error message "Not enough input data" is generated (math_errno = E_NOTENOUGH). If $ndec < 2$, or $ndec > 10$, the error message "Argument 'ndec' must be between 2 and 10" is generated (math_errno = E_DECIMATE).

SEE ALSO

lowpass(), smooth(), interp().

REFERENCE

L. R. Rabiner and R. W. Schafer (1973).

```c
#include "mathlib.h"
#include <stdio.h>
#include <math.h>
#include <stdlib.h>
main()
{
int i, j, k, ndata=21, ndec=3;
double delta;
Real_Vector y, data;

        data = valloc(NULL, ndata);
        delta = twopi_/(double)(ndata-1);  /* generate one cycle of a sinusoid   */
        for (i=0; i < ndata; i++) data[i]=sin(delta*i);

        y = downsamp(data, ndata, ndec); /* lower the sampling rate by 1/3    */

        /* the resampled data should approximate every 3rd point of the input */
        printf("      original            resampled\n");
        printf(" i      data[i]      k    data y[k]\n\n");
        j=-ndec;
        k=0;
        for (i=0; i < ndata; i++) {
            if(i == j+ndec) {
                printf("%2d  %9.6f      %2d   %9.6f\n", i, data[i], k, y[k]);
                j += ndec;
                k++;
            }
            else  printf("%2d  %9.6f\n", i, data[i]);
        }
        return 0;
}
```

PROGRAM OUTPUT

i	original data[i]	k	resampled data y[k]
0	0.000000	0	0.000000
1	0.309017		
2	0.587785		
3	0.809017	1	0.809017
4	0.951057		
5	1.000000		
6	0.951057	2	0.939965
7	0.809017		
8	0.587785		
9	0.309017	3	0.305413
10	0.000000		
11	-0.309017		
12	-0.587785	4	-0.580930
13	-0.809017		
14	-0.951057		
15	-1.000000	5	-1.000000
16	-0.951057		
17	-0.809017		
18	-0.587785	6	-0.587785
19	-0.309017		
20	0.000000		

fact

#include "mathlib.h"
double fact(unsigned *n*);

DESCRIPTION

The *fact* function computes the factorial of n. The factorial represents the number of different ways of arranging or permuting n distinct objects. The factorial of n is often denoted as $n!$, where

$$n! = (n) * (n - 1) * (n - 2) * \ldots * 1$$

The value of $n!$ is always a whole number. Zero factorial is defined to be unity (i. e., $0! = 1$).

Note that $n! = {}_nP_n$ (see the *permut* function).

RETURNS

A double precision whole number representing $n!$.

SEE ALSO

combin(), permut().

```
#include "mathlib.h"
#include <stdio.h>
main()
{
double factorial;
int n=4;
/* find the number of different ways of arranging 4 objects     */
        factorial = fact(n); /* compute 4!                      */
        /* the answer should be factorial= 4! = 24              */
        printf("# of different ways of arranging %d objects = %f\n",
                n, factorial);
        return 0;
}
```

fft2d

```
#include "mathlib.h"
int fft2d(Complex_Matrix m, unsigned rows, unsigned cols, int isign);
```

DESCRIPTION

The *fft2d* function computes the 2-D FFT of the complex matrix m by decomposing the 2-D transform into separable 1-D row and column FFTs.

The *rows* and *columns* arguments specify the number of rows and columns in matrix m. Both of these integer values must be a power of two:

$$rows = 2^i \quad \text{and} \quad cols = 2^j$$

where i and j are integers.

The *isign* argument specifies the direction of the transform. If *isign* $= -1$, then a forward 2-D FFT is performed. If *isign* $= +1$, then an inverse 2-D FFT is performed.

All of the calculations are done "in-place," which means that the resulting transformed matrix is written over the input matrix m.

The *fft2d* function calls the *fft42* function.

RETURNS

0 if successful or -1 if an error occurs. If successful, the 2-D FFT of the complex matrix m is returned in m. If either the *rows* or *cols* argument is not a power of 2, the error message "FFT length must be a power of two" is generated (math_errno = E_FFTPOWER2).

SEE ALSO

fft2d_r(), fftrad2(), fft42().

REFERENCES

J. W. Cooley and J. W. Tukey (1965); A. V. Oppenheim and R. W. Schafer (1975).

```
#include <stdio.h>
#include <stdlib.h>
#include <math.h>
#include "mathlib.h"
main()
{
int i, j, isign=-1;
unsigned rows=8, cols=16;
double w1, w2, i0=2, j0=3;
Real magnitude;
Complex_Matrix m;
Complex f1, f2;

/* Compute the 8 x 16 2-D FFT of a 2-D complex sinusoid */

        m = Cmxalloc(NULL, rows, cols);
        w1 = twopi_*i0/rows;
        w2 = twopi_*j0/cols;
        for(i=0; i<rows; i++) {   /* create 2-D complex sinusoidal data */
            f1 = Complx(cos(w1*i), sin(w1*i));
            for(j=0; j<cols; j++) {
                f2 = Complx(cos(w2*j), sin(w2*j));
                m[i][j] = Cmul(f1, f2);
            }
        }
        fft2d(m, rows, cols, isign);    /* call 2-D FFT */
        for(i=0; i<rows; i++) {
            for(j=0; j<cols; j++) {
                magnitude = Cmag(m[i][j]);
                printf("%4.0f", magnitude);
            }
            printf("\n");
        }
        Cmxfree(m);
        return 0;
}
```

PROGRAM OUTPUT

```
 0   0   0   0   0   0   0   0   0   0   0   0   0   0   0   0
 0   0   0   0   0   0   0   0   0   0   0   0   0   0   0   0
 0   0   0 128   0   0   0   0   0   0   0   0   0   0   0   0
 0   0   0   0   0   0   0   0   0   0   0   0   0   0   0   0
 0   0   0   0   0   0   0   0   0   0   0   0   0   0   0   0
 0   0   0   0   0   0   0   0   0   0   0   0   0   0   0   0
 0   0   0   0   0   0   0   0   0   0   0   0   0   0   0   0
 0   0   0   0   0   0   0   0   0   0   0   0   0   0   0   0
```

fft2d__r

#include "mathlib.h"
Complex__Matrix fft2d__r(Real__Matrix *m*, unsigned *rows*, unsigned *cols*);

DESCRIPTION

The *fft2d_r* function computes the forward 2-D FFT of the real matrix m by decomposing the 2-D transform into separable 1-D row and column FFTs.

The *rows* and *cols* arguments specify the number of rows and columns in matrix m. Both of these integers must be a power of two:

$$rows = 2^i \quad \text{and} \quad cols = 2^j$$

where i and j are integers.

The resulting transformed complex matrix is automatically allocated as the return value of the *fft2d_r* function and should be freed when it is no longer needed.

The *fft2d_r* function calls the *fft2d* function.

RETURNS

A *rows* by *cols* complex matrix representing the 2-D transform of the real matrix m is successful, or NULL if an error occurs. If either the *rows* or *cols* argument is not a power of two, the error message "FFT length must be a power of two" is generated (math_errno = E_FFTPOWER2).

SEE ALSO

fft2d(), fftrad2, fft42().

REFERENCES

J. W. Cooley and J. W. Tukey (1965); A. V. Oppenheim and R. W. Schafer (1975).

```c
#include <stdio.h>
#include <stdlib.h>
#include <math.h>
#include "mathlib.h"
main()
{
int i,j;
unsigned rows=8, cols=8;
double magnitude, i0=2, j0=3;
Complex_Matrix x2d;
Real_Matrix m;

/* Compute the 8 x 8 2-D FFT of the 2-D real sinusoid */

        m = mxalloc(NULL, rows, cols);

        for(i=0; i<rows; i++) {   /* create 2-D real sinusoidal data */
            for(j=0; j<cols; j++) {
                m[i][j] = sin(twopi_*i*i0/rows) * sin(twopi_*j*j0/cols);
            }
        }
        x2d = fft2d_r(m, rows, cols);        /* call real 2-D FFT */
        for(i=0; i<rows; i++) {
            for(j=0; j<cols; j++) {
                magnitude = Cmag(x2d[i][j]);
                printf("%4.0f", magnitude);
            }
            printf("\n");
        }
        Cmxfree(x2d);
        mxfree(m);
        return 0;
}
```

PROGRAM OUTPUT

```
0    0    0    0    0    0    0    0
0    0    0    0    0    0    0    0
0    0    0   16    0   16    0    0
0    0    0    0    0    0    0    0
0    0    0    0    0    0    0    0
0    0    0    0    0    0    0    0
0    0    0   16    0   16    0    0
0    0    0    0    0    0    0    0
```

fft42

```
#include "mathlib.h"
void fft42(Complex__Vector data, unsigned n, int isign);
```

DESCRIPTION

The *fft42* function computes the FFT of the complex vector *data* of length n using a "radix-4 + 2" algorithm. The term "radix-4 + 2" refers to the fact that this algorithm efficiently decomposes the large n-point transform into as many 4-point transforms as possible. If n is not a "power of four," then an additional stage of 2-point transforms is performed. The "radix-4 + 2" computation of the FFT is about twice as fast as the standard Cooley–Tukey "radix-2" approach (see the *fftrad2* function).

The argument *isign* determines the direction of the transform. If *isign* $= -1$, then a forward FFT is performed. If *isign* $= +1$, then an inverse FFT is performed.

All of the calculations are done "in place," which means that the resulting transformed vector is written over the input vector *data*. The length of the transformed vector is n, which must be a power of two:

$$n = 2^m, \quad \text{where } n \text{ and } m \text{ are integers}$$

If the complex sinusoid $A * [\cos(\omega * k * i) + j * \sin(\omega * k * i)]$ is used to generate the *data* vector, the transformed vector will consist of one frequency component at the kth harmonic. If the complex sinusoid $A * [\cos(\omega * k * i) - j * \sin(\omega * k * i)]$ is used, the transformed vector will consist of one frequency component at the $(n - k)$th harmonic, where $k = 1, 2, \ldots, n/2$, $i = 0, 1, 2, \ldots, n - 1$, $\omega = 2 * \pi / n$, and $\pi = 3.141592654\ldots$. Both components are of magnitude $A * n$.

RETURNS

An n point FFT of the complex vector *data* is returned via *data*. If n is not a power of two, the error message "FFT length must be a power of two" is generated (math_errno = E__FFTPOWER2).

SEE ALSO

fftrad2(), Cpowspec(), powspec().

REFERENCES

J. W. Cooley and J. W. Tukey (1965); A. V. Oppenheim and R. W. Schafer (1975).

```c
#include "mathlib.h"
#include <stdio.h>
#include <stdlib.h>
#include <math.h>
main()
{
Complex_Vector data;
unsigned n=16, i;
int isign = -1;
double mag;

    data=Cvalloc(NULL, n);  /* allocate an array for data to be analyzed    */
    for(i=0; i<n; i++) {  /*  generate 4 cycles of a sinusoid               */
        data[i].r = cos(twopi_*4*i/(double) n); /* signal = 4th harmonic    */
        data[i].i = sin(twopi_*4*i/(double) n); /* make the input data complex*/
    }
    fft42(data, n, isign); /* compute forward FFT of data                   */

/*The magnitude of the 4th harmonic should be "n" = 16                      */
/*    All other harmonics should be of zero magnitude                       */

    printf("              magnitude at  \n");
    printf("harmonic #      harmonic\n\n");
    for(i=0; i<n; i++) { /*                                                 */
        mag = sqrt(data[i].r*data[i].r + data[i].i*data[i].i);
        if(mag <= 1.e-14) printf("  %2d           %3.1f\n", i, mag);
        else  printf("  %2d            %9.6f\n", i, mag);
    }
    return 0;
}
```

PROGRAM OUTPUT

```
              magnitude at
harmonic #     harmonic

    0            0.0
    1            0.0
    2            0.0
    3            0.0
    4            16.000000
    5            0.0
    6            0.0
    7            0.0
    8            0.0
    9            0.0
   10            0.0
   11            0.0
   12            0.0
   13            0.0
   14            0.0
   15            0.0
```

fftrad2

```
#include "mathlib.h"
void fftrad2(Complex_Vector data, unsigned n, int isign);
```

DESCRIPTION

The *fftrad2* function computes the FFT of the complex vector *data* of length n using the Cooley–Tukey "radix-2" algorithm. The term "radix-2" refers to the fact that this algorithm efficiently decomposes the large n-point transform into several 2-point transforms.

The argument *isign* determines the direction of the transform. If $isign = -1$, then a forward FFT is performed. If $isign = +1$, then an inverse FFT is performed.

All of the calculations are done "in place," which means that the resulting transformed vector is written over the input vector *data*. The length of the transformed vector is n, which must be a power of two:

$$n = 2^m, \quad \text{where } n \text{ and } m \text{ are integers.}$$

If the real sinusoid $A * \cos(\omega * k * i)$ is used to generate the *data* vector, the transformed vector will consist of two frequency components, one at the kth harmonic and the other at the $(n - k)$th harmonic, where $k = 1, 2, \ldots, n/2$, $i = 0, 1, 2, \ldots, n - 1$, $\omega = 2 * \pi/n$, and $\pi = 3.141592654\ldots$. Both components are of magnitude $A * n/2$.

RETURNS

An n-point FFT of the complex vector *data* is returned via *data*. If n is not a power of two, the error message "FFT length must be a power of two" is generated (math_errno = E_FFTPOWER2).

SEE ALSO

fft42(), Cpowspec(), powspec().

REFERENCES

J. W. Cooley and J. W. Tukey (1965); A. V. Oppenheim and R. W. Schafer (1975).

```c
#include "mathlib.h"
#include <stdio.h>
#include <stdlib.h>
#include <math.h>
main()
{
Complex_Vector data;
unsigned n=16, i;
int isign = -1;
double mag;

    data=Cvalloc(NULL, n);  /* allocate an array for data to be analyzed   */
    for(i=0; i<n; i++) {  /*  generate 4 cycles of a sinusoid              */
        data[i].r = cos(twopi_*4*i/(double) n); /* signal = 4th harmonic   */
        data[i].i = 0.0;  /* make the input data purely real               */
    }
    fftrad2(data, n, isign); /* compute forward FFT of data                */

/*The magnitude of the 4th and 12th harmonic should be n/2 = 8             */
/*       All other harmonics should be of zero magnitude                  */

    printf("             magnitude at  \n");
    printf("harmonic #     harmonic\n\n");
    for(i=0; i<n; i++) { /*                                                */
        mag = sqrt(data[i].r*data[i].r + data[i].i*data[i].i);
        if(mag <= 1.e-14) printf("   %2d            %3.1f\n", i, mag);
        else  printf("   %2d            %9.6f\n", i, mag);
    }
    return 0;
}
```

PROGRAM OUTPUT

harmonic #	magnitude at harmonic
0	0.0
1	0.0
2	0.0
3	0.0
4	8.000000
5	0.0
6	0.0
7	0.0
8	0.0
9	0.0
10	0.0
11	0.0
12	8.000000
13	0.0
14	0.0
15	0.0

fftreal

```
#include "mathlib.h"
Complex__Vector fftreal(Real__Vector data, unsigned n);
```

DESCRIPTION

The *fftreal* function computes the forward FFT of a real data set using the Cooley–Tukey algorithm.

The *data* argument specifies a <u>real</u> vector of length n, where n must be a power of 2.

The transformed result is a <u>complex</u> vector of length $n/2 + 1$. To minimize the use of memory, the *fftreal* function stores the resulting complex vector of length $n/2+1$ in the same memory occupied by the *data* vector of length n. A real vector of length n requires the same amount of storage as a complex vector of length $n/2$. Therefore, in order for the resulting complex vector of length $n/2 + 1$ to fit in the same memory occupied by the *data* vector, the *data* vector must be allocated for $n+2$ elements rather than n elements.

The complex vector returned by the *fftreal* function is actually a pointer to the original *data* vector. Therefore, freeing the resulting complex vector is equivalent to freeing the *data* vector.

If the real sinusoid $A * \cos(\omega * k * i)$ is used to generate the *data* vector, the transformed vector will consist of one frequency component at the kth harmonic, where $k = 1, 2, \ldots, n/2$, $i = 0, 1, 2, \ldots, n - 1$, $\omega = 2 * \pi/n$, and $\pi = 3.141592654\ldots$. This component is of magnitude $A * n/2$.

The *fftreal* function calls the *fftrad2* function.

RETURNS

A complex vector of length $n/2 + 1$ representing the n-point forward FFT of the real vector *data* if successful, or NULL if an error occurs. If n is not a power of 2, the error message "FFT length must be a power of two" is generated (math_errno = E_FFTPOWER2).

SEE ALSO

fftr_inv(), fftrad2(), fft42(), fft2d_r(), fft2d().

REFERENCES

J. W. Cooley, P. A. Lewis, and P. D. Welch (1970) [see DSP-selected papers (1972), pp. 271–293].

```
#include "mathlib.h"
#include <stdio.h>
#include <stdlib.h>
#include <math.h>
main()
{
Real_Vector data;
Complex_Vector cdata;
unsigned n=16, i;
double mag;

/* Compute the real FFT of a cosine waveform */

        data=valloc(NULL, n2); /* allocate an array for data to be analyzed */+
        for(i=0; i<n; i++) { /*  generate 8 cycles of a sinusoid              */
            data[i] = cos(twopi_*2*i/(double) n);   /* signal = 2nd harmonic  */
        }                                     /* make the input data purely real */
        cdata = fftreal(data, n); /* compute forward FFT of data              */

/*The magnitude of the 2nd harmonic should be n/2 = 8                         */
/*        All other harmonics should be of zero magnitude                     */

        printf("              magnitude at  \n");
        printf("harmonic #      harmonic\n\n");
        for(i=0; i<=n/2; i++) {
            mag = sqrt(cdata[i].r*cdata[i].r + cdata[i].i*cdata[i].i);
            if(mag <= 1.e-7) printf("   %2d              %3.1f\n", i, mag);
            else  printf("   %2d           %9.6f\n", i, mag);
        }
        Cvfree(cdata);   /* equivalent to vfree(data) */
        return 0;
}
```

PROGRAM OUTPUT

```
                magnitude at
      harmonic #     harmonic

         0            0.0
         1            0.0
         2            8.000000
         3            0.0
         4            0.0
         5            0.0
         6            0.0
         7            0.0
         8            0.0
```

fftr_inv

```
#include "mathlib.h"
Real_Vector fftr_inv(Complex_Vector cdata, unsigned n);
```

DESCRIPTION

The *fftr_inv* function computes the inverse FFT of a complex data set using the Cooley–Tukey algorithm.

The *cdata* argument specifies a complex vector of length $n/2 + 1$, where n must be a power of 2.

The transformed result is a real vector of length n. To minimize the use of memory, *fftr_inv* function stores the resulting real vector of length n in the same memory occupied by the *cdata* vector of length $n/2 + 1$. A real vector of length n requires the same amount of storage as a complex vector of length $n/2$. Therefore, the memory occupied by the complex vector *cdata* has more than enough space to store the resulting real vector of length n.

The real vector returned by the *fftr_inv* function is actually a pointer to the original *cdata* vector. Therefore, freeing the resulting real vector is equivalent to freeing the *cdata* vector.

The *fftr_inv* function calls the *fftrad2* function.

RETURNS

A real vector of length n representing the inverse FFT of the complex vector *cdata* if successful, or NULL if an error occurs. If n is not a power of 2, the error message "FFT length must be a power of two" is generated (math_errno = E_FFTPOWER2).

SEE ALSO

fftreal(), fftrad2(), fft42(), fft2d_r(), fft2d().

REFERENCES

J. W. Cooley, P. A. Lewis, and P. D. Welch (1970) [see DSP-selected papers (1972), pp. 271–293].

```
#include <stdio.h>
#include <stdlib.h>
#include <math.h>
#include "mathlib.h"
main()
{
Real_Vector data;
Complex_Vector cdata;
unsigned n=16, i;
double ds;

/* Compute the forward real & inverse real FFT of a cosine waveform */

        data=valloc(NULL, n+2);  /* allocate an array for data to be analyzed  */
        for(i=0; i<n; i++) {  /*  generate 8 cycles of a sinusoid              */
            data[i] = cos(twopi_*2*i/(double) n);   /* signal = 2nd harmonic   */
        }                                     /* make the input data purely real */
        cdata = fftreal(data, n); /* compute forward FFT of data                */

        data = fftr_inv(cdata,n); /* compute inverse FFT of data                */

        printf("original data    data passed twice through fft\n");
        for(i=0; i<n; i++) {
            ds = cos(twopi_*2*i/(double) n);
            printf("%2d %9.6f           %9.6f\n", i, ds, data[i]);
        }
        vfree(data);            /* equivalent to Cvfree(cdata) */
        return 0;
}
```

PROGRAM OUTPUT

```
original data     data passed twice through fft
 0  1.000000               1.000000
 1  0.707107               0.707107
 2  0.000000              -0.000000
 3 -0.707107              -0.707107
 4 -1.000000              -1.000000
 5 -0.707107              -0.707107
 6 -0.000000               0.000000
 7  0.707107               0.707107
 8  1.000000               1.000000
 9  0.707107               0.707107
10  0.000000               0.000000
11 -0.707107              -0.707107
12 -1.000000              -1.000000
13 -0.707107              -0.707107
14 -0.000000              -0.000000
15  0.707107               0.707107
```

hwdclose/hwdcreate

```
#include <stdio.h>
#include "mathlib.h"
FILE *hwdcreate(char *filename, HwdHeader *header);
int    hwdclose(FILE *fp, HwdHeader * header);
```

DESCRIPTION

The *hwdcreate* function creates a new Hypersignal Waveform Data file (.TIM, .FRQ, .PHZ, .PWR, .REA, or .IMA extension). These are files used by Hyperception's Hypersignal DSP Software. A Hypersignal Waveform Data file is a binary formatted file containing a sequence of 16 bit integer values. The first 10 integer values, and optionally the next 54 integer values are header information. The integer data values are stored immediately following the header information.

The *hwdcreate* function creates a new file and writes header information to it. The file is opened for read/write access. The *filename* argument specifies the name of the file to create. The name should contain eight characters or less followed by an appropriate Hypersignal Waveform Data file extension (.TIM, .FRQ, .PHZ, .PWR, .REA, or .IMA). The *header* argument specifies the address of a structure containing the header information that is written to the file.

The *hwdclose* function closes a file previously opened by the *hwdcreate* or *hwdopen* functions. The *fp* argument specifies the file pointer returned by the *hwdcreate* or *hwdopen* functions. The *header* argument specifies a pointer to the header information that is written to the file before the file is closed.

The **HwdHeader** structure is defined in **mathlib.h** as follows:

```
typedef struct {
    short amplitude;          /* maximum absolute value of data            */
    short frame_size;         /* number of samples in each frame           */
    short frequency_rem;      /* frequency remainder: frequency % 32767    */
    short fft_order;          /* FFT length: 2^(fft_order)                 */
    short samples_rem;        /* samples remainder: samples % 16384        */
    short frame_overlap;      /* overlap of samples between frames         */
    short time_domain:3;      /* 0=arbitrary, 1=autocorrelation, 2=FIR or IIR */
    short complex:1;          /* 0=real data, 1=complex data               */
    short extended:1;         /* 0=standard header, 1=extended header      */
    short lpc_order:4;        /* LPC autocorrelation order: (0..15)        */
    short window_type:3;      /* 0=rectangular, 1=Hamming, 2=Hanning,      */
                              /* 3=Blackman, 4=Bartlett                    */
    short scale_order:4;      /* Frequency domain scale factor: 2^(scale_order) */
    short channels:6;         /* number of data channels - 1               */
    short true_nyquist:1;     /* true Nyquist rate FFT (0=no, 1=yes)       */
    short reserved:9;         /* reserved for future use                   */
    short frequency_div;      /* frequency dividend: frequency / 32767     */
    short samples_div;        /* samples dividend: samples / 16384         */
    short extension[54];      /* optional extended header information       */
} HwdHeader;
```

The **HwdHeader** structure contains 18 members. The first 17 members are required (i. e., always written to the file). Notice that bit fields are used to pack the first 17 members into the space of 10 short integer values. The eighteenth member named **extension** (an array of 54 short integer values) is optional. The **extension** array is written to the file only if the **extended** member of the structure is set to 1.

The use of this header information is more completely described in the Hypersignal DSP Software Manual. For example, the total number of data samples is calculated as **samples_div * 16384L + samples_rem** (the calculated number of data samples is twice the actual number if the data type is complex (i. e., **complex** = 1)). The sampling frequency is calculated as **frequency_div * 32767L + frequency_rem**.

RETURNS

The *hwdcreate* function returns a file pointer if successful, or NULL if an error occurs. If an error occurs, **math_errno** is set to E_OPEN if *hwdcreate* is unable to create the file, or E_WRITE if an error occurs while writing the header information to the file. The *hwdclose* function returns 0 if successful or −1 if an error occurs.

SEE ALSO

hwdopen(), hwdread(), hwdwrite()

```
#include <stdio.h>
#include "mathlib.h"
/*
    Define header for 2000 data samples at 40000 Hz sampling frequency.
    With the complex member set to 0, the data written must be real rather
    than complex.  With the extended member set to 0, no extended header
    information is written to the file.  With the channels member set to
    0, only a single channel of data can be written.
                                                                        */

HwdHeader header = {

            0,    /* maximum absolute value of data                     */
         2000,    /* number of samples in each frame                    */
         7233,    /* frequency remainder: frequency % 32767             */
            0,    /* FFT length: 2^(fft_order)                          */
         2000,    /* samples remainder: samples % 16384                 */
            0,    /* overlap of samples between frames                  */
            0,    /* 0=arbitrary, 1=autocorrelation, 2=FIR or IIR       */
            0,    /* 0=real data, 1=complex data                        */
            0,    /* 0=standard header, 1=extended header               */
            0,    /* LPC autocorrelation coefficients: (0..15)          */
            0,    /* 0=rectangular, 1=Hamming, 2=Hanning,               */
                  /* 3=Blackman, 4=Bartlett                             */
            0,    /* Frequency domain scale factor: 2^(scale_order)     */
            0,    /* number of data channels - 1                        */
            1,    /* true Nyquist rate FFT (0=no, 1=yes)                */
            0,    /* reserved for future use                            */
            1,    /* frequency dividend: frequency / 32767              */
            0     /* samples dividend: samples / 16384                  */
};

main()
{
    FILE *fp;
    fp = hwdcreate("sample.tim", &header);
    puts("SAMPLE.TIM created.");
    /*    .

          .

          .
              */
    hwdclose(fp, &header);
    return 0;
}
```

hwdopen

#include "mathlib.h"
FILE *hwdopen(char *_filename_, HwdHeader *_header_);

DESCRIPTION

The _hwdopen_ function opens an existing Hypersignal Waveform Data file (.TIM, .FRQ, .PHZ, .PWR, .REA, or .IMA extension). These are files used by Hyperception's Hypersignal DSP Software. A Hypersignal Waveform Data file is a binary formatted file containing a sequence of 16 bit integer values. The first 10 integer values, and optionally the next 54 integer values are header information. The integer data values are stored immediately following the header information.

The _hwdopen_ function opens an existing file and reads header information from it. The file is opened for read/write access.

The _filename_ argument specifies the name of the file to open. The name should contain 8 characters or less followed by an appropriate Hypersignal Waveform Data file extension (.TIM, .FRQ, .PHZ, .PWR, .REA, or .IMA).

The _header_ argument specifies the address of a structure that will contain the header information that is read from the file. See the _hwdcreate_ function for a description of the **HwdHeader** structure.

RETURNS

File pointer if successful, or NULL if an error occurs. If an error occurs, **math_errno** is set to E_OPEN if _hwdopen_ is unable to open the file, or E_READ if an error occurs while reading the header information from the file.

SEE ALSO

hwdcreate(), hwdread(), hwdwrite(), hwdclose()

```c
#include <stdio.h>
#include "mathlib.h"

main()
{
    HwdHeader header;                       /* header information */
    char filename[80];
    long samples, frequency;
    FILE *fp;
    int i;
    printf("Enter file name\nwith .TIM, .FRQ, .PHZ, .PWR, "
            ".REA, or .IMA extension: ");
    gets(filename);
    fp = hwdopen(filename, &header);   /* open existing file        */
    printf("\nHeader Information for file %s\n", filename);
    samples = header.samples_div * 16384L + header.samples_rem;
    frequency = header.frequency_div * 32767L + header.frequency_rem;
    printf("samples       = %ld\n", samples);
    printf("frequency     = %ld\n", frequency);
    printf("amplitude     = %d\n",  header.amplitude);
    printf("frame_size    = %d\n",  header.frame_size);
    printf("fft_order     = %d\n",  header.fft_order);
    printf("frame_overlap = %d\n",  header.frame_overlap);
    printf("time_domain   = %d\n",  header.time_domain);
    printf("complex       = %d\n",  header.complex);
    printf("extended      = %d\n",  header.extended);
    printf("lpc_order     = %d\n",  header.lpc_order);
    printf("window_type   = %d\n",  header.window_type);
    printf("scale_order   = %d\n",  header.scale_order);
    printf("channels      = %d\n",  header.channels + 1);
    printf("true_nyquist  = %d\n",  header.true_nyquist);
    printf("reserved      = %d\n",  header.reserved);
    if (header.extended)
        for (i = 0; i < 54; i++)
            printf("extension[%2d] = %d\n", i, header.extension[i]);
    fclose(fp);                             /* close file with no changes */
    return 0;
}
```

hwdread

> #include "mathlib.h"
> int hwdread(FILE *fp, HwdHeader *header, Real__Vector data,
>   unsigned samples, long index, unsigned channel, Real scale);

DESCRIPTION

The *hwdread* function reads data from a Hypersignal Waveform Data file (.TIM, .FRQ, .PHZ, .PWR, .REA, or .IMA extension). These are files used by Hyperception's Hypersignal DSP Software. A Hypersignal Waveform Data file is a binary formatted file containing a sequence of 16 bit integer values. The first 10 integer values, and optionally the next 54 integer values are header information. The integer data values are stored immediately following the header information.

The *hwdread* function reads data from a file that has been opened by the *hwdopen* or *hwdcreate* function.

The *fp* argument specifies the file pointer returned by the *hwdopen* or *hwdcreate* function.

The *header* argument specifies the address of a structure of type **HwdHeader** that must already be initialized. The *header* is typically initialized by calling the *hwdopen* function. See the *hwdcreate* function for a complete description of the **HwdHeader** structure.

The *data* argument specifies the name of a vector where *hwdread* stores the data that is read. The storage space for this **Real_Vector** must be explicitly allocated by calling the *valloc* function prior to calling the *hwdread* function. As the integer data is read from the file, each value is automatically converted to type **Real** before being stored in the vector.

The *samples* argument specifies the number of data samples to read. The value of *samples* must be less than or equal to the length of the *data* vector.

The *index* argument specifies the offset of the data value (in the file) where reading begins. A value of 0 specifies that reading begins at the first data value in the file, a value of 1 specifies the second data value, and so on. This allows the *hwdread* function to read very large data files in a piecemeal fashion.

The *channel* argument specifies the data channel to read. Since some files may contain multiple channels of data, this argument allows you to select a particular channel. If a file contains only a single channel of data, then the value of *channel* should be 1.

The *scale* argument specifies the factor by which each integer data value is multiplied before being stored in the *data* vector. Since all data values in the file are integers, these integers may not represent the true data values. In other words, the original data values may have been scaled to fit in the range $-32768 \ldots 32767$ before being stored in the file. The *scale* argument allows the data items to be scaled back to their original or true values.

The *hwdread* function determines how to read the file by examining the extended, channels, and complex members of the *header* argument (see the *hwdcreate* function for a description of these members). The extended member specifies the type of header, 0 for standard or 1 for extended. The channels member specifies the number of data channels (minus 1) that are stored in the file. The complex member specifies the type of data, 0 for real data or 1 for complex data.

READING COMPLEX DATA

If a file contains complex data, then the *data* argument should be declared as type **Complex_Vector** rather than **Real_Vector**. In this case, the *Cvalloc* function rather than the *valloc* function must be used to allocate storage space for the *data* argument prior to calling the *hwdread* function. If *data* is declared as a **Complex_Vector**, then you should use a type cast such as **(Real_Vector)** *data* as the *data* argument. The type cast will eliminate any compiler warnings. Also, the calculated number of data samples (from the header information) will be twice the actual number, since each sample consists of two values (a real value and an imaginary value).

RETURNS

0 if successful, or −1 if an error occurs. If an error occurs, **math_errno** is set to E_READ. An attempt to read past the end of the file will cause an error. Also, an error will occur if the specified *channel* is greater than the actual number of data channels in the file. It is very important that the *header* information passed to the *hwdread* function accurately reflects the actual contents of the file. Otherwise the data will not be read correctly.

SEE ALSO

hwdwrite(), hwdcreate(), hwdopen(), hwdclose()

```c
#include <stdio.h>
#include <stdlib.h>
#include "mathlib.h"

main()
{
    char filename[80];
    char *prompt = "--- Press ENTER key to view data, "
                   "or any key plus ENTER to Exit ---";
    FILE *fp;
    HwdHeader header;
    int frame, linecount = 0;
    unsigned i, channel;
    unsigned samples;
    long index, total_samples, frequency;
    Real_Vector data;

    printf("Hypersignal Waveform Data File to Read: ");
    gets(filename);                             /* get file name */
    fp = hwdopen(filename, &header);            /* open file     */
    frequency = header.frequency_div * 32767L + header.frequency_rem;
    total_samples = header.samples_div * 16384L + header.samples_rem;
    if (header.complex) total_samples /= 2;
    frame = header.frame_size;
    if (frame <= 0) frame = 1000;
    printf("\nFile: %s\n", filename);
    printf("      frequency    = %ld\n", frequency);
    printf("      samples      = %ld\n", total_samples);
    printf("      amplitude    = %d\n",  header.amplitude);
    printf("      scale_order  = %d\n",  header.scale_order);
    printf("      fft_order    = %d\n",  header.fft_order);
    printf("      lpc_order    = %d\n",  header.lpc_order);
    printf("      channels     = %d\n",  header.channels + 1);
    printf("      frame_size   = %d\n",  header.frame_size);
    printf("      frame_overlap = %d\n", header.frame_overlap);
    printf("      time_domain  = %d\n",  header.time_domain);
    printf("      window_type  = %d\n",  header.window_type);
    printf("      complex      = %d\n",  header.complex);
    printf("      extended     = %d\n",  header.extended);
    printf("      true_nyquist = %d\n",  header.true_nyquist);
    printf("\n%s", prompt);

    if (getchar() != '\n') exit(0);
    if (header.complex)                         /* check for complex */
        /* allocate space for complex data */
        data = (Real_Vector) Cvalloc(NULL, frame);
    else
        /* allocate space for real data */
        data = valloc(NULL, frame);
    for (channel = 1; channel <= header.channels+1; channel++) {
```

```
    printf("\n------------ Channel %d data ------------\n", channel);
    index = 0;
    total_samples = header.samples_div * 16384 + header.samples_rem;
    if (header.complex) total_samples /= 2;
    while (total_samples > 0) {
        /* read file one frame at a time */
        samples = total_samples > frame ?
                    frame: (unsigned) total_samples;
        hwdread(fp, &header, data, samples, index, channel, 1);
        for (i = 0; i < samples; i++) {  /* print data */
            if (header.complex) {
                printf("data[%ld].r = %.0f, data[%ld].i = %.0f\n",
                index + i, ((Complex_Vector) data)[i].r,
                index + i, ((Complex_Vector) data)[i].i);
            }
            else printf("data[%ld] = %.0f\n", index + i, data[i]);
            /* pause every 24 lines */
            if (++linecount % 24 == 0) {
                printf("%s", prompt);
                if (getchar() != '\n') exit(0);
            }
        }
        index += samples;
        total_samples -= samples;
    }
}
return 0;
}
```

hwdwrite

#include "mathlib.h"
int hwdwrite(FILE *fp, HwdHeader *header, Real__Vector data,
    unsigned samples, long index, unsigned channel, Real scale);

DESCRIPTION

The *hwdwrite* function writes data to a Hypersignal Waveform Data file (.TIM, .FRQ, .PHZ, .PWR, .REA, or .IMA extension). These are files used by Hyperception's Hypersignal DSP Software. A Hypersignal Waveform Data file is a binary formatted file containing a sequence of 16 bit integer values. The first 10 integer values, and optionally the next 54 integer values are header information. The integer data values are stored immediately following the header information.

The *hwdwrite* function writes data to a file that has been opened by the *hwdopen* or *hwdcreate* function.

The *fp* argument specifies the file pointer returned by the *hwdopen* or *hwdcreate* function.

The *header* argument specifies the address of a structure of type **HwdHeader** that must already be initialized. The *header* is typically initialized by calling the *hwdopen* function. However, it may also be initialized directly. See the *hwdcreate* function for a complete description of the **HwdHeader** structure.

The *data* argument specifies the name of a vector that contains the data that is written to the file. The storage space for this **Real__Vector** must be explicitly allocated by calling the *valloc* function prior to calling the *hwdwrite* function. Each **Real** value in the *data* vector is automatically converted to a 16-bit integer before being written to the file.

The *samples* argument specifies the number of data samples to write. The value of *samples* should be less than or equal to the length of the *data* vector.

The *index* argument specifies the offset of the data value (in the file) where writing begins. A value of 0 specifies that writing begins at the first data value in the file, a value of 1 specifies the second data value, and so on. This allows the *hwdwrite* function to write very large data files in a piecemeal fashion.

The *channel* argument specifies the data channel to write. Since some files may contain multiple data channels, this argument allows you to select a particular channel. If a file contains only a single data channel, then the value of *channel* should be 1.

The *scale* argument specifies the factor by which each value in the *data* vector is multiplied before being written to the file. Since all data values in the file must be integers, the **Real** values in the *data* vector may need to be scaled up or down before being stored in the file. To minimize loss of resolution, the data values should be scaled so that the maximum absolute value is as large as possible without exceeding 32767. If the maximum absolute value of the *data* is much less than 32767, then the values should be scaled up (i. e., *scale* > 1). If the maximum absolute value of the *data* exceeds 32767, then the values should be scaled down (i. e., *scale* < 1).

The *hwdwrite* function determines how to write data to the file by examining the extended, channels, and complex members of the *header* argument (see the *hwdcreate* function for a description of these members). The extended member specifies the type of header, 0 for standard or 1 for extended. The channels member specifies the number of data channels (minus 1) that are stored in the file. The complex member specifies the type of data, 0 for real data or 1 for complex data.

The *hwdwrite* function updates the **amplitude** member of the structure pointed to by the *header* argument. This member specifies the maximum absolute value of the data stored in the file. The **amplitude** member is updated only if the absolute value of a *data* item being written to the file is greater than the current maximum.

WRITING COMPLEX DATA

If complex data is being written to a file, then the *data* argument should be declared as type **Complex_Vector** rather than **Real_Vector**. In this case, the *Cvalloc* function rather than the *valloc* function must be used to allocate storage space for the *data* argument prior to calling the *hwdwrite* function. If *data* is declared as a **Complex_Vector**, then you should use a type cast such as **(Real_Vector)** *data* as the *data* argument. The type cast will eliminate any compiler warnings. Also, the calculated number of data samples (from the header information) should be twice the actual number, since each sample consists of two values (a real value and an imaginary value).

After you have finished writing a file, the *hwdclose* function may be used to close the file and update the header information that is stored at the beginning of the file.

After running the following example program, try reading the resulting output file, BANDPASS.TIM, into the Hypersignal DSP software. Select FFT GENERATION from the TIME DOMAIN menu to generate a bandpass frequency response.

RETURNS

0 if successful, or −1 if an error occurs. If an error occurs, **math_errno** is set to E_WRITE. An error will occur if the specified *channel* is greater than the actual number of data channels in the file. Also, an error will occur if any scaled value is greater than 32767 or less than −32768. It is very important that the *header* information passed to the *hwdwrite* function accurately reflect the actual contents of the file. Otherwise the data will not be written correctly.

SEE ALSO

hwdread(), hwdcreate(), hwdopen(), hwdclose()

```
#include <stdio.h>
#include "mathlib.h"

#define NWEIGHTS 31

HwdHeader header = {

        0,    /* maximum absolute value of data                      */
  NWEIGHTS,   /* number of samples in each frame                     */
    10000,    /* frequency remainder: frequency % 32767              */
        0,    /* FFT length: 2^(fft_order)                           */
  NWEIGHTS,   /* samples remainder: samples % 16384                  */
        0,    /* overlap of samples between frames                   */
        2,    /* 0=arbitrary, 1=autocorrelation, 2=FIR or IIR        */
        0,    /* 0=real data, 1=complex data                         */
        0,    /* 0=standard header, 1=extended header                */
        0,    /* LPC autocorrelation order: (0..15)                  */
        0,    /* 0=rectangular, 1=Hamming, 2=Hanning,                */
              /* 3=Blackman, 4=Bartlett                              */
        0,    /* frequency domain scale factor: 2^(scale_order)      */
        0,    /* number of data channels - 1                         */
        0,    /* true Nyquist rate FFT (0=no, 1=yes)                 */
        0,    /* reserved for future use                             */
        0,    /* frequency dividend: frequency / 32767               */
        0     /* samples dividend: samples / 16384                   */
};


main()
{
    int i;
    char *filename = "BANDPASS.TIM";
    FILE *fp;
    Real fl = .20, fh = .30, db = 30.0;
    Real weights[NWEIGHTS];
    fp = hwdcreate(filename, &header);   /* create file    */
    printf("%s created.\n", filename);
    /* create bandpass filter time response */
    bandpass(weights, NWEIGHTS, fh, fl, db, 0);
    /* output time response scaled up by a factor of 32768 */
    hwdwrite(fp, &header, (Real_Vector) weights, NWEIGHTS, 0, 1, 32768);
    hwdclose(fp, &header);  /* close file and update header */
    return 0;
}
```

highpass

#include "mathlib.h"
void highpass(Real *weights*[ ], int *nweights*, double *fc*, double *dB*, int *half*);

DESCRIPTION

The *highpass* function designs a digital highpass filter which can then be implemented with the *convolve* function. The basic approach of window design methods is to truncate the infinite length impulse response of an ideal frequency filter by multiplying it by a time-domain window (in this case, a Kaiser–Bessel window). Time-domain windows are "bell-shaped" (see the *tdwindow* function), which tapers the filter impulse response.

It is assumed that the data to be filtered is sampled at uniform intervals of time, T. The sampling frequency is $F_s = 1/T$. In digital sampling, frequencies are often specified normalized to F_s. For example, $F_s/2$, the Nyquist frequency, is the highest frequency that can be represented and has a normalized value of $1/2$.

Digital highpass filters are characterized by a passband, $(F_2, F_s/2)$, over which the filter's response is reasonably close to unity gain. They also often have a stopband, $(0, F_1)$, over which frequencies are attenuated by some specified amount (i. e., the gain in the stopband is less than $d_s < 1$). The maximum attenuation is at zero frequency (DC).

The fc argument specifies the cutoff frequency, which sets the nominal passband (and stopband) of the Kaiser window filter since it is the average of F_1 and F_2:

$$f_c = (F_1 + F_2)/2$$

The cutoff frequency is the frequency at which the magnitude of the filter response is approximately -6 dB.

The transition band, F_t, is the difference between the passband and the stopband frequencies:

$$F_t = F_2 - F_1$$

The dB argument specifies the desired stopband attenuation in *decibels*:

$$dB = 20 \log_{10}(G_s)$$

where G_s is the gain in the stopband.

The *nweights* argument specifies the length of the filter. The length of the filter needed to achieve the desired values of dB and F_t is given by

$$nweights = (|dB| - 8)/(14.357F_t)$$

The *nweights* filter weights representing the impulse response are returned via the *weights* array. The *weights* array should be declared as an array of type **Real** with at least *nweights* elements.

The *half* argument specifies whether to return all of the filter weights, or just half of them. Since the Kaiser–Bessel highpass filter is a symmetric (even) function, the second half of the filter weights is a mirror image of the first half. This symmetry can be exploited to reduce by a factor of 2, the number of multiplications (and array storage locations) required to implement the filter. If *half* is zero, then all of the *nweights* coefficients are returned through the *weights* array. However, if *half* is nonzero, then only the first *nweights*/2 coefficients are returned. The remaining half of the coefficients could be filled into the *weights* array using the following C statement:

```
for (i = 0; i < nweights/2; i++)
    weights[nweights-1-i] = weights[i];
```

If the filter is symmetric, the *convolve* function needs only the first half of the coefficients.

It should be mentioned that the above design formulas are approximations. If an application requires an exact frequency response, then the filter design should be checked to see that it meets the desired specification. It may be necessary to slightly change the filter design parameters to achieve a more exact frequency response.

The response of a 25-point Kaiser highpass filter is shown in the highpass filter section. The -6 dB cutoff of this filter is $fc = 0.10$ and the stopband attenuation is about -32 dB. This filter is designed in the example below. An interference tone is placed at the -32 dB peak in the stopband at $f = 0.043$. The signal is placed in the passband at $f = 0.20$. The 25-point Kaiser highpass filter reduces the interference from 0.500783 (per data point) to 0.000239 (per data point).

RETURNS

The *nweights* (*nweights*/2 if *half* is nonzero) coefficients representing the impulse response of a highpass filter with cutoff frequency fc and stopband attenuation dB is returned via the *weights* array.

SEE ALSO

lowpass(), bandpass(), convolve(), downsamp(), deriv(), interp(), tdwindow().

REFERENCE

A. V. Oppenheim and R. W. Schafer (1989).

```
#include "mathlib.h"
#include <stdio.h>
#include <stdlib.h>
#include <math.h>

#define NWEIGHTS 25
#define NDATA    100

main()
{
Real_Vector data, y;
Real weights[NWEIGHTS];
int i, ny, ndec=1, itype=1, isym=-1;
double truth, interfer, off, init_error=0.0, residual_error=0.0;

      data=valloc(NULL, NDATA);/* allocate a vector for the data */
      for(i=0; i<NDATA; i++) {/* generate data, one cycle of a sinusoid+noise */
          truth = sin(twopi_*i*0.20); /* put signal in passband of filter &   */
          interfer = sin(twopi_*i*0.043); /*interference at peak in stopband  */
          data[i] = truth + interfer;
          init_error += interfer*interfer/(double) NDATA;
      }
      highpass(weights, NWEIGHTS, 0.10, 32.0, 0); /* design highpass filter   */

      y = convolve(data, NDATA, (Real_Vector) weights, NWEIGHTS,
                  ndec, itype, isym, &ny);
      off = (double) (NWEIGHTS-1)/2.; /* align truth with filtered data       */
      for(i=0; i<ny; i++) { /* compute the error after filtering             */
          truth = sin(twopi_*0.20*(i+off));
          residual_error += (truth-y[i])*(truth-y[i])/(double) ny;
      }/* The Kaiser highpass should remove most of the interference:         */
      printf("the error before filtering is %f\n", init_error); /* = 0.500783*/
      printf("the error after filtering is %f\n", residual_error);/*= 0.000239*/
      return 0;
}
```

hyperdst

```
#include "mathlib.h"
double hyperdst(int n, int k0, int r, int r1, Real pb[ ]);
```

DESCRIPTION

The *hyperdst* function computes the cumulative probability that $k0$ or fewer successes will occur after n trials of a hypergeometric distributed experiment. In a hypergeometric experiment, there are two possible mutually exclusive outcomes for any random sampling (e. g., "an ace" or "not an ace").

The r argument specifies the total size of the population (e. g., 52 cards), and the $r1$ argument specifies the size of the population that will produce one of the mutually exclusive outcomes (e. g., 4 aces). Therefore, the size of the population that will produce the opposite outcome is $r - r1$ (e. g., 48 other cards).

If X represents the number of successful outcomes, then the probability that the number of successful outcomes will be less than or equal to $k0$ may be expressed as follows:

$$P(X \leq k0) = \sum_{i=0}^{k0} \frac{\binom{r1}{i}\binom{r-r1}{n-i}}{\binom{r}{n}}$$

The $pb[\ ]$ argument is used to return the probability that exactly i successes will occur (where $i = 0, 1, 2, \ldots, k0$). Therefore, the $pb[\ ]$ argument should be declared as **Real** $pb[k0 + 1]$; in order to be able to store all of the discrete probabilities. If the discrete probabilities are not needed, then the $pb[\ ]$ argument should be specified as NULL.

The hypergeometric distribution is a discrete process in that only integer valued outcomes are possible. This function is useful in a variety of statistical applications. It is frequently used to predict the outcome of a repetitive sampling from a small population. However, unlike the binomial distribution, the samples are not returned to the population. Hence, the sampling is done *without* replacement. The hypergeometric distribution can be used to predict poker hands dealt from a fair deck of cards (see the example below).

RETURNS

A double precision whole number representing the cumulative probability of having $k0$ or fewer successes after n trials of a hypergeometric distributed experiment. The discrete probabilities of having exactly $0, 1, 2, \ldots, k0$ successes are returned via $pb[\ ]$.

SEE ALSO

binomdst(), and poissdst().

REFERENCE

L. L. Lapin (1983, pp. 113–121).

```
#include "mathlib.h"
#include <stdio.h>
main()
{
int k0,i,n,r,r1;
Real pb[5];
double cumulative;
/* compute the odds of being dealt 4 aces in a 5 card poker hand */
        r1 = 4; /* this is the number of aces in a deck           */
        r = 52; /* this is the number of cards in a deck           */
        n = 5;  /* this is the number of cards in a hand           */
        k0 = 4; /* this is the number of aces desired in a hand    */
        cumulative = hyperdst(n, k0, r, r1, pb);
        for (i=0; i <= k0; i++) {
            printf("the odds that exactly %d aces occur is %f\n", i,pb[i]);
        }
        /* the odds of being dealt 4 aces should be 0.000018 */
        printf("the cumulative probability is %f\n", cumulative);
        return 0;
}
```

iirfiltr

```
#include "mathlib.h"
Real_Vector iirfiltr(Real_Vector data, unsigned ndata, Real scale,
        Real_Vector numerator[ ], Real_Vector denominator[ ],
        unsigned numb_coeff[ ], unsigned nstages);
```

DESCRIPTION

An **infinite impulse response (IIR)** filter is a digital filter that theoretically "rings" forever when excited by an impulse. However, in practice the responses of these filters can be considered finite after some nominal time interval. The classical techniques of analog filter design can also be used to design IIR filters. Several standard techniques exist for transforming analog filter designs into IIR designs. **Impulse invariance** and **bilinear transformation** are the most popular.

The *iirfiltr* function filters the real vector *data* of length *ndata* with a general IIR filter.

The *scale* argument specifies the constant multiplier often required in the cascade realizations of IIR filters. Alternatively, *scale* can simply be used as a gain term. The *scale* argument should be set to 1 if no scaling is required.

The *numerator* [ ] argument is an array of vectors that define the numerator terms for each stage of the IIR filter. Each vector in this array contains the coefficients for one of the stages.

The *denominator* [ ] argument is an array of vectors that define the denominator terms for each stage of the IIR filter. Each vector in this array contains the coefficients for one of the stages.

The *numb_coeff* [ ] argument is an array of integers that defines the number of coefficients in each stage of the IIR filter (i. e., the length of the *numerator* and *denominator* vectors for stages $0 \ldots nstages - 1$).

The *nstages* argument specifies the total number of stages in the IIR filter. The number of stages must be the same for both the numerator and the denominator. If the number of coefficients in the numerator needs to be different from the number of coefficients in the denominator for a particular stage, then the excess elements in the smaller vector may be set to 0.

RETURNS

A real vector of length *ndata*, which is the **IIR** filtered version of the input vector *data* if successful, or NULL if an error occurs.

SEE ALSO

iir_read(), bilinear(), convolve(), convofft, lowpass(), highpass(), bandpass(), downsamp(), resample(), deriv(), interp().

REFERENCE

A. V. Oppenheim and R. W. Schafer (1975).

```
#include <stdio.h>
#include <stdlib.h>
#include <math.h>
#include "mathlib.h"
#define NSTAGES 1
main()
{
unsigned i, numb_coeff[NSTAGES], ndata=256;
Real scale;
Real_Vector data, filtered_data;
Real_Vector numerator[NSTAGES], denominator[NSTAGES];

/* Find the impulse response of the following 2nd order Butterworth
   lowpass filter. The sampling frequency = 10 kHz, bandwidth = 1 kHz,
   transition bandwidth = 3 kHz, stopband attenuation = 20 dB    */

    for (i = 0; i < NSTAGES; i++) {
        numb_coeff[i] = 3;
        numerator[i] = valloc(NULL, numb_coeff[i]);
        denominator[i] = valloc(NULL, numb_coeff[i]);
    }
    scale = 1.12049059727971E-0001;
    numerator[0][0]   =   1.00000000000000E+0000;
    numerator[0][1]   =   2.00000000000000E+0000;
    numerator[0][2]   =   1.00000000000000E+0000;
    denominator[0][0] =   1.00000000000000E+0000;
    denominator[0][1] =  -8.56025524352595E-0001;
    denominator[0][2] =   3.04221763264481E-0001;

/*----------- create an impulse for IIR lowpass filtering --------------*/

    data = valloc(NULL, ndata);
    data[0] = 1.0;
    for(i=1; i<ndata; i++) data[i] = 0.0;

    filtered_data = iirfiltr(data, ndata, scale, numerator, denominator,
                             numb_coeff, NSTAGES);

    printf("The first 20 impulse response coefficients are:\n");
    for(i=0; i<10; i++) {
        printf("%2d %10.6f            %2d %10.6f\n", i, filtered_data[i],
                                      10+i, filtered_data[10+i]);
    }
    mathfree();
    return 0;
}
```

PROGRAM OUTPUT

```
The first 20 impulse response coefficients are:
 0   0.112049        10   0.001088
 1   0.320015        11   0.001545
 2   0.351902        12   0.000991
 3   0.203882        13   0.000379
 4   0.067472        14   0.000023
 5  -0.004268        15  -0.000096
 6  -0.024180        16  -0.000089
 7  -0.019400        17  -0.000047
 8  -0.009251        18  -0.000013
 9  -0.002017        19   0.000003
```

iir_read

```
#include "mathlib.h"
int iir__read(char *filename, IIR__Filter *iir);
```

DESCRIPTION

The *iir_read* function reads an existing Hypersignal FIR or IIR data file (.FIR or .IIR extension). These are files used by Hyperception's Hypersignal DSP Software. An FIR or IIR data file is an ASCII formatted file containing floating point coefficients representing the design of an FIR or IIR filter.

The *iir_read* function opens an existing file and reads all the design coefficients from it.

The *filename* argument specifies the name of the file to open. The name should contain 8 characters or less followed by an appropriate extension (.FIR or IIR).

The *iir* argument specifies the address of a structure where the *iir_read* function stores the design coefficients that are read from the file. The **IIR__Filter** structure is defined in **mathlib.h** as follows:

```
typedef struct {
    Real scale;
    unsigned nstages;
    unsigned numb_coeff[MAX_IIR_STAGES];
    Real_Vector numerator[MAX_IIR_STAGES];
    Real_Vector denominator[MAX_IIR_STAGES];
} IIR_Filter;
```

The *iir_read* function stores the following information in the **IIR__Filter** structure. The **scale** member is the constant multiplier of the transfer function for IIR filters. For FIR filters, **scale** will be 1 unless the filter impulse response is antisymmetric (e. g., Hilbert transform), in which case **scale** will be −1. The **nstages** member is the number of stages for IIR filters. For FIR filters, **nstages** will always be 1. The **numb__coeff** member is an array containing the number of coefficients in each stage. For IIR filters, each of the **nstages** may have a different number of coefficients. Since FIR filters are single stage, only the **numb__coeff[0]** member is used for FIR filters. The **numerator** member is an array of vectors, with each vector containing the numerator coefficients of the transfer function for one of the stages. Only the **numerator[0]** vector is used for FIR filters. The **denominator** member is also an array of vectors, with each vector containing the denominator coefficients of the transfer function for one of the stages. The **denominator** member is not used for FIR filters.

The information stored in the **IIR__Filter** structure may be used by the *iirfiltr* function.

RETURNS

0 if successful, or −1 if an error occurs. If an error occurs, **math_errno** is set to E__OPEN if *iir_read* is unable to open the file, or E__MALLOC if unable to allocate

space for the **numerator** or **denominator** vectors, or E_STAGES if the number of stages for the numerator and denominator are not equal or if the order of a numerator stage is not equal to the order of the corresponding denominator stage, or E_READ if a read error occurs.

SEE ALSO

iirfiltr()

```c
#include <stdio.h>
#include <stdlib.h>
#include "mathlib.h"

main()
{
    char filename[80];
    char *prompt = "--- Press ENTER key to view data, "
                   "or any key plus ENTER to Exit ---";
    FILE *fp;
    IIR_Filter iir;
    int i, j, fir_flag = 0;

    printf("Hypersignal IIR or FIR Data File to Read: ");
    gets(filename);                                 /* get file name */
    iir_read(filename, &iir);                       /* read file     */
    if (iir.nstages == 1 && iir.denominator[0] == NULL) fir_flag++;
    printf("File        : %s\n", filename);
    printf("Filter Type : ");
    if (fir_flag) {                                 /* FIR Filter */
        printf("FIR\n");
        printf("Filter Weights: %d\n\n", iir.numb_coeff[0]);
        printf("%s", prompt);
        if (getchar() != '\n') exit(0);
        printf("\n");
        for (i = 0; i < iir.numb_coeff[0]; i++)
            printf("A[%2d] = %15e\n", i, iir.numerator[0][i]);
    }
    else {                                          /* IIR Filter */
        printf("IIR\n");
        printf("Stages      : %d\n\n", iir.nstages);
        printf("%s", prompt);
        if (getchar() != '\n') exit(0);
        printf("\n");
        for (i = 0; i < iir.nstages; i++) {
            printf("\nStage %d Coefficients\n", i+1);
            for (j = 0; j < iir.numb_coeff[i]; j++)
                printf("A[%2d] = %15e, B[%2d] = %15e\n",
                    j, iir.numerator[i][j], j, iir.denominator[i][j]);
        }
    }
    for (i = 0; i < iir.nstages; i++) {             /* free vectors */
        vfree(iir.numerator[i]);
        if (!fir_flag) vfree(iir.denominator[i]);
    }
    return 0;
}
```

integrat

#include "mathlib.h"
Real__Vector integrat(Real__Vector *fx*, double *dx*, int *n*);

DESCRIPTION
Many important integrals that arise in mathematical problems cannot be expressed in a closed form. The most common examples are definite integrals of the form:

$$y(x) = \int_{x_0}^{x_1} f(x)\, dx$$

where the limits x_0 and x_1 are fixed, and $f(x)$ is any analytical expression, not necessarily integrable.

The *integrat* function uses Simpson's rule to numerically integrate the real vector fx, representing the values of a function $f(x)$ over a range of equally spaced x.

The n argument specifies the number of elements in the vector fx, which must contain at least six real values.

The dx argument specifies the fixed increment between samples:

$$dx = (x_1 - x_0)/(n - 1)$$

where x_0 is the lower limit and x_1 is the upper limit of the integration.

This function approximates the definite integral, $y(x)$, from the point x_0 to the point x_1 with:

$$y(x) = dx * [fx_0 + 4 * fx_1 + 2 * fx_2 + 4 * fx_3 + \dots \\ + 2 * fx_{n-3} + 4 * fx_{n-2} + fx_{n-1}]/3$$

The real vector fx must be allocated by *valloc* prior to calling the *integrat* function. The resulting real vector representing the integral of the data is automatically allocated by the *integrat* function.

The user may select the option of whether or not to write the output vector over the input vector fx. If memory is a limitation, and fx is no longer needed, this may prove useful. This option can be selected by setting the global variable **useinput__ = 1** prior to calling *integrat*. The default option is to not destroy the input vector fx (i. e., **useinput__ = 0**).

It is interesting to note that if the input data is a parabola, then Simpson's formula provides the exact integral.

The error in estimating the integral, $R(x)$, over the region (x_0, x_1) is approximately

$$R(x) = -(dx)^5 * |y''''(x)|/90$$

RETURNS
A real vector of length n representing the integral of fx if successful, or NULL if an error occurs. If $n < 6$, the error message "Not enough input data" is generated (math_errno = E__NOTENOUGH).

SEE ALSO
romberg(), deriv(), deriv1(), interp().

REFERENCE
Abramowitz and Stegun (1964, p. 886, eq. 25.4.5).

```
#include "mathlib.h"
#include <stdio.h>
#include <math.h>
#include <stdlib.h>
main()
{
int i,ndata=17;
Real dx;
Real_Vector y, fx;

        fx = valloc(NULL, ndata);
        dx = twopi_/(double)(ndata-1);  /* generate one cycle of a sinusoid   */
        for (i=0; i < ndata; i++) fx[i]=sin(dx*i);

        y = integrat(fx, dx,  ndata);   /* numerically integrate sin(x) over  */
                                        /* the interval (0, 2*pi)             */
        /* if y = sin(x) the integral should be: Y = 1-cos(x)                 */
        printf("                                numerical      true\n");
        printf("                  integrand:    integral:    integral:\n");
        printf(" i       x          sin(x)       (0, x)      1-cos(x)\n\n");
        for (i=0; i < ndata ; i++) {
            printf("%2d   %9.6f    %9.6f    %9.6f    %9.6f\n",
                            i,dx*i,fx[i], y[i], 1-cos(dx*i));
        }
        return 0;
}
```

PROGRAM OUTPUT

i	x	integrand: sin(x)	numerical integral: (0, x)	true integral: 1-cos(x)
0	0.000000	0.000000	0.000000	0.000000
1	0.392699	0.382683	0.075970	0.076120
2	0.785398	0.707107	0.292933	0.292893
3	1.178097	0.923880	0.617239	0.617317
4	1.570796	1.000000	1.000135	1.000000
5	1.963495	0.923880	1.382709	1.382683
6	2.356194	0.707107	1.707337	1.707107
7	2.748894	0.382683	1.923978	1.923880
8	3.141593	0.000000	2.000269	2.000000
9	3.534292	-0.382683	1.923978	1.923880
10	3.926991	-0.707107	1.707337	1.707107
11	4.319690	-0.923880	1.382709	1.382683
12	4.712389	-1.000000	1.000135	1.000000
13	5.105088	-0.923880	0.617239	0.617317
14	5.497787	-0.707107	0.292933	0.292893
15	5.890486	-0.382683	0.075970	0.076120
16	6.283185	0.000000	0.000000	0.000000

interp

#include "mathlib.h"
Real_Vector interp(Real_Vector *y*, int *ny*, int *nout*, Real_Vector *points*);

DESCRIPTION

In numerical analysis, tables of equidistant data must frequently be interpolated. For example, although many mathematical functions are accurately tabulated, in practice one is often interested in the ("in-between") points. One approach is to connect neighboring points with straight lines. This method is called linear interpolation and is effective if the table has high resolution (i. e., the data is greatly oversampled). However, in many instances the table is coarsely sampled. For these cases, more accuracy can be obtained by interpolating with a higher order polynomial (a line is a first order polynomial).

One of the most versatile approaches is the classical Lagrange interpolation method. At each data point, this technique computes the $(n-1)$th order polynomial that passes through the ordinates of each group of n neighboring points. Lagrange n-point interpolation has the following attractive features:

1. If the data is an mth order polynomial, and $m < n$, the interpolation is exact everywhere.

2. The interpolation is always exact at the input data samples (many interpolation techniques do not have this property).

3. The technique includes interpolation at the edges of data sets.

4. The Lagrange polynomial provides a closed-form expression for the interpolation anywhere over the interval spanned by the n data points.

5. If n is even, and the interpolation is evaluated only over the interval between the two central points, then the interpolation has a linear phase response (and minimum error). This means that the interpolation remains perfectly time-aligned with the input signal, even if the frequency content of the input is changed.

The *interp* function interpolates the real vector y of length ny using a fifth order Lagrange polynomial. The y vector represents a function $f(x)$ over an interval of ny equally spaced x values.

The *nout* argument specifies the length of the resulting real vector that is created by the interpolation. Typically, *nout* should be greater than ny. It is also typical to choose *nout* so that the ratio $(nout - 1)/(ny - 1)$ is an integer. This causes each of the original values of y to be reproduced.

The *points* argument is a real vector of length *nout* that specifies the points between 0 and ny at which interpolated values are desired. If NULL is specified for the *points* argument, then the resulting real vector will contain *nout* equally spaced points. In this case, *nout* must be greater than or equal to ny. Otherwise, the *points* vector specifies the *nout* points between 0 and *nout*, which need not be equally spaced. The values specified by the *points* vector do not have to be integers, nor do they have to be ordered.

The *interp* function calls the *lagrange_* and *xinterp_* functions.

RETURNS

A real vector of length *nout* representing the fifth order Lagrange interpolation of the vector y if successful, or NULL if an error occurs. If *points* = NULL and *nout* < *ny*, the error message "Interpolated length must be >= input length" is generated (math_errno = E_INTERP).

SEE ALSO

lowpass(), smooth(), interp1(), spline().

REFERENCE

L. R. Rabiner and R. W. Schafer (1973).

```
#include "mathlib.h"
#include <stdio.h>
#include <math.h>
#include <stdlib.h>
main()
{
int i, j, k, nfac, ny=7, nout=19;
double delta, truth;
Real_Vector yin, yout;

    yin = valloc(NULL, ny);
    delta = twopi_/(double)(ny-1);   /* generate one cycle of a sinusoid   */
    for (i=0; i < ny; i++) yin[i]=sin(delta*i);

    yout = interp(yin, ny, nout, NULL);/* interpolate by a factor of 3    */

    /* every 3rd point of the interpolation should equal the input       */
    printf("   interpolated        original     exact\n");
    printf(" i    yout[i]      k      yin[k]       value\n\n");
    nfac = (nout-1)/(ny-1);
    j=-nfac;
    k=0;
    for (i=0; i < nout; i++) {
        truth = sin(twopi_*i/(double) (nout-1));
        if(i == j+nfac) {
            printf("%2d   %9.6f    %2d    %9.6f   %9.6f\n",
                        i, yout[i], k, yin[k], truth);
            j += nfac;
            k++;
        }
        else  printf("%2d   %9.6f                        %9.6f\n",
                                i, yout[i], truth);
    }
    return 0;
}
```

PROGRAM OUTPUT

i	interpolated yout[i]	k	original yin[k]	exact value
0	0.000000	0	0.000000	0.000000
1	0.323126			0.342020
2	0.631997			0.642788
3	0.866025	1	0.866025	0.866025
4	0.989574			0.984808
5	0.988386			0.984808
6	0.866025	2	0.866025	0.866025
7	0.640312			0.642788
8	0.339758			0.342020
9	0.000000	3	0.000000	0.000000
10	-0.339758			-0.342020
11	-0.640312			-0.642788
12	-0.866025	4	-0.866025	-0.866025
13	-0.988386			-0.984808
14	-0.989574			-0.984808
15	-0.866025	5	-0.866025	-0.866025
16	-0.631997			-0.642788
17	-0.323126			-0.342020
18	0.000000	6	0.000000	0.000000

interp1

```
#include "mathlib.h"
Real interp1(Real x0, Real_Vector x, Real_Vector y, unsigned n, unsigned order);
```

DESCRIPTION

Many numerical interpolation applications involve data sets that are not equally spaced. For these cases, Lagrange interpolation can provide accurate results as long as the data is sufficiently oversampled and not corrupted with noise. For a brief list of some of the advantages of Lagrange interpolation, see the description section of the *interp* function.

Given a real value $x0$, the *interp1* function uses Lagrange interpolation to calculate a value $y0$. Unlike the *interp* function, the *interp1* function does not require the input data to be equally spaced.

The real vectors x and y specify a set of n x-y data points to be interpolated. The n values in the real vector x must be ordered; that is,

$$x[0] < x[1] < x[2] \ldots < x[n-1]$$

but do not necessarily have to be equally spaced; for example,

$$(x[1] - x[0])! = (x[2] - x[1])$$

The value specified by the $x0$ argument must be within the range of the values specified by the x vector; that is,

$$x[0] \le x0 \le x[n-1]$$

The *order* argument specifies the degree or order of the polynomial that is used to perform the interpolation, where

$$(1 \le order \le 9) \quad \text{and} \quad (order < n)$$

RETURNS

A real value $y0$, which is determined by interpolating the data points specified by the x and y vectors at the point $x0$ if successful, or 0 if an error occurs. If $order < 1$, $order > 9$, or $order \ge n$, the error message "The polynomial order is not in range" is generated (math_errno = E_ORDER). If any of the values in the vector x are equal, the error message "The input table has two identical x values" is generated (math_errno = E_SAMEX).

SEE ALSO

lowpass(), smooth(), interp(), spline().

REFERENCES

Abramowitz and Stegun (1964, p. 878, eqs. 25.2.1 and 25.2.2); R. W. Hornbeck (1975, pp. 43–45).

```c
#include "mathlib.h"
#include <stdio.h>
#include <math.h>
#include <stdlib.h>

main()
{
int i, k, n=7, nout=19;
double delta, delta3, truth;
Real x0, y0;
Real_Vector x, y;
unsigned order = 5;
      y = valloc(NULL, n);
      x = valloc(NULL, n);
      delta = twopi_/(double)(n-1);
      delta3 = delta/3;
   /* define  unequally spaced x values                              */
      x[0]=delta*0; x[1]=delta/3;
      x[2]=delta;   x[3]=delta*2.33333333;
      x[4]=delta*3; x[5]=delta*4.33333333;
      x[6]=delta*6;
   /* generate one cycle of a sinusoid; with nonequally spaced samples   */
      for (i=0; i < n; i++) y[i]=sin(x[i]);
      printf("            interpolated        original    truth:\n");
      printf("     x0          y0        k       y[k]       sin(x)\n\n");
      k=0;
      for (i=0; i < nout; i++) {
            truth = sin(twopi_*i/(double) (nout-1));
            x0 = i*delta3;
            y0 = interp1(x0, x, y, n, order);
            if(fabs(x0-x[k]) < 10.e-6) {
                  printf("%9.6f    %9.6f     %2d    %9.6f    %9.6f\n",
                           x0, y0, k, y[k], truth);
                  k++;
            }
            else  printf("%9.6f    %9.6f                    %9.6f\n",
                                    x0, y0, truth);
      }
      return 0;
}
```

PROGRAM OUTPUT

x0	interpolated y0	k	original y[k]	truth: sin(x)
0.000000	0.000000	0	0.000000	0.000000
0.349066	0.342020	1	0.342020	0.342020
0.698132	0.644430			0.642788
1.047198	0.866025	2	0.866025	0.866025
1.396263	0.981397			0.984808
1.745329	0.979491			0.984808
2.094395	0.862172			0.866025
2.443461	0.642788	3	0.642788	0.642788
2.792527	0.341279			0.342020
3.141593	0.000000	4	0.000000	0.000000
3.490659	-0.341050			-0.342020
3.839724	-0.641814			-0.642788
4.188790	-0.865985			-0.866025
4.537856	-0.984808	5	-0.984808	-0.984808
4.886922	-0.980879			-0.984808
5.235988	-0.851955			-0.866025
5.585054	-0.614752			-0.642788
5.934119	-0.308751			-0.342020
6.283185	-0.000000	6	-0.000000	-0.000000

invprob

```
#include "mathlib.h"
double invprob(double p0);
```

DESCRIPTION

The standard normal distribution (the bell-curve) is the most frequently used statistic in applied probability. The use of the normal distribution usually requires consulting tables of values. Although this may be adequate for a few hand calculations, these tables are cumbersome and are difficult to incorporate into computer programs. The *invprob* function provides a closed-form alternative to the standard normal tables.

The $p0$ argument specifies a cumulative normal probability (i. e., the probability that a random variable X is less than or equal to a specific value $x0$). This is expressed by the following formula:

$$p0 = P(X \le x_0) = \frac{1}{\sqrt{2\pi}} \int_{-\infty}^{x_0} e^{-x^2/2} dx$$

Given $p0$, the *invprob* function calculates the value of $x0$ in the previous formula. Since this value cannot be easily calculated, it is approximated for $p0 \le 0.5$ with a truncated power series:

$$x_0 = -t + (c_0 t + c_1 t^2 + c_2 t^3)/(1 + d_1 t + d_2 t^2 + d_3 t^3)$$

where $t = (-2 * \ln(p0))^{0.5}$
and $c_0 = 2.515517, \quad d_1 = 1.432788$
$c_1 = 0.802853, \quad d_2 = 0.189269$
$c_2 = 0.010328, \quad d_3 = 0.001308$
If $p0 > 0.5$, then the complementary power series is used:

$$x_0 = t - (c_0 t + c_1 t^2 + c_2 t^3)/(1 + d_1 t + d_2 t^2 + d_3 t^3)$$

where $t = (-2 * \ln(1 - p0))^{0.5}$.
The error of this approximation is less than $4.5e - 4$.

The *invprob* function calls the *pseries* function.

RETURNS

A double precision value $x0$ corresponding to the given cumulative normal probability $p0$ if successful, or $-$HUGE_VAL if an error occurs. If $p0 < 0$, the error message "Negative probability undefined" is generated (math_errno = E_NEGPROB).

SEE ALSO

nprob().

REFERENCE

Abramowitz and Stegun (1964, p. 933, eq. 26.2.23).

```
#include "mathlib.h"
#include <stdio.h>
main()
{
double p0=0.5;
/* if P(X<=x0) =.5  and x is a normal random variable find x0      */
    printf("x0 = %f\n", invprob(p0));   /* the answer is x0 = 0     */
    return 0;
}
```

least_sq

#include "mathlib.h"

Real_Vector least_sq(int (*basis)(Real x, Real_Vector fx, unsigned terms), Real_Vector vx, Real_Vector vy, unsigned n, unsigned terms, Real *mse);

DESCRIPTION

One of the most powerful curve-fitting techniques is the method of generalized least squares. The approach is similar to linear regression and polynomial regression, except that the basis functions, $f_i(x)$ for $i = 0, \ldots, t$, are completely general. The least-squares method approximates over the entire data interval with an expression of the form:

$$f(x) = f_0(x) * \text{coef}[0] + f_1(x) * \text{coef}[1] + \ldots + f_{k-1}(x) * \text{coef}[k-1]$$

where $f_0(x) \ldots f_{k-1}(x)$ are the basis functions and $\text{coef}[0] \ldots \text{coef}[k-1]$ are the coefficients. The *least_sq* function calculates these coefficients to minimize the mean-square error between the actual data and the expression.

The vx and vy arguments are real vectors of length n, representing the data as pairs of x-y values.

The *basis* argument specifies a user-defined function that is called by the *least_sq* function to obtain the values of the basis functions. Notice that the user-defined *basis* function requires three arguments. The x argument specifies a value for x in the interval $[vx[0], vx[n-1]]$. The fx argument specifies the real vector in which the user-defined function is expected to store a value for each of the basis functions $f_0(x) \ldots f_{\text{terms}-1}(x)$. The *terms* argument specifies the number of basis functions. The *basis* function should return an integer value of 0 if no errors occur or a nonzero integer value if an error does occur. A nonzero return value causes the *least_sq* function to terminate.

The *terms* argument specifies the number of basis functions (i. e., the number of terms in the expression that is used to approximate the data). When the *least_sq* function calls the user-defined *basis* function, this value is passed as the *terms* argument.

The *mse* argument specifies the address of a variable of type **Real**, where the *least_sq* function returns the mean-square error between the data and the expression. The mean-square error is calculated as follows:

$$mse = (f(vx[0]) - vy[0]^2 + (f(vx[1]) - vy[1])^2 + \ldots \\ + (f(vx[n-1]) - vy[n-1])^2$$

For certain sets of data, the least-squares solution may be ill-conditioned. This is more likely to occur when the number of basis functions is large. Since the least-squares solution is obtained by solving a linear set of equations (the normal equations), the determinant of the normal equations can become very small if too many basis functions are used. The *least_sq* function will fail if the determinant becomes 0.

The *least_sq* function calls the *lineqn* function.

RETURNS

A real vector of length *terms*, representing the coefficients (coef[0] ... coef[terms −1])
that minimize the mean-square error between the data and the expression if successful, or NULL if an error occurs. If the determinant of the normal equations used to
calculate the coefficients becomes 0, the error message "Singular Matrix" is generated
(math_errno = E_MSING).

SEE ALSO

curvreg(), linreg(), pseudinv(), pseries().

REFERENCE

R. W. Hamming (1973).

```
#include "mathlib.h"
#include <stdio.h>


int basis(Real x, Real_Vector fx, unsigned terms)
{  /* this function calculates values for the 3 basis functions
          representing the equation: f(x) = x^2 + x + 1
               f0(x) = 1.0; f1(x) = x; f2(x) = x^2;                    */
int i;
    fx[0] = 1.0;
    for (i=1; i<terms; i++) fx[i] = x*fx[i-1];
    return 0;
}


Real VX[7] = { -3, -2, -1, 0, 1, 2, 3 };
Real VY[7] = { 4, 2, 3, 0, -1, -2, -5 };


main()
{
Real_Vector vx,vy,coeff;
Real mse;
int n=7,terms=3;

/* fit the above "x" and "y" points with the best quadratic polynomial      */

        vx = valloc(VX, n); /* allocate and initialize the "x" values       */
        vy = valloc(VY, n); /* allocate and initialize the "y" values       */
        coeff = least_sq(basis, vx, vy, n, terms, &mse);/*find best fit to data*/

        printf("The best fit quadratic equation is: ");
        printf("f(x) = %f  %f*x  %f*x*x\n", coeff[0], coeff[1], coeff[2]);
        printf("The mean square error = %f\n", mse);
        return 0;
}  /* The answer is: f(x) = 0.666667  -1.392857*x  -0.130952*x*x  */
```

PROGRAM OUTPUT

```
The best fit quadratic equation is: f(x) = 0.666667  -1.392857*x  -0.130952*x*x
The mean square error = 3.095238
```

levinson

#include "mathlib.h"
Real__Vector levinson(Real__Vector *r*, Real__Vector *b*, unsigned *n*);

DESCRIPTION

One of the most important problems in linear algebra is the solution of simultaneous linear equations. Often these equations exhibit a symmetric structure, which can be exploited to reduce the number of computations required to solve the equations. A **Toeplitz** matrix M is a real symmetric matrix with the additional row and column symmetry shown below:

$$M = \begin{bmatrix} m_1 & m_2 & m_3 & \cdots & m_n \\ m_2 & m_1 & m_2 & \cdots & m_{n-1} \\ m_3 & m_2 & m_1 & \cdots & m_{n-2} \\ \vdots & \vdots & \vdots & & \vdots \\ m_n & m_{n-1} & m_{n-2} & \cdots & m_1 \end{bmatrix}$$

Note that M is completely specified by any one of its row (or column) vectors.

The *levinson* function finds the solution vector for a Toeplitz system of linear equations:

$$M[n][n] * v[n] = b[n]$$

where M is the n by n system Toeplitz matrix, b is a constraining vector of length n, and v is the solution vector of length n.

The r argument specifies a real vector of length n corresponding to the first row of matrix M that completely determines the Toeplitz matrix.

The real vectors r and b must be allocated by the *valloc* function prior to calling the *levinson* function. The resulting solution vector of length n is automatically allocated by the *levinson* function and should be freed when it is no longer needed.

RETURNS

A real vector v of length n, which is the solution of a Toeplitz system of linear equations if successful, or NULL if an error occurs.

SEE ALSO

lineqn().

REFERENCES

N. Levinson (1946); E. A. Robinson (1983).

```
#include <stdio.h>
#include <stdlib.h>
#include <math.h>
#include "mathlib.h"

Real M[3][3] = { 1, 2, 3,      /* M[][] is a Toeplitz matrix */
                 2, 1, 2,
                 3, 2, 1 };
main()
{
Real_Vector b, r, v;
int i;
unsigned n = 3;

/* Solve the Toeplitz matrix equation:  M[][]*v[] = b[]     */

        b = valloc(NULL, n);
        r = valloc(NULL, n);

        r[0] = M[0][0];  r[1] = M[0][1];    r[2] = M[0][2];
        b[0] = 1.0;      b[1] = 1.0;        b[2] = 1.0;

        v = levinson(r, b, n);

        printf("The answer is:\n");
        for (i=0; i<n; i++) {
            printf("  %8.4f\n", v[i]);
        }
        mathfree();
        return 0;
}
```

PROGRAM OUTPUT

```
The answer is:
    0.2500
    0.0000
    0.2500
```

lineqn

```
#include "mathlib.h"
Real__Vector lineqn(Real__Matrix m, Real__Vector v, unsigned n);
```

DESCRIPTION

One of the most important problems in Linear Algebra is the solution of simultaneous linear equations. These equations arise in a variety of applications in applied and theoretical math. For example, consider the simple geometrical problem of two lines in a plane (see the example below). If the lines are nonparallel, then they intersect at a unique point. This point can be found as the solution of two simultaneous linear equations. However, if the lines are parallel, then they do not intersect and there is no solution. The case of no solution occurs whenever the determinant of the system matrix is zero (i. e., the system matrix is singular, or noninvertible).

The *lineqn* function finds the solution vector for a system of linear equations:

$$m[n][n] * x[n] = v[n]$$

where m is the n by n system matrix, v is a constraining vector of length n, and x is the solution vector of length n. Since m is a square matrix, the number of equations must be equal to the number of unknowns.

The real matrix m and the real vector v must be allocated by *mxalloc* and *valloc* respectively prior to calling the *lineqn* function. The resulting solution vector of length n is automatically allocated by the *lineqn* function.

The user may select the option of whether or not to write over the input matrix m and vector v with in-place computations. If memory is a limitation, and m and v are no longer needed, this may prove useful. This option can be selected by setting the global variable **useinput__ = 1** prior to calling *lineqn*. If **useinput__ = 1**, then the solution vector is returned in vector v. The default option is to not destroy the input arguments (i. e., **useinput__ = 0**).

RETURNS

The real vector x of length n, representing the solution vector for m and v if successful, or NULL if an error occurs. The determinant of matrix m is also returned in the global variable **determ__**. If no solution exists, the error message "Singular matrix" is generated (math_errno = E__MSING).

SEE ALSO

Clineqn(), levinson().

REFERENCE

J. H. Wilkinson (1965).

```
#include "mathlib.h"
#include <stdio.h>

main()
{
Real_Matrix m;
Real_Vector z,v;
/* find the intersection of the two lines:                            */
/*          2x - 3y = 4                                                */
/*          5x - 7y = -2    the answer is x = -34, y = -24            */

        v = valloc(NULL,2);      /*  allocate the constraining vector     */
        m = mxalloc(NULL,2,2);  /*  allocate the system matrix           */
        v[0] = 4; v[1] = -2;     /* define constraining vector & system matrix */
        m[0][0]=2; m[0][1]=-3; m[1][0]=5; m[1][1]=-7;

        z = lineqn(m,v,2); /* now solve linear system of equations for z[ ] */
        printf(" x = %f, y = %f\n", z[0], z[1]);

        mxfree(m);    /* free matrix m[ ][ ] */
        vfree(v);     /* free vector v[ ]    */
        vfree(z);     /* free vector z[ ]    */
        return 0;
}
```

linreg

```
#include "mathlib.h"
void linreg(Real_Vector x, Real_Vector y, int n, Real *m, Real *b, Real *rsq);
```

DESCRIPTION

The *linreg* function determines the equation for a straight line that minimizes the mean-square error between the line and a set of data points.

The real vectors x and y of length n represent the set of x-y data points. The m and b arguments specify the addresses of two variables of type **Real**, where the *linreg* function returns the two coefficients that define the linear equation as follows:

$$y = mx + b$$

where m is the slope of the line, and b is the y-intercept.

The rsq argument also specifies the address of a variable of type **Real**, where the *linreg* function returns the correlation coefficient. The correlation coefficient ($0 \leq rsq \leq 1$) indicates how closely the data matches the equation for the straight line. The closer rsq is to 1, the better the match between the data and the equation (i. e., the more linear the data).

RETURNS

Three real values are returned via the m, b, and rsq arguments, where m and b are coefficients of a linear equation, and rsq is the correlation coefficient.

SEE ALSO

curvreg(), least_sq(), pseudinv().

REFERENCE

R. W. Hamming (1973).

```
#include "mathlib.h"
#include <stdio.h>
Real X[7] = { -2.01, -1.02, -0.09, 1.01, 1.99, 2.99, 4.01 };
Real Y[7] = { -4.99, -3.01, -.99, 1.01, 2.99, 4.99, 7.01 };
main()
{
Real_Vector x,y;
Real m, b, rsq;
int n=7;

/* fit the above "x" and "y" points with the best line                     */

        x = valloc(X, n); /* allocate and initialize the "x" values        */
        y = valloc(Y, n); /* allocate and initialize the "y" values        */

        linreg(x, y, n, &m, &b, &rsq);/* fit the data with the "best" line  */

        printf("The best fit linear equation is: ");
        printf("y = %f + %f*x\n", b, m);
        printf("The coefficient of determination is %f\n", rsq);

        /* The best fit linear equation is: y = -0.951999 + 1.987499*x */
        /* The coefficient of determination is 0.999777               */
        return 0;
}
```

lmsadapt

```
#include "mathlib.h"
Real_Vector lmsadapt(Real_Vector noise, Real_Vector corrupted,unsigned n,
                     unsigned ndelays, double mu, double rho, double noise_power);
```

DESCRIPTION

Adaptive filters automatically change their coefficients based on the signal and noise statistics to enhance the signal. An adaptive filter can be efficiently implemented with the **LMS** (least mean square) algorithm developed by Widrow.

The *lmsadapt* function implements an LMS adaptive filter. In the context of interference cancelling, this function returns the best estimate of the desired signal component in a signal that has been corrupted with noise. It is assumed that the corrupting noise is known to be correlated to a given noise source. It is further assumed that the true (desired) signal is uncorrelated with this noise source.

The *noise* argument is a vector representing the correlated noise source. This input vector is overwritten with the best LMS estimate of the correlated noise component corrupting the desired signal. The resulting vector is converted to a zero mean signal.

The *corrupted* argument is a vector representing the corrupted signal. This vector contains **both** the desired signal and the undesired noise. It is assumed that the noise in the *corrupted* vector is correlated to the input *noise* vector and uncorrelated to the desired signal. This input vector is also overwritten, the result of converting it to a zero mean signal.

The *n* argument specifies the length of both the *noise* and *corrupted* vectors.

The *ndelays* argument specifies the number of delays used in the LMS filter (i. e., the filter order).

The *mu* argument specifies the convergence step size, where $0 \leq mu \leq 1$. A value of 0 causes the adaptation process to "coast," whereas a value of 1 results in the fastest convergence. The stepsize is determined iteratively as follows:

$$\text{stepsize}[k] = \frac{2 * mu * \text{error}[k]}{(ndelays + 1) * noise\_power[k]}$$

The *rho* argument allows the function to internally adapt its step size to any correlated noise power fluctuations according to the following formula:

$$noise\_power[k] = rho * noise[k]^2 + (1 - rho) * noise\_power[k - 1]$$

where $0 \leq rho \leq 1$. A value of 0 causes no adaptation of the step size to the input noise power, whereas a value of 1 results in the fastest adaptation of the step size to the noise power.

The *noise_power* argument specifies the best estimate of the power in the correlated *noise* source, excluding the DC term. The value must be greater than or equal to 0.

The return value of this function is the desired filtered output vector of length n. This vector represents the best estimate of the true signal. The output vector is automatically allocated by the *lmsadapt* function and should be freed when it is no longer needed.

For an application of this function to **adaptive line enhancement**, see that topic in the Adaptive Signal Processing section 8.5.4.

RETURNS

A real vector of length n representing the LMS filtered estimate of the signal component in the *corrupted* vector if successful, or NULL if an error occurs.

SEE ALSO

smooth().

REFERENCES

B. Widrow and S. D. Stearns (1985); S. D. Stearns and R. A. David (1988).

```
#include <stdio.h>
#include <stdlib.h>
#include <math.h>
#include "mathlib.h"
main()
{
Real_Vector desired, noise_source, corrupted, filtered_output, correlated_noise;
int k, ndelays=20, ndata=250;
double err_before, err_after, freq, tmp, rms_desired;
double mu = 0.05, noise_power = 0.50, rho = 0.2;

/* Adaptive Noise Cancelling with the LMS algorithm : Given a corrupted signal
   and the noise_source, cancel the (unknown) correlated_noise and enhance the
   (unknown) desired signal */

    noise_source = valloc(NULL, ndata);
    corrupted = valloc(NULL, ndata);
    correlated_noise = valloc(NULL, ndata);
    desired = valloc(NULL, ndata);

    freq = 2*pi_/20.; /* for this example, the desired signal is white noise */
    for(k = 0; k<ndata; k++) {
        noise_source[k] = sin(freq*k);
        correlated_noise[k] = (1 + 0.005*k*(0.005*k + 1.0))*sin(freq*k+pi_/3.);
        desired[k] = cos(2*pi_*k/31.0 + pi_/5.0);
        corrupted[k] = desired[k] + correlated_noise[k];
    }
        /* filter the data with the LMS algorithm: */

    filtered_output = lmsadapt(noise_source, corrupted, ndata,
                                        ndelays, mu, rho, noise_power);

    err_after = err_before = rms_desired = 0.0;
    for(k = 0; k<ndata; k++) {
        tmp = desired[k] - corrupted[k];
        err_before += tmp*tmp;
        tmp = desired[k] - filtered_output[k];
        err_after += tmp*tmp;
        rms_desired += desired[k]*desired[k];
    }
    printf("error before LMS filtering = %7.4f\n", err_before/rms_desired);
    printf(" error after LMS filtering = %7.4f\n", err_after/rms_desired);
    mathfree();
    return 0;
}
```

PROGRAM OUTPUT

```
error before LMS filtering =  5.1580
 error after LMS filtering =  0.2548
```

logn

#include "mathlib.h"
double logn(int *n*, double *x*);

DESCRIPTION

Sometimes it is necessary to evaluate the logarithm of a base other than than 10 (the common log) or e (the natural log).

The *logn* function calculates the base n logarithm of the double precision value x (i. e., $\log_n(x)$). The n argument may be any integer value. The x argument must be greater than 0.

The computation utilizes the following relationship:

$$\log_n(x) = \log_e(x) / \log_e(n)$$

The *logn* function is plotted in Figure 9.6 for $n = 2, 6,$ and 10.

LOGARITHM OF INTEGER BASE

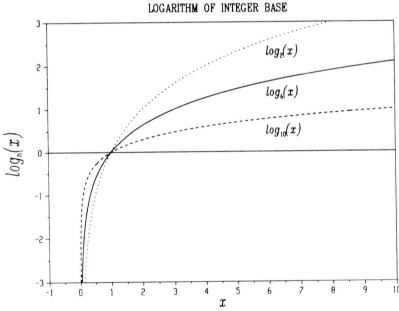

Figure 9.6   Logarithm of Integer Base

The *logn* function calls the natural logarithm *log* function.

RETURNS

A double precision value representing the base n logarithm of x if successful, or -HUGE_VAL if an error occurs. If $x \leq 0$, the error message "Argument is outside the domain of the function" is generated (math_errno = E_DOMERR).

```
#include "mathlib.h"
#include <stdio.h>
main()
{
double xstart, xstop, x, dx;
int i, npts=20;

        xstart = 0.001;
        xstop = 10.;
        dx = (xstop-xstart)/(double) npts;
        x=xstart;
        printf("     x         log2(x)      log6(x)      log10(x)\n\n");
        for (i=0; i<npts; i++) {
           printf("%9.6f  %10.7f  %10.7f  %10.7f\n",
                                      x, logn(2,x), logn(6,x), logn(10,x));
           x += dx;
        }
        return 0;
}
```

PROGRAM OUTPUT

x	log2(x)	log6(x)	log10(x)
0.001000	-9.9657843	-3.8552916	-3.0000000
0.500950	-0.9972615	-0.3857934	-0.3002056
1.000900	0.0012978	0.0005021	0.0003907
1.500850	0.5857798	0.2266106	0.1763373
2.000800	1.0005770	0.3870760	0.3012037
2.500750	1.3223608	0.5115590	0.3980703
3.000700	1.5852991	0.6132774	0.4772226
3.500650	1.8076228	0.6992840	0.5441487
4.000600	2.0002164	0.7737893	0.6021251
4.500550	2.1701013	0.8395098	0.6532656
5.000500	2.3220724	0.8983002	0.6990134
5.500450	2.4595497	0.9514837	0.7403982
6.000400	2.5850587	1.0000372	0.7781802
6.500350	2.7005174	1.0447027	0.8129367
7.000300	2.8074168	1.0860571	0.8451167
7.500250	2.9069387	1.1245574	0.8750757
8.000200	3.0000361	1.1605724	0.9031008
8.500150	3.0874883	1.1944035	0.9294266
9.000100	3.1699410	1.2263006	0.9542473
9.500050	3.2479351	1.2564728	0.9777259

lowpass

```
#include "mathlib.h"
void lowpass(Real weights[ ], int nweights, double fc, double dB, int half);
```

DESCRIPTION

The *lowpass* function designs a digital lowpass filter, which can then be implemented
with the *convolve* function. The basic approach of window design methods is to truncate the infinite length impulse response of an ideal frequency filter by multiplying
it by a time-domain window (in this case, a Kaiser–Bessel window). Time-domain
windows are bell-shaped (see the *tdwindow* function), which tapers the filter impulse
response.

It is assumed that the data to be filtered is sampled at uniform intervals of time,
T. The sampling frequency is $F_s = 1/T$. In digital sampling, frequencies are often
specified normalized to F_s. For example, $F_s/2$, the Nyquist frequency, is the highest
frequency that can be represented and has a normalized value of 1/2.

Digital lowpass filters are characterized by a passband, $(0, F_1)$, over which the
filter's response is reasonably close to unity gain. They also often have a stopband,
$(F_2, F_s/2)$, over which frequencies are attenuated by some specified amount (i. e., the
gain in the stopband is less than $d_s < 1$). The maximum attenuation is at the highest
frequency, $F_s/2$.

The *fc* argument specifies the cutoff frequency, which sets the nominal passband
(and stopband) of the Kaiser window filter, since it is the average of F_1 and F_2:

$$fc = (F_1 + F_2)/2$$

The cutoff frequency is the frequency at which the magnitude of the filter response is
approximately −6 dB.

The transition band, F_t, is the difference between the passband and the stopband
frequencies:

$$F_t = F_2 - F_1$$

The *dB* argument specifies the desired stopband attenuation in *decibels*:

$$dB = 20 \log_{10}(G_s)$$

where G_s is the gain in the stopband.

The *nweights* argument specifies the length of the filter. The length of the filter
needed to achieve the desired values of dB and F_t is given by:

$$nweights = (|dB| - 8)/(14.357 F_t)$$

The *nweights* filter weights representing the impulse response are returned via the
weights array. The *weights* array should be declared as an array of type **Real** with at
least *nweights* elements.

The *half* argument specifies whether to return all of the filter weights, or just
half of them. Since the Kaiser–Bessel lowpass filter is a symmetric (even) function,

the second half of the filter weights is a mirror image of the first half. This symmetry can be exploited to reduce, by a factor of 2, the number of multiplications (and array storage locations) required to implement the filter. If *half* is zero, then all of the *nweights* coefficients are returned through the *weights* array. However, if *half* is nonzero, then only the first *nweights*/2 coefficients are returned. The remaining half of the coefficients could be filled into the *weights* array using the following C statement:

```
for (i = 0; i < nweights/2; i++)
    weights[nweights-1-i] = weights[i];
```

If the filter is symmetric, the *convolve* function needs only the first half of the coefficients.

It should be mentioned that the above design formulas are approximations. If an application requires an exact frequency response, then the filter design should be checked to see that it meets the desired specification. It may be necessary to slightly change the filter design parameters to achieve a more exact frequency response.

It is instructive to compare the "*n*-point" averager (see the *convolve* function) with a Kaiser–Bessel lowpass filter. The responses of a 4-point averager and a 10-point Kaiser lowpass filter are shown in the lowpass filter section. In the example below, this 10-point Kaiser–Bessel lowpass filter reduces the error to about 38 percent less than that of the "4-point" averager (0.045437 per data point, versus 0.073728 per data point).

RETURNS

The *nweights* (*nweights*/2 if *half* is nonzero) coefficients representing the impulse response of a lowpass filter with cutoff frequency fc and stopband attenuation dB is returned via the *weights* array.

SEE ALSO

highpass(), bandpass(), convolve(), downsamp(), deriv(), interp(), tdwindow().

REFERENCE

A. V. Oppenheim and R. W. Schafer (1989).

```c
#include "mathlib.h"
#include <stdio.h>
#include <stdlib.h>
#include <math.h>
#define NWEIGHTS 10
#define NDATA   100


main()
{
Real_Vector data, y;
Real weights[NWEIGHTS];
int i, ny, ndec=1, itype=1, isym=1;
double truth, noise, off, init_error=0.0, filter_error=0.0;
    data=valloc(NULL, NDATA);     /*   allocate a vector for the data      */
    for(i=0; i<NDATA; i++) {/* generate data, one cycle of a sinusoid+noise */
        truth = sin(twopi_*i/(double) NDATA);
        noise = 2*((double) rand()/RAND_MAX - 0.5); /* uniform from +1 to -1*/
        data[i] = truth + noise;
        init_error += noise*noise/(double) NDATA; /* noise = truth - data   */
    }
    lowpass(weights, NWEIGHTS, .4445/4, 45.0, 0); /* design lowpass filter  */
    y = convolve(data, NDATA, (Real_Vector) weights, NWEIGHTS,
                 ndec, itype, isym, &ny);
    off= (double) (NWEIGHTS-1)/2.;  /* align truth with filtered data       */
    for(i=0; i<ny; i++) { /* compute the error after filtering             */
        truth = sin(twopi_*(i+off)/(double) NDATA);
        filter_error += (truth - y[i])*(truth - y[i])/(double) ny;
    }/* The Kaiser lowpass is about 38% better than the 4-pt averager       */
    printf("the error before filtering is %f\n", init_error); /* = 0.345443 */
    printf("the error after filtering is %f\n", filter_error);/* = 0.045437 */
    return 0;
}/* the error for the 4-point averager in the convolve() example was 0.073728 */
```

math_err

#include "mathlib.h"
void math__err(void);

DESCRIPTION

The *math_err* function prints an error message to **stderr**. The error message corresponds to the last error detected by one of the mathlib functions. The global variable **math_errno** determines which error message is printed. If the value of **math_errno** is zero, then the message "No errors detected" is printed. Otherwise, one of the following error messages is printed:

```
value of math_errno|  error message
-------------------|-------------------------------------------------------------
   E_MALLOC      1 |  Insufficent dynamic memory available
   E_MSIZE       2 |  Array too large
   E_NULLPTR     3 |  Invalid NULL pointer
   E_MSING       4 |  Singular matrix
   E_NEQNS       5 |  More unknowns than equations
   E_WINDOW      6 |  Invalid window type specified
   E_DOMERR      7 |  Argument is outside the domain of the function
   E_FACTOR      8 |  Argument 'factor' must be between 2 and 10
   E_DECIMATE    9 |  Argument 'ndec' must be between 2 and 10
   E_NOTENOUGH  10 |  Not enough input data
   E_INTERP     11 |  Interpolated length must be >= input length
   E_LAGRANGE   12 |  Interpolation not defined at input value 't'
   E_NEGPROB    13 |  Negative probability undefined
   E_FFTSIZE    14 |  Length of data must be greater than or equal to FFT size
   E_FFTPOWER2  15 |  FFT length must be a power of two
   E_LIMITS     16 |  Limits on definite integral must be distinct
   E_ROUNDOFF   17 |  Rounding errors prohibit required accuracy
   E_STEPSIZE   18 |  Step size must be nonzero to estimate derivative
   E_DISCRETE   19 |  k0 must be less than total number of objects n
   E_SAMEX      20 |  The input table has two identical x values
   E_ORDER      21 |  The polynomial order is not in range
   E_OPEN       22 |  Could not open file
   E_WRITE      23 |  Error writing to file
   E_READ       24 |  Error reading from file
   E_CURVES     25 |  Argument 'curves' must be between 1 and CURVES_MAX
   E_HEADER     26 |  Invalid file header
   E_RSIZE      27 |  Sizeof(Real) in file != sizeof(Real) in program
   E_NO_MIN     28 |  Minimum probably does not exist
   E_DIVERGE    29 |  No convergence in LIMIT iterations
   E_GRAD_ERR   30 |  Encountered errors in calculating the gradient
   E_USERFUNC   31 |  User function called by least_sq function failed
   E_TRANSFER   32 |  Order of denominator cannot be less than numerator
   E_RANGE      33 |  Value must be between -32768 and 32767
   E_HERMITIAN  34 |  Input matrix is non-Hermitian
   E_STAGES     35 |  Numerator/Denominator must have same stages/order}
```

By default, an error detected by one of the mathlib functions automatically causes one of the above error messages to be printed; then the program is immediately aborted (i. e., terminated). Therefore the *math_err* function is useful only after this default error processing is overridden.

The global variable **math_errmsg** controls whether or not an error message is automatically printed when an error occurs. The global variable **math_abort** controls whether or not the program is automatically aborted when an error occurs. Each of these global variables has a default value of ON. By setting the **math_errmsg** and

math_abort variables to OFF, the default error processing is overridden (i. e., an error message will not be printed and the program will not abort).

After overriding the default error processing, it is essential that the program test for the occurrence of errors. Most functions return a value such as NULL to indicate that an error has occurred. For some functions, it is necessary to test the value of the **math_errno** variable to determine whether or not an error occurred.

RETURNS
None.

```c
#include "mathlib.h"
#include <stdio.h>
#include <stdlib.h>
main()
{
Real_Vector v;

        math_errmsg = OFF; /* turn off printing of error messages */
        math_abort = OFF;  /* turn off program abort */
        v = valloc(NULL, 65535U);
        if (v == NULL) {
            printf("Unable to allocate space for vector v\n");
            math_err();
        }

        logn(2, -1.0);  /* this will definitely cause an error */
        if (math_errno != 0) {
            printf("Can't take log of negative number\n");
            math_err();
            exit(1);
        }
        printf("this should not be printed\n");
        return 0;
}
```

mathfree

#include "mathlib.h"
int mathfree(*void*);

DESCRIPTION
The *mathfree* function frees all vectors and matrices that have been allocated since the start of program execution. Whether allocated explicitly by direct calls to the *valloc*, *Cvalloc*, *mxalloc*, and *Cmxalloc* functions, or implicitly by one of the other library functions, all vectors and matrices are freed.

After calling the *mathfree* function, any variable of type **Real_Vector**, **Complex_Vector**, **Real_Matrix**, or **Complex_Matrix** should be reallocated prior to being used again.

RETURNS
An integer value representing the number of vectors and matrices that were freed by the *mathfree* function.

SEE ALSO
Cmxfree(), mxfree(), vfree(), Cvfree().

```
#include "mathlib.h"
#include <stdio.h>
#include <stdlib.h>
main()
{
Complex_Vector v;
Real_Vector x[6];
Real_Matrix m;
Complex_Matrix c;
int i;

        v = Cvalloc(NULL, 2); /* allocate dynamic memory for a Complex vector */
        m = mxalloc(NULL,2,3);/* allocate dynamic memory for a Real matrix     */
        c = Cmxalloc(NULL,4,5);/* allocate dynamic memory for a Complex matrix */
        for (i=0; i<6; i++) { /* allocate dynamic memory for an array of       */
                            /* 6 Real_Vectors  of length 10                  */
            x[i] = valloc(NULL, 10);
            if (x[i] == NULL) {         /* check for allocation errors        */
                math_err(); exit(1);
            }
        }
        mathfree();    /* free all the dynamically allocated memory  for       */
                    /* v[ ], m[ ][ ], c[ ][ ], and x[i], i=0,1,...5          */
        return 0;
}
```

median

#include "mathlib.h"
Real__Vector median(Real__Vector *x*, unsigned *nx*, unsigned *nfilt*);

DESCRIPTION

Median filtering is a nonlinear technique that is ideal for removing shot-noise or noise from digital dropout sometimes found in images. In its simplest form the median filter is a sliding window that extends over an odd number of data points. At each new window position, the central data point is replaced by the median of the window. The effect of this processing is to filter out large (high frequency) noise spikes, without significantly degrading the desired data. A fast sorting algorithm (e. g., Heapsort) can be used to speed up the filtering.

The *median* function filters the vector x of length nx using a median filter of length *nfilt*. The value of *nfilt* must be odd, where $2 \le nfilt \le nx$.

RETURNS

A real vector of length nx representing the median filtered vector x if successful, or NULL if an error occurs. If *nfilt* is even, less than 2, or greater than nx, the error message "Argument is outside domain of the function" is generated (math_errno = E_DOMERR).

SEE ALSO

lowpass(), highpass(), bandpass(), smooth().

REFERENCE

J. W. Tukey (1971).

```
#include <stdio.h>
#include <stdlib.h>
#include <math.h>
#include "mathlib.h"

Real UNFILTERED[20] = { 0,1,2,3,101, 5,6,7,101,9, 10,11,101,13,14,
                                                   15,16,17,101,19};
main()
{
Real_Vector filtered, x;
int i;
unsigned nx = 20, nfilt = 3;

/* Filter the array UNFILTERED[] with a 3-point median filter */

    x = valloc(UNFILTERED, nx);

    filtered = median(x, nx, nfilt);/* remove the noise with a median filter */

    printf("unfiltered input sequence:\n");
    for(i=0; i<nx-1; i++) {
        printf("%4.0f", x[i]);
    }
    printf("%4.0f\n", x[nx-1]);
    printf("%d-point median-filtered output sequence:\n", nfilt);
    for(i=0; i<nx; i++) {
        printf("%4.0f", filtered[i]);
    }
    printf("\n");
    vfree(x);
    return 0;
}
```

PROGRAM OUTPUT

```
unfiltered input sequence:
   0    1    2    3  101    5    6    7  101    9   10   11  101   13   14   15   16   17  101   19
3-point median-filtered output sequence:
   0    1    2    3    5    6    6    7    9   10   10   11   13   14   14   15   16   17  101   19
```

mxadd

#include "mathlib.h"
Real_Matrix mxadd(Real_Matrix *m1*, Real_Matrix *m2*, unsigned *rows*, unsigned *cols*);

DESCRIPTION

The *mxadd* function adds the two real matrices $m1$ and $m2$. The *rows* and *cols* arguments specify the number of rows and columns in each matrix.

Both $m1$ and $m2$ must be allocated by the *mxalloc* function prior to calling the *mxadd* function. The resulting real matrix, representing the sum of $m1$ and $m2$, is automatically allocated by the *mxadd* function.

The user may select the option of whether or not to write over the input matrix $m1$ with in-place computations. If memory is a limitation, and $m1$ is no longer needed, this may prove useful. This option can be selected by setting the global variable **useinput_ = 1** prior to calling *mxadd*. If this option is selected, the resulting matrix overwrites the input matrix $m1$. The default option is to not destroy the input matrix $m1$ (i. e., **useinput_ = 0**).

RETURNS

A real matrix m of size *rows* by *cols*, where $m = m1 + m2$, or NULL if an error occurs.

SEE ALSO

Cmxsub(), Cmxadd(), mxsub(), Cvadd(), Cvsub(), vadd(), vsub().

```
#include "mathlib.h"
#include <stdio.h>
main()
{
int i,j;
Real_Matrix m,m1,m2;
/* subtract the following two matrices:                              */
/*                                                                   */
/*          m[][] = m1[][] + m2[][]                                  */
/*                                                                   */
/* where m1[][] and m2[] are real and are given below:              */
     m1 = mxalloc(NULL, 2, 2);  /* allocate a 2 by 2 matrix         */
     m2 = mxalloc(NULL, 2, 2); /* allocate another 2 by 2 matrix    */

     m1[0][0] = 2;   m1[0][1] = 2;
     m1[1][0] = 2;   m1[1][1] = 2;
     m2[0][0] = 3;   m2[0][1] = 3;
     m2[1][0] = 3;   m2[1][1] = 3;

     m = mxadd(m1, m2, 2, 2); /* add m1[][] and m2[][] */

     /* the answer should be : m[0][0] = 5;  m[0][0] = 5;
                               m[0][1] = 5;  m[0][1] = 5;   */
     for (i=0; i<2; i++) {
         for (j=0; j<2; j++) {
              printf("m[%d][%d] = %f\n", i, j, m[i][j]);
          }
     }
     mxfree(m);     /* free matrix m[][]  */
     mxfree(m1);    /* free matrix m1[][] */
     mxfree(m2);    /* free matrix m2[][] */
     return 0;
}
```

mxalloc

#include "mathlib.h"
Real__Matrix mxalloc(Real__Ptr *address*, unsigned *rows*, unsigned *cols*);

DESCRIPTION

The *mxalloc* function allocates memory for a real matrix. The *rows* and *cols* arguments specify the number of rows and columns in the matrix.

The *address* argument is a pointer to a **Real** value. If the *address* argument is NULL, the *mxalloc* function allocates space for the matrix from the available dynamic memory. If the *address* argument is not NULL, the *mxalloc* function assumes that it is the address of a two-dimensional array of **Real** values. In this case, only an array of pointers is allocated and initialized to point to the beginning of each row of the two-dimensional array. This provides a mechanism for using preinitialized arrays.

RETURNS

A *rows* by *cols* real matrix if successful, or NULL if an error occurs. If the matrix is too large, an "Array too large" error message is generated (math_errno = E_MSIZE). If there is not enough dynamic memory, an "Insufficient dynamic memory available" error message is generated (math_errno = E_MALLOC).

SEE ALSO

Cmxalloc(), valloc(), Cvalloc().

```
#include "mathlib.h"
#include <stdio.h>
Real MAT1[2][2] = { 1, 2, 3, 4 } ;
main()
{
Real_Matrix  mat1,mat2;
unsigned i,j,rows=2,cols=2;

        mat2 = mxalloc(NULL, rows, cols);  /* allocate real matrix mat2    */
        mat1 = mxalloc(MAT1, rows, cols);  /* allocate and initialize mat1 */

                /* now initialize real matrix mat2 */
        mat2[0][0] = 1; mat2[0][1] = 2; mat2[1][0] = 3; mat2[1][1] = 4;

/* the matrices mat1[][] and mat2[][] should be equal  */

        for (i=0; i<rows; i++) {
            for (j=0; j<cols; j++) {
                printf("mat1[%d][%d] = %3.1f, mat2[%d][%d] = %3.1f\n",
                        i, j, mat1[i][j], i, j, mat2[i][j]);
            }
        }
        mxfree(mat1);    /* free matrix mat1[][]  */
        mxfree(mat2);    /* free matrix mat2[][]  */
        return 0;
}
```

mxcentro

```
#include "mathlib.h"
void mxcentro(Real_Matrix x, unsigned rows, unsigned cols, Real xvalue, Real yvalue);
```

DESCRIPTION
Sometimes the indices of the points in a matrix are more important than the actual values stored at those points. For example, in image processing a point process may occupy less than one pixel, or it may be in between pixels. One is often more interested in the coordinates of the centroid than the actual value of the centroid (e. g., a pixel tracker). Centroiding techniques are ideal for these applications. Because of the focus on the row and column locations, centroiding is sometimes referred to as inverse interpolation.

The *mxcentro* function centroids the unimodal matrix x. The assumption is that the true data represents a unimodal or point process and that x is a quantized version of the process. That is, it is assumed that x has a single maximum and is monotonically decreasing below this value. This function uses a 5×5 modified rectangular centroid around the maximum value of the matrix. An assumption of circular symmetry is made, which excludes the bins on each corner of the rectangle (i. e., only the central 21 bins are used).

The *rows* and *cols* arguments specify the number of rows and columns in matrix x. The *xvalue* and *yvalue* arguments specify the addresses of variables of type **Real**, where *mxcentro* stores the interpolated row (i. e., *xvalue*) and column (i. e., *yvalue*) values, where the actual maximum occurs in the data represented by matrix x.

The *mxcentro* function is ideal for improving the frequency resolution of the 2-D power spectrum in single-tone frequency tests. Similarly, it could be used for improving estimates of the amount of spatial misregistration between images as measured by a 2-D correlation procedure.

RETURNS
The interpolated row and column representing the 2-D centroid of matrix x is returned by *xvalue* and *yvalue*. These are best estimates of the row and column values at which the true maximum of the unimodal process represented by matrix x occurs.

SEE ALSO
v_centro()

REFERENCE
F. B. Hildebrand (1974).

```c
#include <stdio.h>
#include <stdlib.h>
#include <math.h>
#include "mathlib.h"


main()
{
unsigned i, j, imax, jmax, rows=8, cols=8;
double sigma=1.0, r0=4.3, c0=3.1;
double numerator,denominator,improvement;
Real xval,yval,maximum;
Real_Matrix m;
/* Find 2-D centroid of 2-D Gaussian pulse centered at (r0,c0) */
    m = mxalloc(NULL, rows, cols);
    for(i=0; i<rows; i++) {      /* create a 2-dimensional Gaussian pulse  */
        xval = (i-r0)*(i-r0);    /* with center at (r0,c0) and with a      */
        for(j=0; j<cols; j++) { /* variance of sigma                      */
            yval = (j-c0)*(j-c0);
            m[i][j] = exp(-(xval+yval)/(2*sigma));
        }
    }
    maximum = mxmaxval(m, rows, cols, &imax, &jmax); /* find max value     */
    printf("max value in matrix is %f\n", maximum);
    mxcentro(m, rows, cols, &xval, &yval);          /* find centroid of m[][] */
    printf("max occurs at row =%2d col =%2d\n", imax, jmax);
    printf("true center of pulse is at bin x =%6.3f y = %6.3f\n",r0,c0);
    printf("centroid estimate of max bin is x =%6.3f y = %6.3f\n",xval,yval);
    denominator=(r0-xval)*(r0-xval) + (c0-yval)*(c0-yval);
    numerator=(r0-imax)*(r0-imax) + (c0-jmax)*(c0-jmax);
    improvement = sqrt(numerator/denominator);
    printf("factor of improvement over pixel resolution=%5.2f\n",improvement);
    mxfree(m);
    return 0;
}
```

PROGRAM OUTPUT

```
max occurs at row = 4 col = 3
true center of pulse is at bin x = 4.300 y =  3.100
centroid estimate of max bin is x = 4.305 y =  3.049
factor of improvement over pixel resolution= 6.20
```

mxcopy

> #include "mathlib.h"
> void mxcopy(Real__Matrix *dest*, Real__Matrix *src*, unsigned *rows*, unsigned *cols*);

DESCRIPTION

The *mxcopy* function copies the real matrix *src* to the real matrix *dest*. The *rows* and *cols* arguments specify the number of rows and columns in each matrix.

The *dest* and *src* matrices must be allocated by the *mxalloc* function prior to calling the *mxcopy* function.

RETURNS

None.

SEE ALSO

Cmxcopy(), vcopy(), Cvcopy(), Cmxdup(), mxdup(), Cvdup(), vdup().

```
#include "mathlib.h"
#include <stdio.h>
Real SRC[2][2] = { 1, 2, 3, 4 } ;
main()
{
Real_Matrix  src,dest;
unsigned i,j,rows=2,cols=2;

        dest = mxalloc(NULL, rows, cols); /* allocate real matrix dest[][]   */
        src = mxalloc(SRC, rows, cols);   /* allocate and initialize src[][] */

        mxcopy(dest, src, rows, cols);   /* now copy the real matrix src[][] */

/* the matrices src[][] and dest[][] should be equal  */

        for (i=0; i<rows; i++) {
            for (j=0; j<cols; j++) {
                printf("src[%d][%d] = %3.1f, dest[%d][%d] = %3.1f\n",
                       i, j, src[i][j], i, j, dest[i][j]);
            }
        }
        mxfree(src);     /* free matrix src[][]   */
        mxfree(dest);    /* free matrix dest[][]  */
        return 0;
}
```

mxdeterm

#include "mathlib.h"
Real mxdeterm(Real__Matrix *m*, unsigned *n*);

DESCRIPTION

The *mxdeterm* function computes the determinant of the real matrix m, where m contains n rows and n columns.

The real matrix m must be allocated by *mxalloc* prior to calling the *mxdeterm* function.

The user may select the option of whether or not to write over the input matrix m with in-place computations. If memory is a limitation, and m is no longer needed, this may prove useful. This option can be selected by setting the global variable **useinput_ = 1** prior to calling *mxdeterm*. The default option is to not destroy m (i. e., **useinput_ = 0**).

The *mxdeterm* function can be used to determine if a system of simultaneous linear equations has a solution. A solution to the system of equations exists only if the determinant of the system matrix is nonzero. Thus, the case of no solution occurs when the system determinant is zero (i. e., the system matrix is singular, or noninvertible).

Determinants of real matrices have an interesting geometrical interpretation. If m is two-dimensional, the absolute value of its determinant is equal to the area of the parallelogram formed by its row (or column) vectors. Similarly, if m is three-dimensional, the absolute value of its determinant is equal to the volume of the parallelepiped formed by its row (or column) vectors.

RETURNS

A real value representing the determinant of the n by n real matrix m if successful (also stored in global variable **determ_**), or 0 if an error occurs. If the determinant is 0, then a "Singular matrix" error is generated (math_errno = E__MSING).

SEE ALSO

Cmxdeter().

REFERENCE

J. H. Wilkinson (1965).

```
#include "mathlib.h"
#include <stdio.h>
#include <math.h>
Real M[3][3] = { 1,0,0,  0,1,0,  0,0,1 } ;
main()
{
Real det;
Real_Matrix m;

/* use the determinant to compute the volume of the cube formed
   by unit vectors along the x, y, and z axes. i.e., Find the
   determinant of the 3 by 3 identity matrix                        */

       m = mxalloc(M, 3, 3);  /* allocate & initialize system matrix  */

       det = mxdeterm(m, 3);    /* now find the determinant of m[][]   */

       /* the answer should be : det = 1                              */

       printf("det = %f\n", det);
       mxfree(m);    /* free matrix m[][] */
       return 0;
}
```

mxdup

#include "mathlib.h"
Real_Matrix mxdup(Real_Matrix m, unsigned *rows*, unsigned *cols*);

DESCRIPTION

The *mxdup* function makes a duplicate copy of the real matrix m. The *rows* and *cols* arguments specify the number of rows and columns in the matrix.

The real matrix m must be allocated by *mxalloc* prior to calling the *mxdup* function. The *mxdup* function automatically allocates memory for the duplicate copy.

RETURNS

A *rows* by *cols* real matrix that is a duplicate copy of m if successful, or NULL if an error occurs. If there is not enough dynamic memory, an "Insufficient dynamic memory available" error message is generated (math_errno = E_MALLOC).

SEE ALSO

Cmxdup(), vdup(), Cvdup(), Cmxcopy, mxcopy(), vcopy(), Cvcopy().

```
#include "mathlib.h"
#include <stdio.h>
Real M1[2][2] = { 1, 2, 3, 4 } ;
main()
{
Real_Matrix  m1,m2;
unsigned i,j,rows=2,cols=2;
        m1 = mxalloc(M1, rows, cols);  /* allocate and initialize m1[][]  */
        m2 = mxdup(m1, rows, cols);    /* now duplicate the matrix m1[][]  */
/* the matrices m1[][] and m2[][] should be equal  */
        for (i=0; i<rows; i++) {
            for (j=0; j<cols; j++) {
                printf("m1[%d][%d] = %3.1f, m2[%d][%d] = %3.1f\n",
                        i, j, m1[i][j], i, j, m2[i][j]);
            }
        }
        mxfree(m1);    /* free matrix m1[][] */
        mxfree(m2);    /* free matrix m2[][] */
        return 0;
}
```

mxeigen

```
#include "mathlib.h"
Real_Vector mxeigen(Real_Matrix m, unsigned n, unsigned vflag);
```

DESCRIPTION

Eigensystems play a crucial role in the theory of electrical and mechanical resonance, and also in the theory of statics. Eigenvalues and eigenvectors can be used to determine the stability of feedback systems and to control the convergence of associated tracking algorithms. The concepts of similarity transforms and matrix diagonalization are of fundamental importance to eigenanalysis, because the eigenvalues are invariant to such transforms.

A nonsingular $n \times n$ real symmetric matrix has n real eigenvalues and n distinct eigenvectors. The cyclic Jacobi method consists of a sequence of similarity transforms designed to diagonalize a real symmetric matrix. This technique derives an orthogonal transformation matrix whose columns are the desired n eigenvectors. The n eigenvalues are the diagonal elements of the diagonalized matrix.

The *mxeigen* function computes the eigenvalues, and optionally the eigenvectors of a **real symmetric** matrix.

The n argument specifies the number of rows and columns in the square symmetric matrix m.

The *vflag* argument specifies whether or not to compute the eigenvectors in addition to the eigenvalues. If *vflag* is 0, the eigenvectors are not computed. If *vflag* is nonzero, the eigenvectors are computed and returned in matrix m. In this case, each column of matrix m will contain one of the eigenvectors. The eigenvalues and associated eigenvectors are sorted in descending order.

RETURNS

A real vector of length n representing the ordered eigenvalues of the real symmetric matrix m if successful, or NULL if an error occurs.

SEE ALSO

Cmxeigen().

REFERENCE

J. H. Wilkinson (1965).

```
#include <math.h>
#include <stdio.h>
#include "mathlib.h"

Real M[4][4] = {  120,  80,  40,  -16,
                   80, 120,  16,  -40,
                   40,  16, 120,  -80,
                  -16, -40, -80,  120 };
main()
{
Real_Matrix eigenvectors;
Real_Vector eigenvalues;
unsigned i, vflag = 1, n=4;

/* Find the eigenvalues and eigenvectors of A[][] */

    eigenvectors = mxalloc(M, n, n);
    eigenvalues = mxeigen(eigenvectors, n, vflag);

    printf("eigenvalues        eigenvectors \n");
    for (i=0;i<n; i++){
        printf("%8.3f    <%7.4f, %7.4f, %7.4f, %7.4f>\n",
                eigenvalues[i],eigenvectors[i][0],
                    eigenvectors[i][1],eigenvectors[i][2],eigenvectors[i][3]);
    }
    vfree(eigenvalues);
    mxfree(eigenvectors);
    return 0;
}
```

PROGRAM OUTPUT

```
    eigenvalues        eigenvectors
     256.000   < 0.5000, -0.5000, -0.5000, -0.5000>
     144.000   < 0.5000, -0.5000,  0.5000,  0.5000>
      64.000   < 0.5000,  0.5000, -0.5000,  0.5000>
      16.000   <-0.5000, -0.5000, -0.5000,  0.5000>
```

mxfree

#include "mathlib.h"
void mxfree(Real__Matrix m);

DESCRIPTION

The *mxfree* function frees the memory allocated to the real matrix m. The memory for the real matrix m must have been allocated by the *mxalloc* function prior to calling the *mxfree* function.

RETURNS

None.

SEE ALSO

Cmxfree(), vfree(), Cvfree(), mathfree().

```
#include "mathlib.h"
#include <stdio.h>
Real M1[2][2] = { 1,2,3,4 };
main()
{
Real_Matrix m,m1;
int i,j;

        m = mxalloc(NULL, 2, 2); /* allocate dynamic memory for m[][]     */
        m1 = mxalloc(M1, 2, 2);  /* initialize static array m1[][] and
                                    allocate dynamic memory for the 1-D
                                    array of pointers to its rows         */

        for (i=0; i<2; i++) {
                for (j=0; j<2; j++) {
                        m[i][j] = m1[i][j]; /* set m[][] equal to m1[][] */
                        printf("m[%d][%d] = %f\n", i, j, m[i][j]);
                }
        }
          /* now free all the dynamic memory occupied by m[][] and m1[][] */
        mxfree(m);  /*  free the dynamic memory occupied by m[][]         */
        mxfree(m1); /*  free the dynamic memory used by the 1-D array of  */
                /*  pointers to the rows of m1[][]                        */
        return 0;
}
```

mxhisto

```
#include "mathlib.h"
Real__Vector mxhisto(Real__Matrix m, unsigned rows, unsigned cols,
                     unsigned nhist, double *dx, double *x);
```

DESCRIPTION

A histogram is a plot of the frequency of occurrence of events. The bar charts used by financial analysts are the most common examples of histograms. Histograms are also very useful in image processing for categorizing the distribution of grey-scale information.

The *mxhisto* function computes the histogram of matrix m. The *rows* and *cols* arguments specify the number of rows and columns in matrix m.

The *nhist* argument specifies the number of desired bins in the resulting histogram (i. e., the length of the returned vector).

The dx argument specifies the address of a double precision variable where *mxhisto* stores the change in x between each bin of the histogram. The dx value is calculated as $(\text{MAX}(m) - \text{MIN}(m))/nhist$.

The x argument specifies the address of a double precision variable where *mxhisto* stores the starting x value for the histogram. The starting x value corresponds to the center of the first bin in the histogram and is calculated as $\text{MIN}(m) + dx/2$.

The histogram vector is automatically allocated as the return value of this function and should be freed when it is no longer needed.

RETURNS

A real vector of length *nhist* representing the histogram of the matrix m if successful, or NULL if an error occurs. If all the values in m are identical or if *nhist* is 0, the error message "Argument is outside the domain of the function" is generated (math__errno = E__DOMERR).

```
#include<stdio.h>
#include <stdlib.h>
#include <math.h>
#include "mathlib.h"

Real M[2][6] = { 1, 2, 2, 3, 3, 3,
                 4, 4, 4, 4, 5, 5 };
main()
{
unsigned i,rows=2,cols=6, nhist = 13;
double dx, xstart;
Real_Vector hist;
Real_Matrix m;

/* Find the histogram of array M[][] */

    m = mxalloc(M, rows, cols);
    hist = mxhisto(m, rows, cols, nhist, &dx, &xstart);
    printf("bin #   x-value of bin   # of bin occurrences\n");
    for(i=0; i<nhist; i++) {
        printf("%2d          %7.5f              %4.0f\n", i,xstart+i*dx,hist[i]);
    }
    vfree(hist);
    mxfree(m);
    return 0;
}
```

PROGRAM OUTPUT

bin #	x-value of bin	# of bin occurrences
0	1.15385	1
1	1.46154	0
2	1.76923	0
3	2.07692	2
4	2.38462	0
5	2.69231	0
6	3.00000	3
7	3.30769	0
8	3.61538	0
9	3.92308	4
10	4.23077	0
11	4.53846	0
12	4.84615	2

mxident

#include "mathlib.h"

Real_Matrix mxident(unsigned n);

DESCRIPTION

The *mxident* function generates a square, n by n, identity matrix. An identity matrix has elements that are unity along the diagonal and zero elsewhere, for example:

$$m[5][5] = \begin{vmatrix} 1 & 0 & 0 & 0 & 0 \\ 0 & 1 & 0 & 0 & 0 \\ 0 & 0 & 1 & 0 & 0 \\ 0 & 0 & 0 & 1 & 0 \\ 0 & 0 & 0 & 0 & 1 \end{vmatrix}$$

The identity matrix is sometimes called the unit matrix and is often denoted by $I_{n \times n}$. Multiplication of a matrix by the identity matrix produces the original matrix.

The *mxident* function calls the *mxinit* function.

RETURNS

A real matrix of n rows and n columns representing the identity matrix $I_{n \times n}$ if successful, or NULL if an error occurs.

SEE ALSO

mxinit().

```
#include "mathlib.h"
#include <stdio.h>
main()
{
Real_Matrix m;
unsigned i, j, n=3;
    m = mxident(n);    /*   now generate the identity matrix m[][]      */
    printf("the following is the 3x3 identity matrix:\n\n");
    for(i=0; i < n; i++) {
        for(j=0; j < n; j++) printf("  %3.1f ", m[i][j]);
        printf("\n");
    }
    mxfree(m); /*  free m[][]  */
    return 0;
}
```

mxinit

```
#include "mathlib.h"
void mxinit(Real_Matrix m, unsigned rows, unsigned cols, Real value);
```

DESCRIPTION
The *mxinit* function initializes each element of the real matrix *m* with the real scalar specified by *value*.

The *rows* and *cols* arguments specify the number of rows and columns in the matrix. The real matrix *m* must be allocated by *mxalloc* prior to calling the *mxinit* function.

RETURNS
None.

SEE ALSO
Cmxinit(), Cvinit(), vinit().

```
#include "mathlib.h"
#include <stdio.h>
main()
{
Real_Matrix m;
int i,j;

        m = mxalloc(NULL, 2, 2);        /* allocate matrix m[][]          */
        mxinit(m, 2, 2, 0);             /* set matrix m[][] to zero       */
        m[0][0] = m[1][1] = 1;          /* set diagonal terms to 1        */

        printf("the following is the real 2 by 2 identity matrix:\n");
        for (i=0; i<2; i++) {
                for (j=0; j<2; j++) {
                        printf(" %f    ", m[i][j]);
                }
                printf("\n");
        }
        return 0;
}
```

mxinv

#include "mathlib.h"
Real__Matrix mxinv(Real__Matrix m, unsigned n);

DESCRIPTION

The *mxinv* function computes the inverse of the real matrix m.

The n argument specifies the number of rows and columns in the square matrix m. The real matrix m must be allocated by the *mxalloc* function prior to calling the *mxinv* function. The resulting real matrix representing the inverse of m is automatically allocated by the *mxinv* function.

The user may select the option of whether or not to write over the input matrix m with in-place computations. If memory is a limitation, and m is no longer needed, this may prove useful. This option can be selected by setting the global variable **useinput__ = 1** prior to calling *mxinv*. The default option is to not destroy the input matrix m (i. e., **useinput__ = 0**).

Matrix inversion can be used to solve sets of linear equations. Given n equations with n unknowns, the linear set of equations can be written as:

$$M * v = b$$

If M^{-1} exists, then the solution vector v can be found by premultiplying both sides of the above equation by M^{-1}. Since $M^{-1} * M$ is equal to the identity matrix, the solution is

$$v = M^{-1} * b$$

Although matrix inversion is a convenient method for solving systems of linear equations, it is not the most efficient, nor the most accurate. Direct matrix reduction techniques such as LU decomposition (see the *Clineqn* and *lineqn* functions) and Gaussian elimination often provide superior speed and accuracy. This is especially true for large matrices that are nearly singular. However, for small matrices, matrix inversion is often adequate.

RETURNS

An n by n real matrix representing the inverse of the real matrix m if successful, or NULL if an error occurs. If the inverse of m exists, the global variable **determ__** is set equal to the determinant of the resulting matrix. Otherwise, a "Singular matrix" error is generated (math_errno = E__MSING).

SEE ALSO

Cmxinv(), Cmxinv22(), Cmxinv33(), mxinv22(), mxinv33(), Clineqn(), lineqn().

```
#include "mathlib.h"
#include <stdio.h>
Real M[2][2] = { 1,2, 3,4 };
main()
{
int i,j;
Real_Matrix m,minv;

        m = mxalloc(M, 2, 2); /* allocate & initialize 2 by 2 system matrix */
        minv = mxinv(m, 2);            /* now find the inverse of m[][]     */
        /* the answer should be :  */

        /*   minv[0][0] = -2.000000  minv[0][1] = 1.000000        */
        /*   minv[1][0] = 1.500000   minv[1][1] = -0.500000       */

        for (i=0; i<2; i++) {
                for (j=0; j<2; j++) {
                        printf("minv[%d][%d] = %f\n", i, j, minv[i][j]);
                }
        }
        mxfree(m);        /* free matrix m[][]     */
        mxfree(minv);     /* free matrix minv[][] */
        return 0;
}
```

mxinv22

> #include "mathlib.h"
>
> Real__Matrix mxinv22(Real__Matrix m);

DESCRIPTION

The *mxinv22* function computes the inverse of the two by two square real matrix m.

The real matrix m must be allocated by the *mxalloc* function prior to calling the *mxinv22* function. The resulting real matrix representing the inverse of m is automatically allocated by the *mxinv22* function.

The user may select the option of whether or not to write over the input matrix m with in-place computations. If memory is a limitation, and m is no longer needed, this may prove useful. This option can be selected by setting the global variable **useinput__ = 1** prior to calling *mxinv22*. The default option is to not destroy the input matrix m (i. e., **useinput__ = 0**).

The *mxinv22* function is specific to two by two matrices and has some size and speed advantages over the more general *mxinv* function.

RETURNS

A two by two real matrix representing the inverse of the real matrix m if successful, or NULL if an error occurs. If the inverse of m exists, the global variable **determ__** is set equal to the determinant of the resulting matrix. Otherwise, a "Singular matrix" error is generated (math_errno = E__MSING).

SEE ALSO

Cmxinv(), Cmxinv22(), Cmxinv33(), mxinv(), mxinv33(), Clineqn(), lineqn().

```
#include "mathlib.h"
#include <stdio.h>
Real M[2][2] = { 1, 2, 3,4 };
main()
{
int i,j;
Real_Matrix m,minv;

        m = mxalloc(M, 2, 2); /* allocate & initialize 2 by 2 system matrix */
        minv = mxinv22(m);                      /* now find the inverse of m[][]   */


        /* the answer should be :  */

        /*  minv[0][0] = -2.000000  minv[0][1] = 1.000000     */
        /*  minv[1][0] = 1.500000   minv[1][1] = -0.500000    */

        for (i=0; i<2; i++) {
                for (j=0; j<2; j++) {
                        printf("minv[%d][%d] = %f\n", i, j, minv[i][j]);
                }
        }
        mxfree(m);        /* free matrix m[][]    */
        mxfree(minv);     /* free matrix minv[][] */
        return 0;
}
```

mxinv33

#include "mathlib.h"
Real_Matrix mxinv33(Real_Matrix m);

DESCRIPTION

The *mxinv33* function computes the inverse of the three by three square real matrix m.

The real matrix m must be allocated by the *mxalloc* function prior to calling the *mxinv33* function. The resulting real matrix representing the inverse of m is automatically allocated by the *mxinv33* function.

The user may select the option of whether or not to write over the input matrix m with in-place computations. If memory is a limitation, and m is no longer needed, this may prove useful. This option can be selected by setting the global variable **useinput_** = 1) prior to calling *mxinv33*. The default option is to not destroy the input matrix m (i. e., **useinput_** = 0).

The *mxinv33* function is specific to three by three matrices and has some size and speed advantages over the more general *mxinv* function.

RETURNS

A three by three real matrix representing the inverse of the real matrix m if successful, or NULL if an error occurs. If the inverse of m exists, the global variable **determ_** is set equal to the determinant of the resulting matrix. Otherwise, a "Singular matrix" error is generated (math_errno = E_MSING).

SEE ALSO

Cmxinv(), Cmxinv22(), Cmxinv33(), mxinv(), mxinv22(), Clineqn(), lineqn().

```
#include "mathlib.h"
#include <stdio.h>
Real M[3][3] = { 1,0,3, 4,5,0, 0,8,9 };
main()
{
int i,j;
Real_Matrix m,minv;

        m = mxalloc(M, 3, 3); /* allocate & initialize 3 by 3 system matrix */

        minv = mxinv33(m);                    /* now find the inverse of m[][]     */

        /* the answer should be :  */

/* minv[0][0] = 0.319149  minv[0][1] = 0.170213  minv[0][2] = -0.106383 */
/* minv[1][0] = -0.255319 minv[1][1] = 0.063830  minv[1][2] = 0.085106  */
/* minv[2][0] = 0.226950  minv[2][1] = -0.056738 minv[2][2] = 0.035461  */

        for (i=0; i<3; i++) {
                for (j=0; j<3; j++) {
                        printf("minv[%d][%d] = %f\n", i, j, minv[i][j]);
                }
        }
        mxfree(m);        /* free matrix m[][]     */
        mxfree(minv);     /* free matrix minv[][] */
        return 0;
}
```

mxmaxval

#include "mathlib.h"
Real mxmaxval(Real__Matrix *m*, **unsigned** *rows*,
        unsigned *cols*, **unsigned** *imx, **unsigned** *jmx);

DESCRIPTION

The *mxmaxval* function finds the largest element in the real matrix *m*.

The *rows* and *cols* arguments specify the number of rows and columns in the matrix.

The *imx argument is the address of an unsigned integer where *mxmaxval* stores the row number of the largest value. The *jmx argument is the address of an unsigned integer where *mxmaxval* stores the column number of the largest value.

RETURNS

The largest real value in the real matrix *m*. The row and column of the largest value is returned in *imx* and *jmx*, respectively.

SEE ALSO

Cmxmaxvl(), Cmxminvl(), Cvmaxval(), Cvminval(), mxminval(), vmaxval(), vminval().

```
#include "mathlib.h"
#include <stdio.h>
Real M[2][2] = { 1,2, 3,4 };
main()
{
Real_Matrix m;
unsigned imax,jmax;
Real biggest;

        m = mxalloc(M, 2, 2);        /* allocate and initialized matrix m[][] */

        biggest = mxmaxval(m, 2, 2, &imax, &jmax); /* find largest value      */

        printf("the maximum value is = %f\n", biggest);
        printf("it is in row %d and column %d\n", imax, jmax);
        mxfree(m); /* free matrix m[][] */
        return 0;
}
```

mxminval

> #include "mathlib.h"
> Real mxminval(Real__Matrix m, unsigned *rows*, unsigned *cols*, unsigned $*imn$,
>   unsigned $*jmn$);

DESCRIPTION

The *mxminval* function finds the smallest element in the real matrix m.

The *rows* and *cols* arguments specify the number of rows and columns in the matrix.

The $*imn$ argument is the address of an unsigned integer where *mxminval* stores the row number of the smallest value. The $*jmn$ argument is the address of an unsigned integer where *mxminval* stores the column number of the smallest value.

RETURNS

The smallest real value in the real matrix m. The row and column of the smallest value is returned in *imn* and *jmn*, respectively.

SEE ALSO

Cmxmaxvl(), Cmxminvl(), Cvmaxval(), Cvminval(), mxmaxval(), vmaxval(), vminval().

```
#include "mathlib.h"
#include <stdio.h>
Real M[2][2] = { 1,2, 3,4 };
main()
{
Real_Matrix m;
unsigned imin,jmin;
Real smallest;

        m = mxalloc(M, 2, 2);        /* allocate and initialized matrix m[][] */

        smallest = mxminval(m, 2, 2, &imin, &jmin); /* find smallest value    */

        printf("the minimum value is = %f\n", smallest);
        printf("it is in row %d and column %d\n", imin, jmin);
        mxfree(m); /* free matrix m[][] */
        return 0;
}
```

mxmul

#include "mathlib.h"
Real__Matrix mxmul(Real__Matrix $m1$**, Real__Matrix** $m2$**,**
           unsigned $n1$**, unsigned** $n2$**, unsigned** $n3$**);**

DESCRIPTION

The *mxmul* function multiplies the two real matrices $m1$ and $m2$.

The number of rows and columns in $m1$ is specified by $n1$ and $n2$ respectively (i. e., $m1[n1][n2]$). The number of rows and columns in $m2$ is specified by $n2$ and $n3$ respectively (i. e., $m2[n2][n3]$). The number of columns in matrix $m1$ must be equal to the number of rows in matrix $m2$.

The real matrices $m1$ and $m2$ must be allocated by *mxalloc* prior to calling the *mxmul* function. The resulting real matrix representing the product of $m1$ and $m2$ is automatically allocated by the *mxmul* function. The resulting matrix contains $n1$ rows and $n3$ columns.

RETURNS

An $n1$ by $n3$ real matrix representing the product of $m1$ and $m2$ if successful, or NULL if an error occurs.

SEE ALSO

Cmxmul(), Cmxmul1(), Cmxmul2(), mxmul1(), mxmul2().

```
#include "mathlib.h"
#include <stdio.h>
Real M1[2][2] = { 2,2, 2,2 };
Real M2[2][2] = { 3,3, 3,3 };
main()
{
int i,j;
Real_Matrix m,m1,m2;
/* multiply the following two matrices:                            */
/*        m[][] = m1[][] * m2[][]                                   */
/* where m1[][] and m2[][] are real and are given below:           */
      m1 = mxalloc(M1, 2, 2);  /* allocate a 2 by 2 matrix         */
      m2 = mxalloc(M2, 2, 2); /* allocate another 2 by 2 matrix    */
      m = mxmul(m1, m2, 2, 2, 2); /* multiply m1[][] by m2[][]     */

      /* the answer should be : m[i][j] = 12;  for all i and j     */

      for (i=0; i<2; i++) {
          for (j=0; j<2; j++) {
              printf("m[%d][%d] = %f\n", i, j, m[i][j]);
          }
      }
      mxfree(m);     /* free matrix m[][]  */
      mxfree(m1);    /* free matrix m1[][] */
      mxfree(m2);    /* free matrix m2[][] */
      return 0;
}
```

mxmul1

#include "mathlib.h"

Real__Matrix mxmul1(Real__Matrix $m1$, Real__Matrix $m2$, unsigned $n1$, unsigned $n2$, unsigned $n3$);

DESCRIPTION

The *mxmul1* function multiplies the transpose of the real matrix $m1$ by the real matrix $m2$ (i. e., $m1^T * m2$).

The number of rows and columns in $m1$ is specified by $n2$ and $n1$ respectively (i. e., $m1[n2][n1]$). The number of rows and columns in $m2$ is specified by $n2$ and $n3$ respectively (i. e., $m2[n2][n3]$). The number of rows in matrix $m1$ must be equal to the number of rows in matrix $m2$.

The real matrices $m1$ and $m2$ must be allocated by *mxalloc* prior to calling the *mxmul1* function. The resulting real matrix representing the product of $m1^T$ and $m2$ is automatically allocated by the *mxmul1* function. The resulting matrix contains $n1$ rows and $n3$ columns.

RETURNS

An $n1$ by $n3$ real matrix representing the product of $m1^T$ and $m2$ if successful, or NULL if an error occurs.

SEE ALSO

Cmxmul(), Cmxmul1(), Cmxmul2(), mxmul(), mxmul2().

```
#include "mathlib.h"
#include <stdio.h>
Real M1[2][3] = { 1,1,1,   2,2,2 };
Real M2[2][2] = { 1,1,   2,2 };
main()
{
int i,j;
Real_Matrix m,m1,m2;
/* multiply the following two matrices:                               */
/*        m[][] = TRANSPOSE { m1[][] } * m2[][]                        */
/* where m1[][] and m2[][] are real and are given below:              */
      m1 = mxalloc(M1, 2, 3);  /* allocate a 2 by 3 matrix            */
      m2 = mxalloc(M2, 2, 2); /* allocate a 2 by 2 matrix             */
      m = mxmul1(m1, m2, 3, 2, 2);/*multiply TRANSPOSE{m1[][]} by m2[][] */

/* the answer is :  m[i][j] = 5, for all i and j                      */

      for (i=0; i<3; i++) {
          for (j=0; j<2; j++) {
              printf("m[%d][%d] = %f\n", i, j, m[i][j]);
          }
      }
      mxfree(m);     /* free matrix m[][]  */
      mxfree(m1);    /* free matrix m1[][] */
      mxfree(m2);    /* free matrix m2[][] */
      return 0;
}
```

mxmul2

```
#include "mathlib.h"
Real_Matrix mxmul2(Real_Matrix m1, Real_Matrix m2, unsigned n1, unsigned n2,
                unsigned n3);
```

DESCRIPTION

The *mxmul2* function multiplies the real matrix $m1$ by the transpose of the real matrix $m2$ (i. e., $m1 * m2^T$).

The number of rows and columns in $m1$ is specified by $n1$ and $n2$ respectively (i. e., $m1[n1][n2]$). The number of rows and columns in $m2$ is specified by $n3$ and $n2$ respectively (i. e., $m2[n3][n2]$). The number of columns in matrix $m1$ must be equal to the number of columns in matrix $m2$.

The real matrices $m1$ and $m2$ must be allocated by *mxalloc* prior to calling the *mxmul2* function. The resulting real matrix representing the product of $m1$ and $m2^T$ is automatically allocated by the *mxmul2* function. The resulting matrix contains $n1$ rows and $n3$ columns.

RETURNS

An $n1$ by $n3$ real matrix representing the product of $m1$ and $m2^T$ if successful, or NULL if an error occurs.

SEE ALSO

Cmxmul(), Cmxmul1(), Cmxmul2(), mxmul(), mxmul1().

```
#include "mathlib.h"
#include <stdio.h>
Real M1[2][2] = { 1,1, 2,2 };
Real M2[3][2] = { 1,1, 2,2, 3,3 };
main()
{
int i,j;
Real_Matrix m,m1,m2;
/* multiply the following two matrices:                          */
/*        m[][] = m1[][] * TRANSPOSE { m2[][] }                   */
/* where m[][] and m1[][] are real and are given below:          */
       m1 = mxalloc(M1, 2, 2);  /* allocate a 2 by 2 matrix      */
       m2 = mxalloc(M2, 3, 2); /* allocate a 3 by 2 matrix       */
       m = mxmul2(m1, m2, 2, 2, 3);/*multiply m1[][] by TRANSPOSE{ m2[][]} */
/* the answer is :   m[0][0] = 2, m[0][1] = 4, m[0][2] = 6
                     m[1][0] = 4, m[1][1] = 8, m[1][2] = 12      */
       for (i=0; i<2; i++) {
           for (j=0; j<3; j++) {
                printf("m[%d][%d] = %f\n", i,j, m[i][j]);
             }
         }
       mxfree(m);     /* free matrix m[][]  */
       mxfree(m1);    /* free matrix m1[][] */
       mxfree(m2);    /* free matrix m2[][] */
       return 0;
}
```

mxscale

#include "mathlib.h"
Real__Matrix mxscale(Real__Matrix m, unsigned *rows*, unsigned *cols*, Real *value*);

DESCRIPTION

The *mxscale* function multiplies the real matrix m by the real scalar *value*.

The *rows* and *cols* arguments specify the number of rows and columns in the matrix. The real matrix m must be allocated by *mxalloc* prior to calling the *mxscale* function. The resulting real matrix representing the product of m and *value* is automatically allocated by the *mxscale* function.

The user may select the option of whether or not to write over the input matrix m with in-place computations. If memory is a limitation, and m is no longer needed, this may prove useful. This option can be selected by setting the global variable (**useinput__ = 1**) prior to calling *mxscale*. The default option is to not destroy the input matrix m (i. e., **useinput__ = 0**).

RETURNS

A *rows* by *cols* real matrix representing the product of the real matrix m with the real *value* if successful, or NULL if an error occurs.

SEE ALSO

CmxscalC(), CmxscalR(), CvscalC(), CvscalR(), vscale().

```
#include "mathlib.h"
#include <stdio.h>
Real M[2][2] = { 1,1, 1,1 };
main()
{
Real_Matrix m,m1;
unsigned rows=2,cols=2;
int i,j;
Real value;

    value = 2;   /* define a real scalar                            */
    m=mxalloc(M, rows, cols);  /* allocate and initialize m[][]     */
    m1 = mxscale(m, rows, cols, value);/* now scale real matrix m[][]    */
    /* the answer should be : m1[0][0] = 2;   m1[0][1] = 2;
                              m1[1][0] = 2;   m1[1][1] = 2;          */
    for (i=0; i<2; i++) {
        for (j=0; j<2; j++) {
            printf("m1[%d][%d] = %f\n", i, j, m1[i][j]);
        }
    }
    mxfree(m);    /* free matrix m[][]  */
    mxfree(m1);   /* free matrix m1[][] */
    return 0;
}
```

mxsub

#include "mathlib.h"
Real__Matrix mxsub(Real__Matrix $m1$**, Real__Matrix** $m2$**, unsigned** *rows*, **unsigned** *cols*);

DESCRIPTION

The *mxsub* function subtracts the real matrix $m2$ from the real matrix $m1$.

The *rows* and *cols* arguments specify the number of rows and columns in each matrix. The real matrices $m1$ and $m2$ must be allocated by *mxalloc* prior to calling the *mxsub* function. The resulting real matrix representing the difference between $m1$ and $m2$ is automatically allocated by the *mxsub* function.

The user may select the option of whether or not to write over the input matrix $m1$ with in-place computations. If memory is a limitation, and $m1$ is no longer needed, this may prove useful. This option can be selected by setting the global variable **useinput_ = 1** prior to calling *mxsub*. If this option is selected, the resulting matrix overwrites the input matrix $m1$. The default option is to not destroy the input matrix $m1$ (i. e., **useinput_ = 0**).

RETURNS

A *rows* by *cols* real matrix representing the difference between $m1$ and $m2$ if successful, or NULL if an error occurs.

SEE ALSO

Cmxsub(), Cmxadd(), mxadd(), Cvadd(), Cvsub(), vadd(), vsub().

```c
#include "mathlib.h"
#include <stdio.h>
#include <stdlib.h>
main()
{
int i,j;
Real_Matrix m,m1,m2;
/* subtract the following two matrices:                               */
/*                                                                    */
/*          m[][] = m1[][] - m2[][]                                   */
/*                                                                    */
/* where m1[][] and m2[][] are real and are given below:             */
      m1 = mxalloc(NULL, 2, 2);  /* allocate a 2 by 2 matrix         */
      m2 = mxalloc(NULL, 2, 2); /* allocate another 2 by 2 matrix    */
      m1[0][0] = 2;   m1[0][1] = 2;
      m1[1][0] = 2;   m1[1][1] = 2;
      m2[0][0] = 3;   m2[0][1] = 3;
      m2[1][0] = 3;   m2[1][1] = 3;

      m = mxsub(m1, m2, 2, 2); /* subtract m2[][] from m1[][]         */

      /* the answer should be : m[0][0] = -1;   m[0][0] = -1;
                                m[0][1] = -1;   m[0][1] = -1;         */
      for (i=0; i<2; i++) {
          for (j=0; j<2; j++) {
              printf("m[%d][%d] = %f\n", i, j, m[i][j]);
          }
      }
      mxfree(m);     /* free matrix m[][]  */
      mxfree(m1);    /* free matrix m1[][] */
      mxfree(m2);    /* free matrix m2[][] */
      return 0;
}
```

mxtrace

#include "mathlib.h"
Real mxtrace(Real__Matrix m, unsigned n);

DESCRIPTION

The *mxtrace* function computes the trace of the real matrix m. The trace of a matrix is the sum of its diagonal terms:

$$m[0][0] + m[1][1] + \ldots + m[n][n]$$

The n argument specifies the number of rows and columns in the square matrix m.

The real matrix m must be allocated by the *mxalloc* function prior to calling the *mxtrace* function.

The trace of a matrix can be used to normalize the matrix coefficients. If the matrix is nonsingular (i. e., the determinant is nonzero), then it can be diagonalized. It is interesting to note that the trace of the diagonalized matrix is always equal to the trace of the original matrix. That is, the trace of a matrix is invariant to matrix diagonalization.

RETURNS

A real value representing the sum of the diagonal elements of the real matrix m.

SEE ALSO

Cmxtrace().

```
#include "mathlib.h"
#include <stdio.h>
Real M[2][2] = { 1, 0, 0, 1};
main()
{
Real trace;
Real_Matrix m;
unsigned n=2;
        m = mxalloc(M, n, n);    /* allocate and initialize m[][]      */
        trace = mxtrace(m, n);  /* now compute the trace of m[][]      */
/*   the answer should be:   trace = 2;                                */
        printf("trace = %f\n", trace);
        mxfree(m); /*  free m[][]  */
        return 0;
}
```

mxtransp

#include "mathlib.h"
Real_Matrix mxtransp(Real_Matrix *m*, unsigned *rows*, unsigned *cols*);

DESCRIPTION
The *mxtransp* function forms the transpose of the real matrix m.

The *rows* and *cols* arguments specify the number of rows and columns in the matrix. The transpose is obtained by interchanging the rows and columns of matrix m. Therefore, the number of rows in the tranposed matrix is equal to *cols*, and the number of columns is equal to *rows*.

The real matrix m must be allocated by the *mxalloc* function prior to calling the *mxtransp* function. The resulting real matrix representing the transpose of m is automatically allocated by the *mxtransp* function.

If m is square, the user may select the option of whether or not to write the resulting output over the input matrix m. If memory is a limitation, and m is no longer needed, this may prove useful. This option can be selected by setting the global variable **useinput_ = 1** prior to calling *mxtransp*. The default option is to not destroy the input matrix m (i. e., **useinput_ = 0**).

RETURNS
A *cols* by *rows* real matrix representing the transpose of matrix m (i. e., m^T) if successful, or NULL if an error occurs.

SEE ALSO
Cmxtrans().

```
#include "mathlib.h"
#include <stdio.h>
Real M[2][3] = { 0,   1,   2,
                 3,   4,   5 };
main()
{
Real_Matrix m,mt;
unsigned i, j, rows=2, cols=3;

        m=mxalloc(M, rows, cols);  /* allocate and initialize matrix m[][] */
        mt = mxtransp(m, rows, cols); /* form the matrix transpose of m[][] */
/* the answer should be: mt[i][j] = m[j][i]; i=0,1,... cols; j=0,1,...,rows*/
        for(i=0; i<cols; i++) {
                for(j=0; j<rows; j++) {
                        printf("mt[%d][%d] = %5.2f, m[%d][%d] = %5.2f\n",
                                i, j, mt[i][j], j, i, m[j][i]);
                }
        }
        mxfree(m);   /* free matrix m[][]   */
        mxfree(mt);  /* free matrix mt[][]  */
        return 0;
}
```

newton

#include "mathlib.h"
double newton(double (*fx)(double x), double *x, double dx, unsigned *limit*, double *err*);

DESCRIPTION

One of the most common and difficult problems in mathematical analysis is the solution of nonlinear equations. For example, consider the following equation:

$$\sin(x) = 3 * x + 0.1$$

Although this is a very simple equation, there is no analytical method of solving for x. All attempts at a straightforward solution are of no avail. Numerical root-finding algorithms are ideal for problems such as these. The Newton–Raphson technique is a simple iterative approach for finding the x where $f(x) = 0$. The formula results from the first two terms of the Taylor series expansion at x_i:

$$f(x) = f(x_i) + f'(x_i) * (x - x_i) + \dots$$

Setting $f(x) = 0$ and solving for $x = x_{i+1}$ yields:

$$x_{i+1} = x_i - f(x_i)/f'(x_i)$$

In this form, the derivative $f'(x)$ is explicitly required. It is often more convenient to approximate the derivative:

$$f'(x_i) = [f(x_i) - f(x_i + dx)]/dx$$

where a good choice for dx is

$$dx = x_i * 1.e - 5$$

Including this approximation, the new algorithm becomes

$$x_{i+1} = x_i - dx * [-1 + f(x_i + dx)/f(x_i)]^{-1}$$

One very nice feature of the Newton–Raphson technique is that the number of required iterations is usually insensitive to the initial guess. This is because the step size is weighted by the derivative of the function. The worse the initial guess, the larger the initial weighting. The number of iterations depends more on the behavior of $f'(x)$ as $f(x)$ approaches zero. For example, finding the zeros of $\sin(x)$ requires very few iterations, since the slope of the sine function is maximum at the zero-crossings. However, finding the zeros of $\cos(x) - 1$ requires comparatively more iterations, since its derivative (and the weighted step size) vanishes at the zero-crossings. In either case, the number of required iterations is fairly independent of the initial guess.

The *newton* function finds a value x for which the user-defined function fx is equal to 0.

The user-defined function fx requires a single double precision argument x. The fx function must return a double precision value representing $f(x)$.

The x argument specifies the address of a double precision variable, the value of which is used as the initial guess. If the function fx has more than one zero crossing, the initial value of x will affect which zero crossing the *newton* function finds. When the *newton* function terminates, the value of x for which $fx = 0$ is also returned through this argument.

The dx argument specifies the change in x to use in approximating the derivative at each point during the search. A value of $1.0e-5$ is typically a reasonable value to use for dx.

The *limit* argument specifies the maximum number of iterations. The *newton* function automatically terminates after *limit* iterations. Since some functions may not have a value of x at which $fx = 0$, this provides a means to terminate the search. If *limit* iterations occur before a zero is found, the *newton* function returns the value of x for which fx is closest to zero.

The *err* argument specifies the absolute error tolerance. If at some point x, the absolute value of fx is less than *err*, then a zero crossing is considered to have been found.

SEE ALSO

p_roots().

RETURNS

A double precision value representing fx, which will typically be within *err* of 0. The value of x for which $|fx| < err$ is returned via the x argument.

REFERENCE

D. Kahaner, C. Moler, and S. Nash (1989).

```
#include "mathlib.h"
#include <stdio.h>
#include <math.h>
#include <stdlib.h>
/* example subroutine to evaluate f(x) = 0 */
double fx(double x)
{                                       /* zero occurs at x=-0.049989591... */
        return sin(x)-3*x-0.1;
}
main()
{
int limit;
double y, x, dx, err;

        x = -0.5;           /* inital guess for x */
        limit = 1000;       /* maximum number of iterations */
        dx = 1.e-5;         /* change in x for determining derivative */
        err = 1.e-15;       /* absolute error tolerance */
        y = newton(fx, &x, dx, limit, err);
        printf("x = %.9f, and y = f(x) = %.9f\n", x, y);
        return 0;

}
```

normal

$$\text{\#include "mathlib.h"}$$
$$\text{double normal(double } sigma, \text{ double } u);$$

DESCRIPTION

Random numbers are frequently used to predict expected outcomes of statistical events. Many random events are too complex to evaluate directly and must be characterized via Monte Carlo simulations. These simulations usually require a large number of random trials. The specific outcome of any one trial is relatively insignificant. The investigator is more interested in the average statistical behavior across all of the trials. The generation of independent pseudorandom number sequences is essential to these statistical simulations.

The *normal* function generates a random number from a Normal (Gaussian) distribution with mean u and standard deviation *sigma*. The method sums a sequence of 12 independent random variables X_i, which are uniformly distributed over the interval $(0, 1)$:

$$Y = X_1 + X_2 + X_3 + \ldots + X_{12}$$

The mean and variance of the X_i are given by

$$u_x = E[X_i] = 1/2, \quad \text{and} \quad VAR[X_i] = E[(X_i - u_x)^2] = 1/12$$

From the central limit theorem, the standard normal random variable, Z, with zero mean and unity variance can be approximated by

$$Z = Y - 6$$

The desired normal random variable, W, is obtained with the simple transformation:

$$W = Z * sigma + u = (Y - 6) * sigma + u$$

Different random number sequences may be generated by calling the standard C function *srand* prior to calling the *normal* function. The *srand* function requires a single unsigned integer argument which seeds the random number generator. A different random number sequence is generated for each different seed value.

RETURNS

A double precision value representing a random number generated from a normal (Gaussian) distribution with a mean of u and a standard deviation of *sigma*.

SEE ALSO

poisson(), urand(), nprob(), invprob().

REFERENCES

L. R. Rabiner and B. Gold (1975, pp. 570–571); P. L. Meyer (1970); D. E. Knuth (1981).

```
#include "mathlib.h"
#include <stdio.h>
#include <math.h>
#include <stdlib.h>
main()
{
int i,n;
double stdev,amean,mean,var,vout;

        n=100;
        amean=8.0;
        stdev=3.0;
        srand(1); /* do'nt seed the uniform random generator with the clock */
        mean=var=0.0;
        for (i=0; i < n; i++) {
            vout=normal(stdev,amean);
            mean += vout;
            var += vout*vout;
        }
        var = var/(n-1) - (mean/n) * mean/(n-1);
        mean /= n;

        printf("            Normal Random Variable Test (100 trials):\n\n");
        printf("    true mean = %f,      true standard deviation = %f\n",
                                                        amean, stdev);
        printf("observed mean = %f,  observed standard deviation = %f\n",
                                                        mean, sqrt(var));

        return 0;
}
```

nprob

#include "mathlib.h"
double nprob(double $x0$);

DESCRIPTION

The standard normal distribution (the bell-curve) is the most frequently used statistic in applied probability. The use of the normal distribution usually requires consulting tables of values. Although this may be adequate for a few hand calculations, these tables are cumbersome and are difficult to incorporate into computer programs.

The *nprob* function provides a closed form alternative to the standard normal tables by approximating the cumulative normal distribution.

Given $x0$, the *nprob* function determines the probability that a random variable X will be less than or equal to $x0$. This probability may be expressed as follows:

$$P(X \leq x0) = \frac{1}{\sqrt{2\pi}} \int_{-\infty}^{x0} e^{-x^2/2} \, dx$$

Since this integral cannot be easily evaluated, it is approximated for $x0 \geq 0$ with a truncated power series:

$$P(X \leq x0) = 1 - f(x0) * \left(b_1 t + b_2 t^2 + b_3 t^3 + b_4 t^4 + b_5 t^5\right)$$

where $t = (1 + 0.2316419 * x0)^{-1}$ and $f(x0) = 0.39894228 * e^{-x0^2/2}$
and $b_1 = 0.31938153$ $b_2 = -0.356563782$
    $b_3 = 1.781477937$ $b_4 = -1.821255978$
    $b_5 = 1.330274429$
If $x0 < 0$, then the complementary power series is used:

$$P(X \leq x0) = f(x0) * b_1 t + b_2 t^2 + b_3 t^3 + b_4 t^4 + b_5 t^5)$$

The error of the approximation is less than $7.5e - 8$.

The *nprob* function calls the *pseries* function.

RETURNS

A double precision value representing the cumulative probability that a normally distributed random variable X is less than or equal to $x0$.

SEE ALSO

invprob().

REFERENCE

Abramowitz and Stegun (1964, p. 932, eq. 26.2.17).

```
#include "mathlib.h"
#include <stdio.h>
main()
{
double x0=0;
/* if x is a normal random variable, and x0=0, find the probability P(X <= x0)*/
    printf("P(X <= x0) = %f\n", nprob(x0)); /* the answer is P(X <= 0) = 0.5  */
    return 0;
}
```

permut

#include "mathlib.h"
double permut(int n, int r);

DESCRIPTION

The *permut* function computes the number of different ways of choosing r objects from a total of n objects, where $r \le n$.

The order of selection is important, since selecting object 1 followed by object 2 is considered distinct from selecting object 2 followed by object 1. The number of such permutations, $_nP_r$, is given by

$$_nP_r = n!/(n-r)!$$

RETURNS

A double precision whole value representing the number of permutations of n objects taken r at a time.

SEE ALSO

combin(), fact().

```
#include "mathlib.h"
#include <stdio.h>
main()
{
double nPr;
int n=4,r=2;
/* find the number of different ways of choosing 2 things from 4      */
      nPr = permut(n, r);
      /* the answer should be nPr= 4!/(4-2)! = 12 */
      printf("# of permutations of %d things taken %d at a time = %f\n",
            n, r, nPr);
      return 0;
}
```

poissdst

#include "mathlib.h"
double poissdst(int $k0$, double λ_T, Real $pb[\,]$);

DESCRIPTION

The Poisson distribution is useful in a large number of statistical applications. It is a discrete process in that only integer valued outcomes are possible. The Poisson distribution is often used to predict the number of discrete random events that will occur in a fixed time interval. This distribution has many applications in queueing theory. For example, the Poisson distribution can be used to predict automobile traffic flow (see the example below).

The *poissdst* function computes the cumulative probability that X or fewer events will occur in a given time interval T.

The $k0$ argument specifies the number of events X. We wish to compute the probability that $k0$ or fewer events will occur.

The λ_T argument specifies the total number of events that should occur in the time interval T. This is the average rate of occurrence of events multiplied by the time interval.

Given $k0$ and λ_T, the probability may be expressed as follows:

$$P(X \le k0) = \sum_{i=0}^{k0} \frac{(\lambda_T)^i e^{-\lambda_T}}{i!}$$

The $pb[\,]$ argument specifies an array of length $k0 + 1$ in which the *poissdst* function stores the probabilities that exactly $0, 1, 2, \ldots, k0$ events will occur. If these individual probabilities are not needed, then specify NULL for the $pb[\,]$ argument.

RETURNS

A double precision value representing the probability that $k0$ or fewer events will occur over a time interval in which the average number of events is λ_T. The probabilities that $0, 1, 2, \ldots, k0$ events will occur are returned via the $pb[\,]$ argument.

SEE ALSO

binomdst(), and hyperdst().

REFERENCE

P. L. Meyer (1970, pp. 159–170).

```
#include "mathlib.h"
#include <stdio.h>
#define RATE 3
#define COUNT 35
#define TIME 10
main()
{
int k0,i;
double lt;
Real pb[COUNT+1];
double cumulative;
/* if cars are arriving at a stop sign at a rate of RATE per minute,   */
/* compute the probability that COUNT or fewer will arrive in          */
/* TIME minutes.  Assume a Poisson process:                            */
     lt = RATE*TIME; /* this is average number arrivals expected       */
     k0 = COUNT;      /* this is the number of arrivals desired         */
     cumulative = poissdst(k0, lt, pb);/* compute Poisson distribution */
     printf("the odds that %d or less cars arrive is %f\n\n",
               k0, cumulative);
     for (i=0; i <= k0; i++) {
          printf("the odds that exactly %2d cars arrive is %f\n", i,pb[i]);
     }
     return 0;
}
```

poisson

```
#include "mathlib.h"
unsigned poisson(int pmean);
```

DESCRIPTION

Random numbers are frequently used to predict expected outcomes of statistical events. Many random events are too complex to evaluate directly and must be characterized via Monte Carlo simulations. These simulations usually require a large number of random trials. The specific outcome of any one trial is relatively insignificant. The investigator is more interested in the average statistical behavior across all the trials. The generation of independent pseudorandom number sequences is essential to these statistical simulations.

The Poisson distribution is very useful in modeling the number of discrete events that occur in a fixed time interval when the time between the events is exponentially distributed. For the Poisson distribution, the variance equals the mean.

The *poisson* function generates a random number from a Poisson distribution with a mean and variance of *pmean*.

The most straightforward method of generating a Poisson random integer is to sum a sequence of K independent random variables Y_i, which are exponentially distributed with unity mean and variance. K is incremented until the sum is equal to or greater than the desired mean *pmean*. At this point, the value of K is the desired Poisson random integer.

The sequence of exponentially distributed random variables, Y_i, is easily constructed from any uniform random number sequence X_i according to

$$Y_i = -\log_e(X_i)$$

It is easy to show that the sum of the exponential random variables, Y, has a mean and variance of K, where K is approximately *pmean*, since

$$VAR(Y) = K * VAR[Y_i] = K$$

and

$$E[Y] = K * E[Y_i] = K$$

where

$$Y = Y_1 + Y_2 + Y_3 + \ldots + Y_K$$

The method used by the *poisson* function is a slight variation of the technique just described. K uniform random variables, X_i, are multiplied together. K is incremented and the product declines until it is less than or equal to e^{-pmean}. At this point, the value of K is the desired Poisson random integer. This approach results from raising both sides of the last equation to the base of e:

$$e^{-Y} = X_1 * X_2 * X_3 * \ldots * X_K \geq e^{-\text{pmean}}$$

where $Y_i = -\log_e(X_i)$.

Different random number sequences may be generated by calling the standard C function *srand* prior to calling the *poisson* function. The *srand* function requires a single unsigned integer argument, which seeds the random number generator. A different random number sequence is generated for each different seed value.

RETURNS

An unsigned integer value representing a random number generated from a Poisson distribution with a mean and variance of *pmean*.

SEE ALSO

normal(), urand(), poissondst().

REFERENCES

P. L. Meyer (1970); D. E. Knuth (1981).

```
#include "mathlib.h"
#include <stdio.h>
#include <math.h>
#include <stdlib.h>
main()
{
int i,n,pmean;
double stdev,amean,mean,var;
unsigned vout;

    n=100;
    pmean=20;
    srand(2); /* seed the uniform random generator with the clock */
    mean=var=0.0;
    for (i=0; i < n; i++) {
        vout=poisson(pmean);
        mean += vout;
        var += vout*vout;
    }
    var = var/(n-1) - (mean/n) * mean/(n-1);
    mean /= n;
    printf("   Poisson Random Variable Test (100 trials):\n\n");
    printf("    true mean = %d            true variance = %d\n",
                                                pmean, pmean);
    printf("observed mean = %4.1f         observed variance = %4.1f\n",
                                                mean, var);
    return 0;
}
```

powspec

#include "mathlib.h"
Real__Vector powspec(Real__Vector v, unsigned nv, unsigned npw, Real__Vector w);

DESCRIPTION

The *powspec* function computes the power spectrum of the real vector v of length nv, using Welch's periodogram approach. Most modern spectrum analyzers use this approach. The signal length nv is divided into several nonoverlapping segments, each of length npw, where npw is a "power of two" and $npw \leq nv$. Each segment is Fourier transformed, and all the transforms are averaged, yielding the desired power spectrum.

The w argument specifies a window of length npw for windowing the data segments. If windowing is specified, a modified periodogram approach is used to compute the power spectrum. The *tdwindow* function provides several different windows for this purpose. If no windowing is desired, then NULL should be specified for the w argument. No windowing is equivalent to specifying a rectangular window.

If the input vector v is representative of a real sinusoid $A * \{\sin(2 * \pi * k * i/n)\}$, where $k = 1, 2, \ldots, npw/2$, $i = 0, 1, 2, \ldots, n - 1$, and $\pi = 3.141592654\ldots$, then the resulting vector representing the power spectrum will consist of two frequency components, one at the $(k + 1)$th element of the power spectrum and the other at the $(n - k + 1)$th element. With no windowing, both components are of magnitude $A^2/4$.

The *powspec* function calls the *fftrad2* function.

RETURNS

A real vector of length nv representing the power spectrum of the real vector v if successful, or NULL if an error occurs. If $nv < npw$, the error message "Length of data must be greater than or equal to FFT size" is generated (math_errno = E__FFTSIZE).

SEE ALSO

Cpowspec(), fftrad2(), fft42(), tdwindow().

REFERENCES

P. D. Welch (1970); A. V. Oppenheim and R. W. Schafer (1975).

```
#include "mathlib.h"
#include <stdio.h>
#include <stdlib.h>
#include <math.h>

main()
{
Real_Vector v;
unsigned npw=16, i, nv=128;
double noise;
Real_Vector w,spectrum1,spectrum2;

    v=valloc(NULL, nv);  /* allocate an array for data to be analyzed     */
    w = tdwindow(npw, 7);/*"7" is a Hamming window, see the tdwindow() routine*/
    for(i=0; i<nv; i++) {  /*  generate nv/npw cycles of a sinusoid+noise   */
        noise = 2*((double) rand()/RAND_MAX - 0.5); /* uniform from +1 to -1 */
        v[i] = sin(twopi_*4*i/(double) npw) + noise; /* signal = 4th harmonic */
    }
    spectrum1 = powspec(v, npw, npw, w); /* compute spectrum of one segment */
    spectrum2 = powspec(v, nv, npw, w);  /* compute spectrum entire data set*/
/*4th harmonic of 2nd spectrum should be higher&noise floor should be smoother*/
    printf("           spectrum of       spectrum of\n");
    printf("harmonic #   1st segment      all segments\n\n");
    for(i=0; i<npw; i++) { /*                                             */
     printf("   %2d         %9.6f         %9.6f\n",
                              i, spectrum1[i],spectrum2[i]);
    }
    vfree(v);          /* free v[]        */
    vfree(spectrum1);  /* free spectrum1[] */
    vfree(spectrum2);  /* free spectrum2[] */
    vfree(w);          /* free w[]        */
    return 0;
}
```

PROGRAM OUTPUT

harmonic #	spectrum of 1st segment	spectrum of all segments
0	0.000006	0.015357
1	0.016551	0.032928
2	0.042415	0.026181
3	0.076729	0.079332
4	0.176842	0.241440
5	0.057615	0.074847
6	0.015346	0.024337
7	0.020153	0.011717
8	0.006704	0.006853
9	0.020153	0.011717
10	0.015346	0.024337
11	0.057615	0.074847
12	0.176842	0.241440
13	0.076729	0.079332
14	0.042415	0.026181
15	0.016551	0.032928

p_roots

#include "mathlib.h"
Complex__Vector p__roots(Real *coeff* **[], int** *n*, **unsigned** *limit*);

DESCRIPTION

The roots of polynomials have been of great interest to mathematicians over the centuries. An nth order polynomial has, in general, n complex roots. Yet closed form solutions exist only for polynomials less than fifth order, and the solutions for third and fourth order polynomials are very cumbersome.

The Newton–Raphson method is a useful tool for the iterative evaluation of roots of differentiable functions (e. g., polynomials). Polynomials with real coefficients are analytic functions and are differentiable everywhere in the complex plane. When used in conjunction with the Cauchy–Riemann equations, the Newton–Raphson technique can be extended to evaluate complex roots of high order polynomials.

The *p_root* function evaluates the real and imaginary roots of polynomials with real coefficients using the Newton–Raphson method.

The *coeff* argument is an array of $n + 1$ real coefficients defining an nth order polynomial $f(x)$, where:

$$f(x) = coef[0] + x * coef[1] + x^2 * coef[2] + \ldots + x^{n-1} * coef[n-1]$$

The n argument specifies the order of the polynomial. The order must be in the range $0 < n \le 36$.

The *limit* argument specifies the maximum number of iterations allowed to evaluate any root (500 is typically a reasonable choice).

RETURNS

A complex vector of length n representing the real and imaginary roots of a polynomial with real coefficients *coeff* if successful, or NULL if an error occurs. The highest order coefficient must be nonzero, otherwise the error message "Argument is outside the domain of the function" is generated (math_errno = E__DOMERR). If the polynomial order is less than 1 or greater than 36, the error message "The polynomial order is not in range" (math_errno = E__ORDER).

SEE ALSO

newton().

REFERENCE

F. B. Hildebrand (1949).

```c
#include "mathlib.h"
#include <stdio.h>
#include <math.h>
#define NORDER 3
main()
{
Complex_Vector root;
int i;
Real coeff[NORDER+1];
unsigned limit=500;

/*Find the roots of this polynomial (x+2)(x-1)(x+1) = -2 -1*x +2*x^2 + 1*x^3  */

        coeff[0] = -2;
        coeff[1] = -1;
        coeff[2] = 2;
        coeff[3] = 1;

        root = p_roots(coeff, NORDER, limit);

        for(i=0; i<NORDER; i++) {
            printf("root[%d].r=%8.4f, root[%d].i=%8.4f\n",i,root[i].r,i,root[i].i);
        }
        Cvfree(root);
        return 0;
}
```

PROGRAM OUTPUT

```
root[0].r= -1.0000, root[0].i=  0.0000
root[1].r=  1.0000, root[1].i=  0.0000
root[2].r= -2.0000, root[2].i=  0.0000
```

pseries

#include "mathlib.h"
double pseries(double x, double $coef$[ ], unsigned n);

DESCRIPTION

The *pseries* function calculates the value of a function represented by a Taylor series:

$$f(x) = coef[0] + x * coef[1] + x^2 * coef[2] + \ldots + x^{n-1} * coef[n-1]$$

The x argument specifies the value of x for which $f(x)$ is calculated, and the *coef* argument specifies the value of the n coefficients of the polynomial expression.

Using the *pseries* function, some care must be taken to insure that the series converges over the domain of interest. For example, the Taylor series expansion for e^x converges rapidly if $|x| < 1$ (see the example below). However, if $|x| \geq 1$, the Taylor series is divergent and truncation errors become infinitely large.

The *pseries* function uses Horner's method, which is numerically more accurate than computing the previous expression directly. Horner's formula for five terms is

$$coef[0] + x * (coef[1] + x * (coef[2] + x * (coef[3] + x * coef[4])))$$

Chebyshev polynomials should be used for more accurate approximations. For the same number of terms, the Chebyshev error in approximating e^x is much less than the truncated Taylor series example below (see the example for the *chebser* function). Moreover, for the same approximation error, the Chebyshev coefficients can be specified with considerably less precision than required for the Taylor series method. Nonetheless, the truncated Taylor series is a straightforward and valuable approach in mathematical approximation.

RETURNS

A double precision value representing $f(x)$ at the point x, calculated using the coefficients for a Taylor series of length n.

SEE ALSO

chebser().

```c
#include "mathlib.h"
#include <stdio.h>
#include <math.h>
double coef[7] = { 1., 1., .5, .166666667,
                   .041666667, .008333333, .001388889};
main()
{
int i,n,m=20;
double dx,x,approx;
/* evaluate the 7-term Taylor series approximation to exp(x), -1 < x < 1   */
      dx = 2.0/(double) m;
      x=-1;     /* evaluate the approximation for -1 < x < 1              */
      n = 7;    /* this is the number of terms in the polynomial series   */
      printf("    x      exp(x)  approximation\n");
      for(i=0; i<=m; i++) {
              approx = pseries(x, coef, n);/* evaluate the polynomial    */
              printf("%f %f %f\n", x, exp(x), approx);
              x += dx;
      }
      return 0;
}
```

pseudinv

#include "mathlib.h"
Real__Vector pseudinv(Real__Matrix m, **Real__Vector** v, **unsigned** *neqns,* **signed** *unknwns*);

DESCRIPTION

Suppose a data set is known to belong to a linear process. We may wish to estimate the parameters that characterize this process, but we find that the data is corrupted with random noise (e. g., measurement errors). For example, the points on a wall all lie in a plane. With no noise, only three points are needed to characterize the plane's equation. However, in measuring the distance to the wall, we introduce random noise, which degrades our estimate of the plane's equation. One way to reduce these random effects is to take several additional data points (more than three) and overconstrain the system. Pseudo-inverse methods can be used to solve this overconstrained system of equations. These approaches are very powerful and are optimal, since they minimize the mean-square error.

The *pseudinv* function finds the solution vector for an overconstrained system of linear equations. A system of equations is considered to be overconstrained when the number of equations is greater than the number of unknowns.

Consider the following system of linear equations:

$$m[\text{neqns}][\text{unknwns}] * x[\text{unknwns}] = v[\text{neqns}]$$

where m is the system matrix, x is the solution vector, v is the constraining vector, *neqns* is the number of equations, and *unknwns* is the number of unknowns ($neqns \geq unknwns$).

If the system matrix m is square (i. e., *neqns* = *unknwns*) and invertible, then solving for x is straightforward (see the *lineqn* function). Otherwise the previous equation must be transformed into an equivalent square system by premultiplying both sides of the equation by m^T (i. e., the transpose of matrix m):

$$[m^T * m] * x = m^T * v$$

Let $[m^T * m] = A[unknwns][unknwns]$ and let $m^T * v = w[unknwns]$ be the new constraining vector. Then the previous expression can be rewritten:

$$A * x = w$$

Since A is a square matrix, it can be inverted (if it is nonsingular) and the desired solution vector, $x[unknwns]$, is given by

$$x = A^{-1} * w$$

A^{-1} is called a pseudo-inverse matrix.

The *pseudinv* function calls the *mxmul1, vmxmul1,* and *lineqn* functions.

RETURNS

A real vector of length *unknwns* representing the solution vector for the system of linear equations m constrained by v if successful, or NULL if an error occurs. If *neqns < unknwns*, the error message "More unknowns than equations" is generated (math_errno = E_NEQNS). If the square matrix A is noninvertible, the error message "Singular matrix" is generated (math_errno = E_MSING).

SEE ALSO

Clineqn(), lineqn().

```c
#include "mathlib.h"
#include <stdio.h>
#include <math.h>
#include <stdlib.h>
Real M[5][2] = { -2,-1, -1,0, 0,1, 1,2, 3,4 };
Real V[5] = { 1,1,1,1,1 };
main()
{
int i,j,neqns=5,unknwns=2;
Real_Matrix m;        /* m[neqns][unknwns] */
Real_Vector z;        /* z[unknwns] */
Real_Vector v;        /* v[neqns] */
double noise;
/*------------------------------------------------------------------*/
/* z0*x + z1*y = 1; find best z0 and z1 for several noisy x and y   */
/*                                                                  */
/*  | x0  y0 |   | z0 |     | 1 |                                   */
/*  | x1  y1 |   | z1 |     | 1 |                                   */
/*  | x2  y2 |          =   | 1 |   or  M*z = V  find best z.       */
/*  | x3  y3 |              | 1 |   With no noise the answer        */
/*  | x4  y4 |              | 1 |     is:   z0 = -1, z1 = 1          */
/*------------------------------------------------------------------*/
        m = mxalloc(M, neqns, unknwns);
        v = valloc(V, neqns);
        for (i=0; i<neqns; i++) {     /* degrade true linear points with noise */
            for (j=0; j<unknwns; j++){/*generate uniform noise from +.1 to -.1*/
                noise = .2*((double) rand()/RAND_MAX - 0.5);
                    m[i][j] += noise;
            }
        }
        z = pseudinv(m, v, neqns, unknwns); /* solve the overconstrained set  */
                                    /*    of linear equations         */
        printf("the best answer with noisy data is: z0=%f, z1=%f\n", z[0], z[1]);
        return 0;
} /* the answer is z0 = -0.999 31; and z1 = 0.982814                          */
```

resample

#include "mathlib.h"

Real__Vector resample(Real__Vector *data*, **unsigned** *ndata*, **unsigned** **nout*, **int** *last*,
   double fs_in, **double** fs_out, **double** *transition*, **double** *dB*);

DESCRIPTION

The *resample* function uses a Kaiser–Bessel weighted lowpass FIR filter to resample digital data to any **integral** new sampling frequency. This routine provides an almost arbitrary accuracy for both integer decimation and integer interpolation. The *resample* function can filter data sets of unlimited length, since it is designed to be called on consecutive data segments.

The *data* argument specifies a real vector of length *ndata* to be resampled. The *nout* argument specifies the address of an unsigned integer variable where the *resample* function stores the length of the resulting resampled vector.

The *last* argument specifies whether or not this function will be called more than once for the same data set. If so, a speed improvement is obtained by retaining internal working storage between function calls. If *last* = 0, then the internally allocated working storage is saved for the next call. If *last* = 1, then the working storage is freed before exiting the function. The final call to the *resample* function should specify *last* as 1 in order to free the allocated memory.

The fs_in and fs_out arguments specify the input and output sampling rates in hertz. The values of these two arguments must be integrally related (i. e., one is an integer multiple of the other).

The *transition* argument specifies the desired transition bandwidth of the resampling filter normalized by MAX$\{fs\_in, fs\_out\}$. The transition bandwidth determines the steepness of the filter and is internally limited by

$$0.02 \leq transition \leq 0.20$$

The *dB* argument is the desired stopband attenuation (in decibels) of the resampling filter, where *dB* is internally limited by

$$50 \leq dB = 20^* \log_{10}(passband\_gain/stopband\_gain) \leq 90$$

The *resample* function automatically designs the resampling filter (just once) based on the acceptable resampling error. The acceptable resampling error is determined from the value of the *dB* and *transition* arguments. The filter length is calculated as

$$filter\_length = (|dB| - 8)/(14.357 * transition)$$

which limits the filter length to

$$15 \leq filter\_length \leq 287$$

The *resample* function calls the *convolve* and *lowpass* functions.

RETURNS

A real vector of length *nout* representing the resampled version of the input vector *data* if successful, or NULL if an error occurs. If *ndata* is less than fs_in/fs_out + filter_length, the error message "Not enough input data" is generated (math_errno = E_NOTENOUGH).

SEE ALSO

downsamp(), interp().

REFERENCE

L. R. Rabiner and R. W. Schafer (1973).

```
#include <math.h>
#include <stdio.h>
#include "mathlib.h"
main()
{
Real_Vector data,out,y,spectrum,w;
double f0,fmax,freq,fs_in,fs_out,dB=50.0,transition=0.2;
int i,j,ny,m=4,ntimes=4,ndata=100,last=0,nfft=32;
unsigned nout;
```

(continued on next page)

(continued)

```
/* Interpolate a sine wave by a factor of two; in four segments */

        fs_in = 1000.;
        fs_out = 2.0*fs_in;
        f0 = 375.0;                /* freq = stimulus frequency in Hz (6th harmonic)  */
        data=valloc(NULL, ndata); /* allocate an array for data to be analyzed  */
        ny = (int) (ntimes*ndata*fs_out/fs_in);
        y=valloc(NULL, ny);
        if(fs_out/m > fs_in/2) fmax = fs_in/2;
        else fmax = fs_out/m;/* fmax is the maximum allowable frequency (in Hz) */
        if(f0 > fmax) f0 = fmax;
        ny = 0;
        for(j=0; j<ntimes; j++) {
            for(i=0; i<ndata; i++) data[i] = sin(twopi_*f0*(i+j*ndata)/fs_in);
            if(j == ntimes-1) last = 1;
            out=resample(data,ndata,&nout,last,fs_in,fs_out,transition,dB);
            for (i=0;i<nout;i++) y[i+ny] = out[i];
            ny += nout;
            vfree(out);
        }
        w = tdwindow(nfft, 7);            /* get a Hamming window */
        spectrum = powspec(y,ny,nfft,w);   /* do a power spectrum of the output */
        freq = fs_out/nfft;
        printf("frequency (Hz)   magnitude (dB) \n");
        for (i=0;i<nfft/m; i++){
            printf(" %8.3f              %f\n", i*freq,spectrum[i]);
        }
        mathfree();
        return 0;
}
```

PROGRAM OUTPUT

frequency (Hz)	magnitude (dB)
0.000	0.000000
62.500	0.000000
125.000	0.000000
187.500	0.000002
250.000	0.000014
312.500	0.030096
375.000	0.152700
437.500	0.030075

romberg

```
#include "mathlib.h"
double romberg(double (*fx)(double x), double x0, double x1, double accuracy);
```

DESCRIPTION

Many important integrals that arise in mathematical problems cannot be expressed in a closed form. The most common examples are definite integrals of the form

$$y(x) = \int_{x_0}^{x_1} f(x) \, dx$$

where the limits x_0 and x_1 are fixed and $f(x)$ is any analytical expression, not necessarily integrable.

Numerical integration provides quick and accurate solutions to difficult integrals or integrals without analytical solutions. Perhaps the fastest such method is the trapezoidal rule in connection with Romberg's principle. This method approximates the definite integral $y(x)$ from the point x_0 to the point x_1 by successively applying the trapezoidal rule. At the kth iteration, the integration interval (x_0, x_1) is divided into 2^{k-1} equidistant subintervals. Each new trapezoidal approximation can be obtained from the preceding approximations without having to recompute $f(x)$ at the overlapping grid locations. This accounts for the extraordinary speed of the Romberg integration approach.

The *romberg* function numerically integrates the user-defined function fx using the Romberg approach.

The user-defined function fx requires a single double precision argument x, and is expected to return a double precision value representing $f(x)$.

The x_0 and x_1 arguments specify the limits of the integration. The fx function is integrated over the range x_0 to x_1, where $x_0 != x_1$.

The *accuracy* argument specifies the absolute error to allow in calculating the integral. If the value of the integral is large, then the *accuracy* should not be too small. For example, if it is known that the value of the integral is approximately $10.e + 15$, a reasonable value for *accuracy* would be 100, whereas an unreasonable value would be $10.e - 13$. The *romberg* function will fail if it is unable to achieve the specified *accuracy*.

One disadvantage of Romberg's approach is that it sometimes fails when periodic functions are integrated over an integer number of periods. If the grid locations of the integration panels nearly align with the zero crossings of the function, the Romberg approach may terminate prematurely with an incorrect answer. These situations can be avoided by segmenting the integration interval into lengths that are not integer multiples of the period.

RETURNS

A double precision value representing the definite integral of the function fx over the interval x_0 to x_1 if successful, or 0 if an error occurs. If $x_0 = x_1$, the error message "Limits on definite integral must be distinct" is generated (math_errno = E_LIMITS) and 0 is returned. If singularities are encountered, or if the value of the integral is very large compared to *accuracy*, or if the length of the integration interval $(x1 - x0)$ is too large, then the error message "Rounding errors prohibit required accuracy" is generated (math_errno = E_ROUNDOFF).

SEE ALSO

integrat(), deriv(), deriv1(), interp().

REFERENCE

Hornbeck, (1975, pp. 150–154).

```
#include "mathlib.h"
#include <stdio.h>
#include <math.h>
#include <stdlib.h>

/* example function to evaluate the value of fx at x*/
double fx(double x)
{
        return sin(x);
}


main()
{
double integral, x0=0, x1=1.50*pi_, accuracy = 1.e-10;

/* numerically integrate sin(x) over the interval: (0, 1.50*pi) */

        integral=romberg(fx, x0, x1, accuracy);/*Romberg integration routine*/

/*  if y = sin(x) the integral should be: Y = 1-cos(x1)                    */

        printf(" Romberg integration of sin(x) over (0, 1.50*pi):\n\n");

        printf("     true integral = %20.16f\n", 1-cos(x1));
        printf("numerical integral = %20.16f\n", integral);
        printf(" the difference is = %20.16f\n", integral-1+cos(x1));
        return 0;
}
```

smooth

#include "mathlib.h"
Real_Vector smooth(Real_Vector *data*, int *ndata*, int *factor*);

DESCRIPTION

The *smooth* function uses a Kaiser lowpass filter (see the *lowpass* function) to smooth functions represented by a set of equally spaced data points. For many data processing applications, desired signals are corrupted with undesired random errors. These errors typically consist of both high and low frequencies, while the signal has mainly low frequencies. Data smoothing is a filtering process that reduces the high frequency noise, thus enhancing the low frequency signals.

The *data* argument specifies a real vector of length *ndata*. The amount of smoothing can be adjusted by selecting different smoothing factors. The *factor* argument sets the cutoff frequency of the smoothing filter. This is an integer factor (between 2 and 10) by which the input bandwidth is reduced. The minimum amount of smoothing occurs when *factor* = 2, and the maximum amount occurs when *factor* = 10. When smoothing by large factors, some care must be taken that the signals of interest are sufficiently oversampled, or they will be filtered out with the noise.

The *smooth* function fixes the stopband attenuation at $d_s = 0.001$, which corresponds to -60 dB. The lowpass filter length *nweights* (see the *lowpass* function) is automatically chosen by the following empirical formula:

$$nweights = [1.08 * |\log_{10}(d_s)| + 1.174] * factor + 0.5$$

or

$$nweights = (\text{int})(4.414 * factor + 0.5)$$

This sets the corner frequencies of the lowpass smoothing filter to the following arbitrary but reasonable values:

$$F_1 = 0.1235/(F_s/factor)$$
$$F_2 = 1.253/(F_s/factor)$$

where F_s is the data sampling rate in hertz.

The Kaiser lowpass filter smooths only the central $ndata - nweights + 1$ points of the data. The upper and lower edges (i. e., $i < nweights/2$ and $i > ndata - nweights/2 - 1$) of the data are smoothed by a 3-point FIR filter using the following three filter weights:

$$h_0 = 1/4, \qquad h_1 = 1/2, \qquad h_2 = 1/4$$

The *smooth* function calls the *convolve* function, and the *lowpass* function.

RETURNS

A real vector of length *ndata* representing a smoothed version of the *data* vector if successful, or NULL if an error occurs. If *ndata* < *nweights*, the error message "Not enough input data" is generated (math_errno = E_NOTENOUGH). If *factor* ≤ 1, or *factor* ≥ 11, the error message "Argument 'factor' must be between 2 and 10" is generated (math_errno = E_FACTOR).

SEE ALSO

lowpass(), downsamp(), interp().

REFERENCE

Abramowitz and Stegun (1964, p. 883, eq. 25.3.6).

```
#include "mathlib.h"
#include <stdio.h>
#include <stdlib.h>
#include <math.h>


main()
{
Real_Vector data,y;
int ndata=20, factor=3, i;
double truth[20], noise;

    data=valloc(NULL, ndata);/* allocate an array for data to be filtered */
    for(i=0; i<ndata; i++) {/* generate data, one cycle of a sinusoid+noise*/
        truth[i] = sin(twopi_*i/(double) ndata);
        noise = ((double) rand()/RAND_MAX - 0.5); /* uniform from +.5 to -.5*/
        data[i] = truth[i] + noise;
    }
    y = smooth(data, ndata, factor); /* now smooth the data           */

    printf("         data[i]      smoothed     data[i]\n");
    printf(" i      + noise         data       (no noise)\n\n");
    for(i=0; i<ndata; i++) {
        printf(" %2d      %9.6f      %9.6f     %9.6f\n",
                            i, data[i], y[i], truth[i]);
    }
    vfree(data);  /* free data vector */
    vfree(y);     /* free smoothed data vector */
    return 0;
}
```

PROGRAM OUTPUT

i	data[i] + noise	smoothed data	data[i] (no noise)
0	-0.500000	-0.500000	0.000000
1	-0.059448	0.056124	0.309017
2	0.843393	0.598754	0.587785
3	0.767680	0.840643	0.809017
4	0.983818	0.863564	0.951057
5	0.718940	0.729946	1.000000
6	0.498086	0.785438	0.951057
7	0.987901	0.799543	0.809017
8	0.767097	0.734372	0.587785
9	0.743738	0.510651	0.309017
10	-0.116504	0.159301	-0.000000
11	-0.289592	-0.248895	-0.309017
12	-0.256797	-0.653282	-0.587785
13	-1.274470	-0.978638	-0.809017
14	-1.397619	-1.259999	-0.951057
15	-0.970290	-1.029523	-1.000000
16	-0.779893	-0.957851	-0.951057
17	-1.301326	-1.021732	-0.809017
18	-0.704381	-0.863068	-0.587785
19	-0.742181	-0.742181	-0.309017

spline0 - spline

#include "mathlib.h"
Real__Vector spline0(Real__Vector x, **Real__Vector** y, **unsigned** n);
Real spline(Real $x0$, **Real__Vector** x, **Real__Vector** y, **unsigned** n, **Real__Vector** *deriv*);

DESCRIPTION

Interpolation using a high order polynomial function can sometimes produce large fluctuations in between the data points. Also, piecewise polynomial interpolation often produces sharp cusps at the data points that are unpleasing to the eye. Both of these difficulties can be minimized by increasing the number of data points, or by decreasing the degree of the polynomial. When this is not an option, then derivative constraints can be added to the polynomials, resulting in much smoother interpolations. Interpolations performed using derivative constraints are commonly referred to as splines.

The cubic spline is one of the most common splines used for interpolation. The cubic spline interpolates using a succession of third order polynomials of the form:

$$f_i(x) = b_{0i} + b_{1i}x + b_{2i}x^2 + b_{3i}x^3$$

where i is the interval over which the interpolation is performed. The smooth appearance of a cubic spline results from matching the slope and curvature of these cubic formulas in adjacent intervals.

The *spline0* function computes the second derivatives required to match the slope and curvature of the cubic formulas across adjacent intervals of the data. The x and y vectors specify a set of n data points, and the *spline0* function returns a real vector of length n, representing the second derivative at each data point. This vector of second derivatives is then passed to the *spline* function as the *deriv* argument.

The *spline* function uses the x, y, and *deriv* vectors of length n to calculate a value $y0$ for a given value $x0$, where $x[0] \le x0 \le x[n-1]$. Cubic spline interpolation is used to determine $y0$.

The x, y, and n arguments are the same for both the *spline0* and *spline* functions. The values in the x vector must be ordered, but not necessarily equally spaced (i. e., $x[0] < x[1] < x[2] < \ldots < x[n-1]$).

RETURNS

spline0: A real vector of length n representing the second derivatives at the data points specified by the x and y vectors if successful, or NULL if an error occurs. If $x[i] = x[j]$, where $i \mathbin{!=} j$, the error message "The input table has two identical x values" is generated (math_errno = E__SAMEX).

spline: A real value $y0$ representing the interpolated value at the point $x0$, where $x[0] \le x0 \le x[n-1]$.

SEE ALSO

lowpass(), smooth(), interp(), interp1().

REFERENCE

R. W. Hornbeck (1975, pp. 47–50).

```
#include "mathlib.h"
#include <stdio.h>
#include <math.h>
#include <stdlib.h>
main()
{
int i, k, n=7, nout=19;
double delta, delta3, truth;
Real x0, y0;
Real_Vector x, y, deriv;
      y = valloc(NULL, n);
      x = valloc(NULL, n);
      delta = twopi_/(double)(n-1);
      delta3 = delta/3;
   /* define  unequally spaced x values                                    */
      x[0]=delta*0; x[1]=delta/3;
      x[2]=delta;   x[3]=delta*2.33333333;
      x[4]=delta*3; x[5]=delta*4.33333333;
      x[6]=delta*6;
   /* generate one cycle of a sinusoid; with nonequally spaced samples      */
      for (i=0; i < n; i++) y[i]=sin(x[i]);
      printf("           interpolated        original     truth:\n");
      printf("    x              y         k      y[k]       sin(x)\n\n");
      k=0;
      deriv = spline0(x, y, n);  /* find 2nd derivative of data             */
      for (i=0; i < nout; i++) {
            truth = sin(twopi_*i/(double) (nout-1));
            x0 = i*delta3;
            y0 = spline(x0, x, y, n, deriv); /*interpolate with spline       */
            if(fabs(x0-x[k]) < 10.e-6) {
                  printf("%9.6f    %9.6f    %2d    %9.6f    %9.6f\n",
                              x0, y0, k, y[k], truth);
                  k++;
            }
            else  printf("%9.6f    %9.6f                      %9.6f\n",
                                    x0, y0, truth);
      }
      return 0;
}
```

PROGRAM OUTPUT

x	interpolated y	k	original y[k]	truth: sin(x)
0.000000	0.000000	0	0.000000	0.000000
0.349066	0.342020	1	0.342020	0.342020
0.698132	0.644242			0.642788
1.047198	0.866025	2	0.866025	0.866025
1.396263	0.974505			0.984808
1.745329	0.968753			0.984808
2.094395	0.855829			0.866025
2.443461	0.642788	3	0.642788	0.642788
2.792527	0.343615			0.342020
3.141593	0.000000	4	0.000000	0.000000
3.490659	-0.341601			-0.342020
3.839724	-0.643372			-0.642788
4.188790	-0.869659			-0.866025
4.537856	-0.984808	5	-0.984808	-0.984808
4.886922	-0.963998			-0.984808
5.235988	-0.825754			-0.866025
5.585054	-0.599434			-0.642788
5.934119	-0.314396			-0.342020
6.283185	-0.000000	6	-0.000000	-0.000000

stats

```
#include "mathlib.h"
double stats(Real__Vector x, int n, double *pop__var, double *samp__var);
```

DESCRIPTION

The two most commonly used measures in statistics are the mean and variance. The mean, u_X, of a random variable, X, is defined as the expected (average) value:

$$u_X = E[X] = [X_1 + X_2 + \ldots + X_n]/n$$

The variance of a population of size n is s_p^2

$$s_p^2 = [(X_1 - u_X)^2 + (X_2 - u_X)^2 + \ldots + (X_n - u_X)^2]/n$$

In practice, the total population can rarely be observed. Often, only a small sample of the population can be studied. Suppose that m samples are drawn from a population of size n, where $m << n$. For these cases, the variance formula should be normalized by a factor of $m - 1$, rather than m. The resulting sample variance s_s^2 is given by

$$s_s^2 = [(X_1 - u_X)^2 + (X_2 - u_X)^2 + \ldots + (X_m - u_X)^2]/(m - 1)$$

s_s^2 is an unbiased estimator of s_p^2. The presence of the factor of $1/(m - 1)$ causes the expected value of the sample variance to be equal to the true population variance, since

$$E[s_s^2] = s_p^2$$

The *stats* function calculates the mean of the vector x of length n.

The *pop__var* and *samp__var* arguments specify the addresses of variables of type **double**. The *stats* function returns the population variance via the *pop__var* argument, and the sample variance via the *samp__var* argument.

RETURNS

A double precision value representing the mean of the data specified by the real vector x. The population and sample variance are returned via the *pop__var* and *samp__var* arguments, respectively.

REFERENCE

P. L. Meyer (1970).

```
#include "mathlib.h"
#include <stdio.h>
#include <math.h>
#include <stdlib.h>
main()
{
int i,n;
double tvar,tmean,xmean,pop_var,samp_var;
Real_Vector x;

    n=100;
    x = valloc(NULL, n); /* allocate vector x[]                    */

    for (i=0; i < n; i++) { /* generate one cycle of a sine wave    */
        x[i]=sin(twopi_*i/n);
    }
/* now find the mean and variance of x[] */

    xmean = stats(x, n, &pop_var, &samp_var);

    tmean=0;    /* this is the true mean of the sine wave          */
    tvar=1/2.;  /* this is the true variance of the sine wave      */

    printf("          Find the Mean and Variance of a Sine Wave:\n\n");
    printf("      true mean = %f,                    true variance = %f\n",
                                                        tmean, tvar);
    printf("estimated mean = %f,  estimated (population) variance = %f\n",
                                                        xmean, pop_var);
    return 0;
}
```

PROGRAM OUTPUT

```
        Find the Mean and Variance of a Sine Wave:

    true mean = 0.000000,                    true variance = 0.500000
 estimated mean = 0.000000,  estimated (population) variance = 0.500000
```

tdwindow

#include "mathlib.h"
Real_Vector tdwindow(unsigned *nw*, unsigned *wtype*);

DESCRIPTION

Window functions are frequently used in digital signal processing. The most common applications are in spectral analysis, antenna design, and digital filtering.

The *tdwindow* function provides the following nine time-domain windows, depending on the value of *wtype*:

wtype	TYPE OF TIME DOMAIN WINDOW
0	approximate Blackman
1	three-term Blackman–Harris, −67 dB
2	three-term Blackman–Harris, −61 dB
3	four-term Blackman–Harris, −92 dB
4	four-term Blackman–Harris, −74 dB
5	exact Blackman
6	Hanning window
7	Hamming window
8	rectangular window

The *nw* argument specifies the length of the window (i. e., the number of time-domain weights).

The *tdwindow* function calls the *wind_* function.

RETURNS

A real vector of length *nw* representing any one of nine different time-domain windows if successful, or NULL if an error occurs. If *wtype* < 0 or *wtype* > 8, the error message "Invalid window type specified" is generated (math_errno = E_WINDOW).

SEE ALSO

bandpass(), highpass(), lowpass().

REFERENCE

F. J. Harris (1978).

```
#include "mathlib.h"
#include <stdio.h>

main()
{
Real_Vector w;
unsigned nw=16, wtype = 6;
int i;

        w = tdwindow(nw, wtype);    /* generate the window weights */

        printf("The Hanning Window weights are:\n\n");
        for(i=0; i<nw; i++) {
            printf("%2d    %f\n", i, w[i]);
        }
        vfree(w); /* free the dynamic memory allocated for w[]          */
        return 0;
}
```

PROGRAM OUTPUT

```
The Hanning Window weights are:

 0     0.000000
 1     0.005764
 2     0.022058
 3     0.046066
 4     0.073635
 5     0.100000
 6     0.120601
 7     0.131877
 8     0.131877
 9     0.120601
10     0.100000
11     0.073635
12     0.046066
13     0.022058
14     0.005764
15     0.000000
```

urand

```
#include "mathlib.h"
double urand();
void uraninit(long seed);
```

DESCRIPTION

Random numbers are frequently used to predict expected outcomes of statistical events. Many random events are too complex to evaluate directly and must be characterized via Monte Carlo simulations. These simulations usually require a large number of random trials. The specific outcome of any one trial is relatively insignificant. The investigator is more interested in the average statistical behavior across all of the trials. The generation of independent pseudorandom number sequences is essential to these statistical simulations.

Uniform random numbers are perhaps the most important pseudorandom sequences. Many statistical distributions can be generated with computations that involve the uniform distribution. For convenience, most uniform random number routines generate numbers between 0 and 1. Other ranges of values are easily obtained through simple arithmetic translations.

If X is a random variable uniformly distributed over the interval $(0, 1)$, its mean and variance are

$$u_X = E[X] = 1/2, \quad \text{and} \quad VAR[X] = E[(X - u_X)^2] = 1/12$$

The *urand* function returns a random number uniformly distributed between 0 and 1. A linear congruential random generator with a full 32-bit seed (see references below) is used to generate the number sequences. Different pseudorandom sequences can be generated by changing the seed value with the *uraninit* function. The default seed value used by the *urand* function is 1.

The *uraninit* function is similar to the standard C function *srand*. The *seed* argument specifies a seed value for the *urand* random number generator. For each value of *seed*, a different pseudorandom sequence is generated. If *seed* < 0, then a random seed value is selected.

RETURNS

urand: A double precision value representing a uniformly distributed random number between 0 and 1.

uraninit: None.

SEE ALSO

normal(), poisson().

REFERENCES

S. K. Park and K. W. Miller (1988); D. E. Knuth (1981).

```c
#include "mathlib.h"
#include <stdio.h>
#include <math.h>
#include <stdlib.h>
main()
{
int i,n;
double mean,var,vout;
long seed=1234;  /* keep the seed value fixed at 1234 */

    n=100;
    mean=var=0.0;
    uraninit(seed);
    for (i=0; i < n; i++) {
        vout=urand();
        mean += vout;
        var += vout*vout;
    }
    var = var/(n-1) - (mean/n) * mean/(n-1);
    mean /= n;

    printf("          Uniform Random Variable Test (100 trials):\n\n");
    printf("    true mean = 1/2 = 0.5,  true variance = 1/12 = 0.0833...\n");
    printf("observed mean = %f,   observed variance = %f\n", mean, var);
    return 0;
}
```

vadd

```
#include "mathlib.h"
Real_Vector vadd(Real_Vector v1, Real_Vector v2, unsigned n);
```

DESCRIPTION
The *vadd* function adds the two real vectors $v1$ and $v2$. The resulting sum is also a real vector.

The n argument specifies the number of elements in vectors $v1$ and $v2$. The real vectors $v1$ and $v2$ must be allocated by the *valloc* function prior to calling the *vadd* function. The resulting real vector representing the sum of $v1$ and $v2$ is automatically allocated by the *vadd* function.

The user may select the option of whether or not to write over the input vector $v1$ with in-place computations. If memory is a limitation, and $v1$ is no longer needed, this may prove useful. This option can be selected by setting the global variable **useinput_ = 1** prior to calling *vadd*. If this option is selected, the resulting sum overwrites the input vector $v1$. The default option is to not destroy the input vector $v1$ (i. e., **useinput_ = 0**).

RETURNS
A real vector v of length n, where $v = v1 + v2$.

SEE ALSO
Cmxadd(), Cmxsub(), mxadd(), mxsub(), Cvsub(), Cvadd(), vsub().

```
#include "mathlib.h"
#include <stdio.h>
main()
{
int i;
Real_Vector v,v1,v2;
unsigned n=2;
/* add the following two vectors:                                         */
/*                                                                        */
/*          v[] = v1[] + v2[]                                             */
/*                                                                        */
/* where v1[] and v2[] are real and are given below:                     */
        v1 = valloc(NULL, n);  /* allocate vector of length 2            */
        v2 = valloc(NULL, n);  /* allocate another vector of length 2    */
        v1[0] = 2;   v1[1] = 2;  v2[0] = 3;   v2[1] = 3;

        v = vadd(v1, v2, n);  /* add v1[] and v2[]                       */

        /* the answer should be : v[0] = 5;   v[1] = 5;                  */

        for (i=0; i<n; i++) {
            printf("v[%d] = %f\n", i, v[i]);
        }
        vfree(v);    /* free vector v[]  */
        vfree(v1);   /* free vector v1[] */
        vfree(v2);   /* free vector v2[] */
        return 0;
}
```

valloc

> #include "mathlib.h"
> Real__Vector valloc(Real__Ptr *address*, unsigned *n*);

DESCRIPTION

The *valloc* function allocates memory for a real vector of length *n*.

The *address* argument is a pointer to a **Real** value. If the *address* argument is NULL, the *valloc* function allocates space for the vector from the available dynamic memory. If the *address* argument is not NULL, the *valloc* function assumes that it is the address of a single-dimensioned array of **Real** values. In this case, no memory is allocated; the *valloc* function simply returns a pointer to the beginning of the single-dimensioned array. This provides a mechanism for using preinitialized arrays.

RETURNS

A real vector of length *n* if successful, or NULL if an error occurs. If the vector is too large, an "Array too large" error message is generated (math_errno = E__MSIZE). If there is not enough dynamic memory, an "Insufficient dynamic memory available" error message is generated (math_errno = E__MALLOC).

SEE ALSO

Cmxalloc(), mxalloc(), Cvalloc().

```
#include "mathlib.h"
#include <stdio.h>
Real VECT1[4] = { 1, 2, 3, 4 } ;
main()
{
Real_Vector  vect1,vect2;
unsigned n=4;
int i;

        vect2 = valloc(NULL, n);   /* allocate real vector vect2      */
        vect1 = valloc(VECT1, n);  /* allocate and initialize vect1   */

        /* now initialize real vector vect2 */
        vect2[0] = 1; vect2[1] = 2; vect2[2] = 3; vect2[3] = 4;

/* the vectors vect1[] and vect2[] should be equal  */

        for (i=0; i<n; i++) {
                printf("vect1[%d] = %3.1f, vect2[%d] = %3.1f\n",
                        i, vect1[i], i, vect2[i]);
        }
        vfree(vect2);    /* free vector vect2[] */
        return 0;
}
```

v__centro

#include "mathlib.h"
Real v__centro(Real__Vector y, unsigned n);

DESCRIPTION

It frequently happens that a one-dimensional variable or function is given in tabular form and that the abscissa values are of more interest than the ordinates. For example, in spectral analysis one may be more interested in frequency content than exact amplitude. Centroiding techniques are ideal for these applications. Because of the focus on the coordinate values, centroiding is sometimes referred to as inverse interpolation.

The *v_centro* function centroids the unimodal vector y of length n. The assumption is that the true data represents a unimodal or point process and that y is a quantized version of the process. That is, it is assumed that y has a single maximum and is monotonically decreasing below this value. This approach uses a 2-bin or a 3-bin centroid, depending on which is the more appropriate fit to the data around the maximum.

The *v_centro* function is ideal for improving the frequency resolution of the power spectrum in single-tone frequency tests. It is also useful for improving the estimate of the time lag that best matches the pure delays of time series in correlation procedures.

RETURNS

A real value representing the inverse interpolation of the real vector y. This is a best estimate of the abscissa (x-value) at which the true maximum of the unimodal process (represented by vector y) occurs.

SEE ALSO

mxcentro().

REFERENCE

F. B. Hildebrand (1974).

```
#include <stdlib.h>
#include <stdio.h>
#include <math.h>
#include "mathlib.h"

main()
{
unsigned i, nfft = 16;
Real freq_estimate,improvement,freq_bin = 6.72;
Real_Vector data,spectrum;
/* Find the centroid of an FFT of a sinusoid that is not a harmonic of the
   fundamental frequency */

    data = valloc(NULL, nfft);
    for(i=0; i<nfft; i++) data[i] = sin(twopi_*freq_bin*i/nfft);
    spectrum = powspec(data, nfft, nfft, NULL);
    for(i=0; i<nfft/2+1; i++) spectrum[i] = sqrt(spectrum[i]);
    freq_estimate = v_centro(spectrum, nfft/2+1);
    improvement=fabs(((int)(freq_bin+0.5)-freq_bin)/(freq_bin-freq_estimate));
    printf("true frequency bin =%6.3f\n",freq_bin);
    printf("centroid estimate of bin =%6.3f\n",freq_estimate);
    printf("factor of improvement in resolution over fft=%5.2f\n",improvement);
    vfree(data);
    vfree(spectrum);
    return 0;
}
```

PROGRAM OUTPUT

```
true frequency bin = 6.720
centroid estimate of bin = 6.694
factor of improvement in resolution over fft=10.71
```

vcopy

```
#include "mathlib.h"
Real_Vector vcopy(Real_Vector dest, Real_Vector src, unsigned n);
```

DESCRIPTION

The *vcopy* function copies the real vector *src* to the real vector *dest*.

The *n* argument specifies the number of elements in each vector.

The *dest* and *src* vectors must be allocated by the *valloc* function prior to calling the *vcopy* function.

RETURNS

None.

SEE ALSO

Cmxcopy(), mxcopy(), Cvcopy(), Cmxdup(), mxdup(), Cvdup(), vdup().

```
#include "mathlib.h"
#include <stdio.h>
Real SRC[4] = { 1, 2, 3, 4 } ;
main()
{
Real_Vector  src,dest;
unsigned n=4;
int i;

        dest = valloc(NULL, n);     /* allocate real vector dest[]      */
        src = valloc(SRC, n);       /* allocate and initialize src[]    */

        vcopy(dest, src, n); /* now copy the real vector src[] into dest[] */

/* the vectors src[] and dest[] should be equal  */

        for (i=0; i<n; i++) {
                printf("src[%d] = %3.1f, dest[%d] = %3.1f\n",
                        i, src[i], i, dest[i]);
        }
        vfree(dest);    /* free vector dest[] */
        return 0;
}
```

vcross

#include "mathlib.h"
Real_Vector vcross(Real_Vector v1, Real_Vector v2);

DESCRIPTION

The *vcross* function computes the cross product of the two real vectors $v1$ and $v2$ of length 3.

The cross product (sometimes called the vector product) between two three-dimensional vectors is given by the following determinant:

$$v1 \times v2 = \begin{vmatrix} i & j & k \\ x1 & y1 & z1 \\ x2 & y2 & z2 \end{vmatrix}$$

where i, j, and k are the unit vectors in the x, y, and z directions, the variables $x1, y1$, and $z1$ are the components of $v1$, and the variables $x2, y2$, and $z2$ are the components of $v2$.

The cross product has an interesting geometrical interpretation. If θ is the angle between the two vectors, then

$$v1 \times v2 = u|v1||v2|\sin(\theta)$$

where u is the unit vector perpendicular to the plane of $v1$ and $v2$, and so directed that a right-handed screw driven in the direction of u would carry $v1$ into $v2$.

The cross product is not commutative, since

$$v1 \times v2 = -v2 \times v1$$

RETURNS

A real vector of length 3 representing the cross product of the two three-dimensional vectors specified by $v1$ and $v2$ if successful, or NULL if an error occurs.

SEE ALSO

vdot().

```
#include "mathlib.h"
#include <stdio.h>
main()
{
int i;
Real_Vector v,v1,v2;
/* form the vector product between the following two vectors:          */

        v1 = valloc(NULL, 3);  /* allocate vector of length 2           */
        v2 = valloc(NULL, 3);  /* allocate another vector of length 2   */
        v1[0] = 1;   v1[1] = 0;  v1[2] = 0;
        v2[0] = 0;   v2[1] = 1;  v2[2] = 0;

        v = vcross(v1, v2);  /* form the cross product between v1[] and v2[]  */

        /* the answer should be : v[0] = 0;  v[1] = 0;  v[2] = 1;        */

        for (i=0; i<3; i++) {
            printf("v[%d] = %f\n", i, v[i]);
        }
        vfree(v);     /* free vector v[]  */
        vfree(v1);    /* free vector v1[] */
        vfree(v2);    /* free vector v2[] */
        return 0;
}
```

vdot

#include "mathlib.h"
Real vdot(Real__Vector $v1$, Real__Vector $v2$, unsigned n);

DESCRIPTION

The *vdot* function computes the dot product of the two real vectors $v1$ and $v2$ of length n.

The dot product (sometimes called the inner product) of two vectors is given by

$$v1 * v2 = v1[0]v2[0] + v1[1]v2[1] + \ldots + v1[n-1]v2[n-1]$$

where n is the length of both vectors.

The dot product has an interesting geometrical interpretation. If θ is the angle between the two vectors, then

$$v1 * v2 = |v1||v2| \cos(\theta)$$

When the dot product is normalized by the magnitude of the larger vector, it represents the projection of the smaller vector onto the larger.

RETURNS

A real value representing the dot product of the two real vectors $v1$ and $v2$ of length n.

SEE ALSO

Cvdot(), vcross().

```
#include "mathlib.h"
#include <stdio.h>
main()
{
Real_Vector v1,v2;
Real dot;
unsigned n=2;
/* form the dot product between the following two vectors:          */
        v1 = valloc(NULL, n);  /* allocate vector of length 2       */
        v2 = valloc(NULL, n); /* allocate another vector of length 2 */
        v1[0] = 2;   v1[1] = 2;  v2[0] = 0;   v2[1] = 1;

        dot = vdot(v1, v2, n); /* form the dot product between v1[] and v2[] */

        /* the answer should be : 2.00                              */

        printf("the dot product = %f\n", dot);

        vfree(v1);   /* free vector v1[] */
        vfree(v2);   /* free vector v2[] */
        return 0;
}
```

vdup

```
#include "mathlib.h"
Real__Vector vdup(Real__Vector v, unsigned n);
```

DESCRIPTION

The *vdup* function makes a duplicate copy of the real vector v.

The n argument specifies the number of elements in the vector. The real vector v must be allocated by *valloc* prior to calling the *vdup* function. The *vdup* function automatically allocates memory for the duplicate copy.

RETURNS

A real vector of length n that is a duplicate copy of v if successful, or NULL if an error occurs. If there is not enough dynamic memory, an "Insufficient dynamic memory available" error message is generated (math_errno = E__MALLOC).

SEE ALSO

Cmxdup(), mxdup(), Cvdup(), Cmxcopy(), mxcopy(), vcopy(), Cvcopy().

```
#include "mathlib.h"
#include <stdio.h>
Real V1[2] = { 1,2 } ;
main()
{
Real_Vector  v1,v2;
unsigned i;
unsigned n=2;

        v1 = valloc(V1, n);         /* allocate and initialize v1[]     */
        v2 = vdup(v1, n);           /* now duplicate the vector v1[]    */

/* the vectors v1[] and v2[] should be equal  */

        for (i=0; i<n; i++) {
                printf("v1[%d] = %3.1f, v2[%d] = %3.1f\n", i, v2[i], i, v2[i]);
        }
        vfree(v2);  /* free vector v2[]  */
        return 0;
}
```

vectomat

#include "mathlib.h"
Real__Matrix vectomat(Real__Vector $v1$, Real__Vector $v2$, unsigned n);

DESCRIPTION
The *vectomat* function multiplies the transpose of the real vector $v1$ by the real vector $v2$.

The n argument specifies the number of elements in vectors $v1$ and $v2$. The resulting product is a real matrix of n rows by n columns. Using row-vector notation, we have

$$v1^{T}[n] * v2[n] = m[n][n]$$

The real vectors $v1$ and $v2$ must be allocated by *valloc* prior to calling the *vectomat* function. If desired, $v1$ and $v2$ can be the same vector. The resulting n by n real matrix is automatically allocated by the *vectomat* function.

RETURNS
An n by n real matrix m, where $m[n][n] = v1^{T}[n] * v2[n]$ if successful, or NULL if an error occurs.

SEE ALSO
Cvectomx(), Cvmxmul(), Cvmxmul1().

```c
#include "mathlib.h"
#include <stdio.h>
#include <math.h>
#include <stdlib.h>
Real V1[5] = { 2,2,2,2,2 };
Real V2[5] = { 3,3,3,3,3 };
main()
{
int i,j,n=5;
Real_Vector v1,v2;
Real_Matrix m;

        v1 = valloc(V1, n);
        v2 = valloc(V2, n);
        m = vectomat(v1, v2, n); /* multiply the two vectors int a matrix m[][]*/
        for (i=0; i<n; i++) {
            for (j=0; j<n; j++) {
                printf("%d %d %f\n", i,j,m[i][j]);
            }
        } /*   the answer is m[i][j] = 6,  for all i and j = 0,1,...4         */
        mxfree(m) ; /* free matrix m[][] */
        return 0;
}
```

vfree

#include "mathlib.h"
void vfree(Real__Vector *v*);

DESCRIPTION

The *vfree* function frees the memory allocated to the real vector *v*. The memory for
the real vector *v* must have been allocated by the *valloc* function prior to calling the
vfree function.

RETURNS

None.

SEE ALSO

Cmxfree(), mxfree(), Cvfree(), mathfree().

```
#include "mathlib.h"
#include <stdio.h>
Real V1[2] = { 1,2 };
main()
{
Real_Vector v,v1;
int i;

        v = valloc(NULL, 2); /* allocate dynamic memory for v[]        */
        v1 = valloc(V1, 2);  /* initialize static array v1[]; no
                                dynamic memory is allocated for v1[]    */

        for (i=0; i<2; i++) {
                v[i] = v1[i];           /* set v[] equal to v1[]        */
                printf("v[%d] = %f\n", i, v[i]);
        }
        vfree(v);  /*  free the dynamic memory occupied by v[]          */
        return 0;
}
```

vinit

> #include "mathlib.h"
> void vinit(Real__Vector v, unsigned n, Real *value*);

DESCRIPTION

The *vinit* function initializes each of the n elements of the real vector v with the real scalar *value*.

The real vector v must be allocated by *valloc* prior to calling the *vinit* function.

RETURNS

None.

SEE ALSO

Cmxinit(), mxinit(), Cvinit().

```
#include "mathlib.h"
#include <stdio.h>
main()
{
Real_Vector v;
int i;
unsigned n=2;

        v = valloc(NULL, n);        /* allocate vector m[][]                */

        vinit(v, n, 0);             /* set vector v[] to zero               */

        printf("the following real vector of length 2 should be zero:\n");
        for (i=0; i<n; i++) {
                printf(" %f\n", v[i]);
        }
        return 0;
}
```

vmag

#include "mathlib.h"
Real vmag(Real__Vector *v*, unsigned *n*);

DESCRIPTION

The *vmag* function computes the magnitude of the real vector *v*. The magnitude is equal to the square root of the sum of the squares of each element in vector *v*.

The *n* argument specifies the number of elements in the vector. The real vector *v* must be allocated by *valloc* prior to calling the *vmag* function.

RETURNS

A real value that represents the magnitude of the vector *v*.

SEE ALSO

Cvmag(), Cmag(), Cmxmag().

```
#include "mathlib.h"
#include <stdio.h>
Real V[2] = { 1, 1 };
main()
{
Real_Vector v;
Real magnitude;
int i;

        v = valloc(V, 2);    /* initialize v[] to static real array V[]    */

        magnitude = vmag(v, 2);        /* compute the magnitude of v[]    */

        /* the answer from the Pythagorean theorem = sqrt(2) = 1.41421...    */

        printf("the magnitude of v[] = %f\n", magnitude);
        return 0;
}
```

vmaxval

> **#include "mathlib.h"**
> **Real vmaxval(Real_Vector v, unsigned n, unsigned $*imx$);**

DESCRIPTION

The *vmaxval* function finds the element in the real vector v that has the largest magnitude.

The n argument specifies the number of elements in the vector. The $*imx$ argument is the address of an unsigned integer where *vmaxval* stores the element (i. e., index) number of the largest value.

RETURNS

The largest real value in the real vector v. The index of the element corresponding to the largest value is returned via the imx argument.

SEE ALSO

Cmxmaxvl(), Cmxminvl(), Cvmaxval(), mxmaxval(), mxminval(), vminval().

```
#include "mathlib.h"
#include <stdio.h>
Real V[2] = { 3,4 };
main()
{
Real_Vector v;
unsigned imax;
Real biggest;

        v = valloc(V, 2);             /* allocate and initialize vector v[]    */

        biggest = vmaxval(v, 2, &imax); /* find biggest value of v[]           */

        printf("the maximum value is = %f\n", biggest);
        printf("it is element # %d of the vector v[]\n", imax);
        return 0;
}
```

vminval

#include "mathlib.h"
Real vminval(Real_Vector v, unsigned n, unsigned *imn);

DESCRIPTION

The *vminval* function finds the element in the real vector v that has the smallest magnitude.

The n argument specifies the number of elements in the vector. The *imn argument is the address of an unsigned integer where *vminval* stores the element (i. e., index) number of the smallest value.

RETURNS

The smallest real value in the real vector v. The index of the element corresponding to the smallest value is returned via the imn argument.

SEE ALSO

Cmxmaxvl(), Cmxminvl(), Cvmaxval(), Cvminval(), mxmaxval(), mxminval(), vmaxval().

```
#include "mathlib.h"
#include <stdio.h>
Real V[2] = { 3,4 };
main()
{
Real_Vector v;
unsigned imin;
Real smallest;

        v = valloc(V, 2);           /* allocate and initialize vector v[]    */

        smallest = vminval(v, 2, &imin); /* find smallest value of v[]        */

        printf("the minimum value is = %f\n", smallest);
        printf("it is element # %d of the vector v[]\n", imin);
        return 0;
}
```

vmxmul

```
#include "mathlib.h"
Real_Vector vmxmul(Real_Matrix m, Real_Vector v, unsigned n1, unsigned n2);
```

DESCRIPTION

The *vmxmul* function multiplies the real matrix m by the real vector v.

The $n1$ argument specifies the number of rows in matrix m. The $n2$ argument specifies the number of columns in matrix m, as well as the number of elements in vector v.

The real matrix m and the real vector v must be allocated by *mxalloc* and *valloc* respectively prior to calling the *vmxmul* function. The resulting real vector of length $n1$, representing the product of m and v, is automatically allocated by the *vmxmul* function.

RETURNS

A real vector $v1$ of length $n1$, where $v1[n1] = m[n1][n2] * v[n2]$ if successful, or NULL if an error occurs.

SEE ALSO

vmxmul1(), Cvmxmul(), Cvmxmul1(), Cvectomx(), vectomat().

```
#include "mathlib.h"
#include <stdio.h>
#include <math.h>
#include <stdlib.h>
Real V[3] = { 4, 4, 4 };
Real M[2][3] = { 1, 3, 5, 2, 4, 6 };
main()
{
int i,n1=2,n2=3;
Real_Vector v,v1;
Real_Matrix m;

        v = valloc(V, n2);      /* allocate and initialize v[]              */
        m = mxalloc(M, n1, n2);/* allocate and initialize m[][]            */
        v1 = vmxmul(m, v, n1, n2);  /* compute the product v1[] = m[][]*v[]   */
        for (i=0; i<n1; i++) {
                printf("%d %f\n", i, v1[i]);
        }
 /* the answer is v1[0] = 36, v1[1] = 48                                     */
     mxfree(m) ; /* free matrix m[][] */
     return 0;
}
```

vmxmul1

#include "mathlib.h"
Real__Vector vmxmul1(Real__Matrix m, **Real__Vector** v, **unsigned** $n1$, **unsigned** $n2$);

DESCRIPTION

The *vmxmul1* function multiplies the transpose of the real matrix m by the real vector v.

The $n1$ argument specifies the number of columns in matrix m. The $n2$ argument specifies the number of rows in matrix m, as well as the number of elements in vector v.

The real matrix m and the real vector v must be allocated by *mxalloc* and *valloc* respectively prior to calling the *vmxmul1* function. The resulting real vector of length $n1$, representing the product of m^T and v is automatically allocated by the *vmxmul1* function.

RETURNS

A real vector $v1$ of length $n1$, where $v1[n1] = m^T[n2][n1] * v[n2]$ if successful, or NULL if an error occurs.

SEE ALSO

vmxmul(), Cvmxmul(), Cvmxmul1(), Cvectomx(), vectomat().

```
#include "mathlib.h"
#include <stdio.h>
#include <math.h>
#include <stdlib.h>
Real V[3] = { 4, 4, 4 };
Real M[3][2] = { 1, 2, 3, 4, 5, 6 };
main()
{
int i,n1=2,n2=3;
Real_Vector v,v1;
Real_Matrix m;

    v = valloc(V, n2);      /* allocate and initialize v[]              */
    m = mxalloc(M, n2, n1);/* allocate and initialize m[][]             */
    v1 = vmxmul1(m, v, n1, n2);/* compute the product v1[] = m'[][]*v[] */
    for (i=0; i<n1; i++) {
            printf("%d %f\n", i, v1[i]);
    }
 /* the answer is v1[0] = 36, v1[1] = 48                                */
    mxfree(m) ; /* free matrix m[][] */
    return 0;
}
```

vread

#include "mathlib.h"
Real__Vector vread(FILE $^*fp$, unsigned n);

DESCRIPTION

The *vread* function reads the next n real values from the file pointed to by fp. The file must have been opened as a binary file (e. g., $fp = $ fopen("filename", "rb")), since the *vread* function assumes that the real values are stored in binary format. The number of bytes read from the file is equal to $n *$ **sizeof(Real)**. The file pointer should be advanced this number of bytes toward the end-of-file if the n real values are read successfully.

The *vread* function automatically allocates a real vector of length n to store the real values that are read from the file.

The *vwrite* function may be used to create a file of binary real values.

RETURNS

A real vector of length n if successful, or NULL if an error occurs. If a read error occurs, an "Error reading from file" error message is generated (math_errno = E_READ).

SEE ALSO

vwrite(), xywrite(), Cvread(), Cvwrite(), xyread().

```
#include "mathlib.h"
#include <stdio.h>
#include <stdlib.h>
Real V1[3] = { 1.1, 2.2, 3.3 };
Real V2[3] = { 4.4, 5.5, 6.6 };
Real V3[3] = { 7.7, 8.8, 9.9 };
main()
{
FILE *fp;
Real_Vector v1,v2,v3,x1,x2,x3;
int i;

    v1 = valloc(V1, 3);
    v2 = valloc(V2, 3);
    v3 = valloc(V3, 3);

    fp = fopen("binary.dat", "w+b");/*open file "binary.dat" for binary write */
    if (fp == NULL) {               /*see if any errors were generated        */
            printf("can't open binary.dat\n");
            exit(1);
    }

    vwrite(fp, v1, 3);              /* write vector v1[] to "binary.dat"    */
    vwrite(fp, v2, 3);              /* write vector v2[] to "binary.dat"    */
    vwrite(fp, v3, 3);              /* write vector v3[] to "binary.dat"    */

    fseek(fp, 2*3*sizeof(Real), SEEK_SET);/* move pointer to 3rd vector     */
    x3 = vread(fp, 3);              /* read vector x3[] from binary.dat     */

    fseek(fp, 3*sizeof(Real), SEEK_SET);/* move pointer to 2nd vector       */
    x2 = vread(fp, 3);              /* read vector x2[] from binary.dat     */

    fseek(fp, 0, SEEK_SET);/* move pointer to start of 1st vector           */
    x1 = vread(fp, 3);              /* read vector x1[] from binary.dat     */

    for( i=0; i<3; i++){
        printf("x3[%d] = %3.1f, x2[%d] = %3.1f, x1[%d] = %3.1f\n",
                i,x3[i], i,x2[i], i,x1[i]);
    }
    fclose(fp);
    return 0;
}
```

PROGRAM OUTPUT

```
x3[0] = 7.7, x2[0] = 4.4, x1[0] = 1.1
x3[1] = 8.8, x2[1] = 5.5, x1[1] = 2.2
x3[2] = 9.9, x2[2] = 6.6, x1[2] = 3.3
```

vscale

#include "mathlib.h"
Real_Vector vscale(Real_Vector *v*, unsigned *n*, Real *value*);

DESCRIPTION

The *vscale* function multiplies the real vector *v* by the real scalar *value*.

The *n* argument specifies the number of elements in the vector. The real vector *v* must be allocated by *valloc* prior to calling the *vscale* function. The resulting real vector representing the product of *v* and *value* is automatically allocated by the *vscale* function.

The user may select the option of writing over the input vector *v* with in-place computations. If memory is a limitation, and *v* is no longer needed, this may prove useful. This option can be selected by setting the global variable **useinput_ = 1** prior to calling *vscale*. The default option is to not destroy the input vector *v* (i. e., **useinput_ = 0**).

RETURNS

An *n* element real vector representing the product of the real vector *v* with the real *value* if successful, or NULL if an error occurs.

SEE ALSO

CmxscalC(), CmxscalR(), mxscale(), CvscalC(), CvscalR().

```
#include "mathlib.h"
#include <stdio.h>
Real V[2] = { 1,1 };
main()
{
Real_Vector v,v1;
unsigned n=2;
int i;
Real value;

        value = 2;                 /* define a real scalar              */
        v=valloc(V, n);            /* allocate and initialize v[]       */
        v1 = vscale(v, n, value);  /* now scale real vector v[] by "value" */
        /* the answer should be : v1[0] = 2;  v1[1]= 2;                 */
        for (i=0; i<2; i++) {
                printf("v1[%d] = %f\n", i, v1[i]);
        }
        vfree(v1);    /* free vector v1[] */
        return 0;

}
```

vsub

#include "mathlib.h"
Real_Vector vsub(Real_Vector $v1$**, Real_Vector** $v2$**, unsigned** n**);**

DESCRIPTION

The *vsub* function subtracts the real vector $v2$ from the real vector $v1$.

The n argument specifies the number of elements in each vector. The real vectors $v1$ and $v2$ must be allocated by *valloc* prior to calling the *vsub* function. The resulting real vector representing the difference between $v1$ and $v2$ is automatically allocated by the *vsub* function.

The user may select the option of writing over the input vector $v1$ with in-place computations. If memory is a limitation, and $v1$ is no longer needed, this may prove useful. This option can be selected by setting the global variable **useinput_ = 1** prior to calling *vsub*. If this option is selected, the resulting vector overwrites the input vector $v1$. The default option is to not destroy the input vector $v1$ (i. e., **useinput_ = 0**).

RETURNS

A real vector of length n representing the difference between $v1$ and $v2$ if successful, or NULL if an error occurs.

SEE ALSO

Cmxadd(), Cmxsub(), mxadd(), mxsub(), Cvsub(), Cvadd(), vsub().

```
#include "mathlib.h"
#include <stdio.h>
main()
{
int i;
Real_Vector v,v1,v2;
unsigned n=2;
/* subtract the following two vectors:                             */
/*                                                                 */
/*          v[] = v1[] - v2[]                                      */
/*                                                                 */
/* where v1[] and v2[] are real and are given below:              */
    v1 = valloc(NULL, n);  /* allocate vector of length 2          */
    v2 = valloc(NULL, n); /* allocate another vector of length 2   */
    v1[0] = 2;   v1[1] = 2;  v2[0] = 3;   v2[1] = 3;

    v = vsub(v1, v2, n); /* subtract v2[] from v1[]                 */

    /* the answer should be : v[0] = -1;  v[1] = -1;               */

    for (i=0; i<n; i++) {
        printf("v2[%d] = %f\n", i, v[i]);
    }
    vfree(v);    /* free vector v[]  */
    vfree(v1);   /* free vector v1[] */
    vfree(v2);   /* free vector v2[] */
    return 0;
}
```

vwrite

#include "mathlib.h"
int vwrite(FILE * fp, Real__Vector v, unsigned n);

DESCRIPTION

The *vwrite* function writes the n real values in the real vector v to the file pointed to by fp. The file must have been opened as a binary file (e. g., fp = fopen("filename", "wb")), since the *vwrite* function outputs the real values in binary format. The number of bytes written to the file is equal to n * **sizeof(Real)**. The file pointer should be advanced this number of bytes toward the end-of-file if the n real values are written successfully.

The real vector v must be allocated by *valloc* prior to calling the *vwrite* function. The *vread* function may be used to read real values from a binary file.

RETURNS

0 if the real vector v is successfully written to the file, or -1 if an error occurs. If a write error occurs, an "Error writing to file" error message is generated (math_errno = E__WRITE).

SEE ALSO

vread(), Cvread(), Cvwrite(), xywrite(), xyread().

```
#include "mathlib.h"
#include <stdio.h>
#include <stdlib.h>
Real V1[3] = { 1.1, 2.2, 3.3 };
Real V2[3] = { 4.4, 5.5, 6.6 };
Real V3[3] = { 7.7, 8.8, 9.9 };
main()
{
FILE *fp;
Real_Vector v1,v2,v3,x1,x2,x3;
int i;

    v1 = valloc(V1, 3);
    v2 = valloc(V2, 3);
    v3 = valloc(V3, 3);

    fp = fopen("binary.dat", "w+b");/*open file "binary.dat" for binary write */
    if (fp == NULL) {               /*see if any errors were generated       */
            printf("can't open binary.dat\n");
            exit(1);
    }

    vwrite(fp, v1, 3);              /* write vector v1[] to "binary.dat"   */
    vwrite(fp, v2, 3);              /* write vector v2[] to "binary.dat"   */
    vwrite(fp, v3, 3);              /* write vector v3[] to "binary.dat"   */
    fseek(fp, 0, SEEK_SET);    /* position file indicator to top of file   */

    x1 = vread(fp, 3);              /* read vector x1[] from "binary.dat"   */
    x2 = vread(fp, 3);              /* read vector x2[] from "binary.dat"   */
    x3 = vread(fp, 3);              /* read vector x3[] from "binary.dat"   */

    for( i=0; i<3; i++){
        printf("x1[%d] = %3.1f, x2[%d] = %3.1f, x3[%d] = %3.1f\n",
                i,x1[i], i,x2[i], i,x3[i]);
    }
    fclose(fp);
    return 0;
}
```

PROGRAM OUTPUT

```
x1[0] = 1.1, x2[0] = 4.4, x3[0] = 7.7
x1[1] = 2.2, x2[1] = 5.5, x3[1] = 8.8
x1[2] = 3.3, x2[2] = 6.6, x3[2] = 9.9
```

whitnois

#include "mathlib.h"
double whitnois(double *sigma*, double *u*);

DESCRIPTION

Random noise generators are frequently used in simulations to predict expected responses of engineering systems. Many engineering systems are too complex to evaluate directly and must be characterized via statistical waveform analysis.

The *whitnois* function generates uniform random white noise with a standard deviation specified by *sigma* and a mean specified by *u*. The power (i. e., variance) of the white noise process is *sigma*^{2}.

The following method uses the uniform random distribution to generate the noise samples:

If X is a random variable uniformly distributed with zero mean and a standard deviation of one, then

$$X' = (b - a)(X - 1/2) + (a + b)/2$$

is also uniformly distributed with a mean u and a standard deviation of *sigma*, where X' is the desired noise sample and

$$u = (a + b)/2$$
$$s = (b - a)/\sqrt{12}$$

Thus, a and b are uniquely determined by the desired mean and standard deviation.

Different white noise sequences may be generated by calling the standard C function *srand* prior to calling the *whitnois* function. The *srand* function requires a single unsigned integer argument, which seeds the random number generator. A different random number sequence is generated for each different seed value.

RETURNS

A double precision value representing a uniform random white noise sample of mean u and standard deviation *sigma*.

SEE ALSO

poisson(), urand(), normal().

```
#include <stdio.h>
#include <stdlib.h>
#include <math.h>
#include "mathlib.h"
main()
{
double tmean=9.0,std_dev = 3.0,mean,var,noise;
int i,n = 500;

/*  Uniform Random White Noise Test (500 trials)   */

    srand(2);
    mean = var = 0.0;
    for(i=0; i<n; i++) {
        noise = whitnois(std_dev, tmean);
        mean += noise;
        var += noise*noise;
    }
    var = var/(n-1) - (mean/n) * mean/(n-1);
    mean /= n;
    printf("   Uniform Random White Noise Test (500 trials):\n\n");
    printf("   true mean = %4.0f         true standard deviation = %4.0f\n",
                                                tmean, std_dev);
    printf("observed mean = %4.1f      observed standard deviation = %4.1f\n",
                                                mean, sqrt(var));

    return 0;
}
```

PROGRAM OUTPUT

```
    Uniform Random White Noise Test (500 trials):

   true mean =    9          true standard deviation =    3
observed mean =  9.1      observed standard deviation =  2.9
```

xyinfo

```
#include "mathlib.h"
int xyinfo(char *filename, int *format, int *curves, unsigned points[ ], Labels *labels,
        int *size);
```

DESCRIPTION

The *xyinfo* function reads only the header information from a data file created by the *xywrite* function, by the GRAFIX program, or by an editor or spreadsheet. The file may contain data stored in either binary or ASCII format (see the *xywrite* function for a description of the data file).

The argument *filename* is a character string that specifies the name of the data file. The argument *format* specifies the address of an integer variable where *xyinfo* returns the file format. The file format is one of the following integer values defined in **mathlib.h**: XY_ASCII, XY_BINARY, XYY_ASCII, or XYY_BINARY.

The argument *curves* specifies the address of an integer variable where *xyinfo* returns the number of curves that the file contains.

The argument *points*[ ] is an array of integers where *xyinfo* returns the number of points for each curve. The number of points for curve 0 is returned in *points*[0], the number of points for curve 1 in *points*[1], and so on. If the format is XYY (i. e., XYY_ASCII or XYY_BINARY), then all the curves will contain the same number of points (i. e., each element of *points*[ ] will have the same value).

The NULL constant may optionally be passed as the *points*[ ] argument. If NULL is passed, then no information regarding the number of points in each curve is returned. If the argument is not NULL, then the *points*[ ] array must be dimensioned with at least as many elements as there are curves in the file. For example, if the file contains only one curve, then a dimension of 1 is sufficient. If the file contains five curves, then a dimension of 5 is sufficient. The maximum number of curves in a file is **CURVES_MAX**, a constant defined in **mathlib.h**. A dimension of CURVES_MAX is sufficient to allow the *xyinfo* function to read the header information of any data file created by the *xywrite* function.

The argument *labels* specifies the address of a structure of type **Labels**, where *xyinfo* returns character strings that describe the title, xaxis, and yaxis, as well as each curve. The **Labels** structure is defined in **mathlib.h** as follows:

```
typedef struct {
        char title[TITLE_MAX+1];
        char xaxis[AXIS_MAX+1];
        char yaxis[AXIS_MAX+1];
        char curve[CURVES_MAX][LABEL_MAX+1];
    } Labels;
```

The NULL constant may optionally be passed as the *labels* argument. If NULL is passed, then no information regarding labels is returned. If the argument is not NULL, then the title is returned in the string *labels*.title, the xaxis in the string *labels*.xaxis, and the yaxis in the string *labels*.yaxis. The label for curve 0 is returned in the string *labels*.curve[0], the label for curve 1 in the string *labels*.curve[1], and so on. If the file contains no labels, then each of these strings will have a length of 0. The title string is limited to TITLE_MAX characters. The xaxis and yaxis strings are limited to AXIS_MAX characters. The number of curves is limited to CURVES_MAX, and the string describing each curve is limited to LABEL_MAX characters. All these constants are defined in **mathlib.h**. Strings that are longer than these constants are truncated before being input from the file.

The argument *size* specifies the address of an integer variable, where *xyinfo* returns the size of the binary floating point values stored in the file. If the format is binary (i. e., XY_BINARY or XYY_BINARY), then the size of the floating point values stored in the file must match the sizeof(Real) in the program. Otherwise the *xyread* function cannot be used to read the data. If the format is ASCII (i. e., XY_ASCII or XYY_ASCII), then the *xyinfo* function returns 0 for the *size*, since the file does not contain any binary values.

RETURNS
An integer value representing the length of the header for binary files, or the number of lines in the header for ASCII files. A value of −1 is returned if an error occurs and **math_errno** is set to E_OPEN if *xyinfo* is unable to open the file, or E_HEADER if the file header is invalid, or E_CURVES if the number of curves is less than one or greater than CURVES_MAX.

SEE ALSO
xywrite(), xyread(), vread(), vwrite().

```
#include "mathlib.h"
#include <stdio.h>
#include <string.h>
char prompt[] = "Enter name of data file to view --> ";

main()
{
    int i, format, size, curves;
    unsigned points[CURVES_MAX];
    Labels labels;
    char type[11], buffer[80];

    math_errmsg = OFF;   /* turn off error messages */
    math_abort = OFF;    /* turn off program abort */
    printf("%s", prompt);

    /* Keep getting file names until only the ENTER key is pressed */
    while (strlen(gets(buffer)) != 0) {
```

(continued on next page)

(continued)

```
        /* read the file */
        if (xyinfo(buffer, &format, &curves, points, &labels, &size) == -1) {
            printf("Attempted to process data file  --> %s\n", buffer);
            math_err();
            printf("\n\n%s", prompt);
            continue;
        }

        /* determine the file format */
        switch (format) {
            case XY_ASCII  : strcpy(type, "XY_ASCII");   break;
            case XY_BINARY : strcpy(type, "XY_BINARY");  break;
            case XYY_ASCII : strcpy(type, "XYY_ASCII");  break;
            case XYY_BINARY: strcpy(type, "XYY_BINARY"); break;
        }

        /* Output information about the file */
        printf("%s contains %d curve(s), %s format\n",
                    buffer, curves, type);
        if (labels.title[0] != '\0') printf("Title = %s\n", labels.title);
        else printf("No text for Title\n");
        if (labels.xaxis[0] != '\0') printf("X axis = %s\n", labels.xaxis);
        else printf("No text for X axis\n");
        if (labels.yaxis[0] != '\0') printf("Y axis = %s\n", labels.yaxis);
        else printf("No text for Y axis\n");

        /* Output the labels for each curve */
        for (i=0; i < curves; i++) {
            if (labels.curve[i][0] != '\0')
                printf("Curve %d = %s\n", i, labels.curve[i]);
            else printf("No text for curve %d\n", i);
        }

        /* Output the size of a floating point value if binary */
        if (format == XY_BINARY || format == XYY_BINARY) {
            printf("Size = %d ", size);
            if (size == sizeof(Real))
                printf("is equal to sizeof(Real)\n");
            else
                printf("is not equal to sizeof(Real) = %d\n", sizeof(Real));
        }

        printf("\n\n%s", prompt);
    }
    return 0;
}
```

xyread

```
#include "mathlib.h"
int xyread(char *filename, int *format, int *curves, unsigned points[],
                Labels *labels, Real_Vector x[], Real_Vector y[]);
```

DESCRIPTION

The *xyread* function reads one or more curves from a data file, where each curve is represented by an x and y vector (i. e., a set of x and y data points). The file may have been created by the *xywrite* function, the GRAFIX program, or manually with an editor or spreadsheet. The file may contain data stored in either binary or ASCII format (see the *xywrite* function for a description of the data file).

The argument *filename* is a character string that specifies the name of the data file to read.

The argument *format* specifies the address of an integer variable where *xyread* returns the file format. The file format is one of the following integer values: XY_ASCII, XY_BINARY, XYY_ASCII, or XYY_BINARY (defined in **mathlib.h**).

The argument *curves* specifies the address of an integer variable where *xyread* returns the number of curves that the file contains. Since array indexing in C starts at 0, it is convenient to think of the curves as being numbered from $0 \ldots curves - 1$. That is, the first curve is 0, the second 1, and so on with the last curve being *curves* -1.

The argument *points*[ ] specifies an array of integers where *xyread* returns the number of points for each curve. The number of points for curve 0 is returned in *points*[0], the number of points for curve 1 in *points*[1], and so on. If the format is XYY (i. e., XYY_ASCII or XYY_BINARY), then all of the curves will contain the same number of points (i. e., each element of *points*[ ] will have the same value).

The argument *labels* specifies the address of a structure of type **Labels** where *xyread* returns character strings that describe the <u>title</u>, <u>xaxis</u>, and <u>yaxis</u>, as well as each <u>curve</u>. The **Labels** structure is defined in **mathlib.h** as follows:

```
typedef struct {
        char title[TITLE_MAX+1];
        char xaxis[AXIS_MAX+1];
        char yaxis[AXIS_MAX+1];
        char curve[CURVES_MAX][LABEL_MAX+1];
} Labels;
```

The NULL constant may optionally be passed as the *labels* argument. If NULL is passed, then no information regarding labels is returned. If the argument is not NULL, then the title is returned in the string *labels*.title, the xaxis in the string *labels*.xaxis, and the yaxis in the string *labels*.yaxis. The label for curve 0 is returned in the string *labels*.curve[0], the label for curve 1 in the string *labels*.curve[1], and so on. If the file contains no labels, then each of these strings will have a length of 0. The title string is limited to TITLE_MAX characters. The xaxis and yaxis strings are limited to AXIS_MAX characters. The number of curves is limited to CURVES_MAX, and the string describing each curve is limited to LABEL_MAX characters. All these constants are defined in **mathlib.h**. Strings that are longer than these constants are truncated before being input from the file.

The arguments $x[]$ and $y[]$ are **Real_Vector** arrays where *xyread* returns the x and y vectors for each curve in the file. The x and y vectors for curve 0 are returned in $x[0]$ and $y[0]$ respectively, the x and y vectors for curve 1 in $x[1]$ and $y[1]$ respectively, and so on. If the format is XYY (i. e., XYY_ASCII or XYY_BINARY), then each element of the $x[]$ array will reference the same x vector (i. e., each element of $x[]$ will have the same value).

The individual data points in a vector are referenced by specifying a second dimension. For example, the first point in curve 0 is referenced by $x[0][0]$ and $y[0][0]$, the second point by $x[0][1]$ and $y[0][1]$, and so on. The first point in curve 1 is referenced by $x[1][0]$ and $y[1][0]$, the second point by $x[1][1]$ and $y[1][1]$, and so on.

The *points*$[]$, $x[]$, and $y[]$ arrays must be dimensioned with at least as many elements as there are curves in the file being read. For example, if the file contains only one curve, then a dimension of 1 is sufficient. If the file contains five curves, then a dimension of 5 is sufficient. The maximum number of curves that the *xyread* function will read is **CURVES_MAX**, a constant defined in **mathlib.h**. A dimension of CURVES_MAX is sufficient to allow the *xyread* function to read any data file created by the *xywrite* function.

The *xyread* function automatically allocates memory for the x and y vectors using the *valloc* function, then returns the starting address of each vector in the $x[]$ and $y[]$ arrays. Therefore, before calling the *xyread* function a second time, the *vfree* function should be called to free each of the vectors stored in the $x[]$ and $y[]$ arrays. Otherwise the memory occupied by these vectors will be lost.

DATA FILE FORMAT
The first line in the file must be a header that defines information about the data. The header must contain definitions for the following keywords: **Format=**, **Curves=**, and **Points=**. For binary formats, the keyword **Size=** must also be defined. Optionally, the keywords **Title=**, **Xaxis=**, **Yaxis=**, and **Label=** may be defined. See the *xywrite* function for further details.

The format must be XY_ASCII, XY_BINARY, XYY_ASCII, or XYY_BINARY (e. g., Format = XY_ASCII). The file must contain at least one curve (e. g., Curves = 1), but not more than CURVES_MAX curves. The number of points in each curve must be specified (e. g., Points = 5 or Points = 5, 10, 15). If the format is binary (i. e., Format = XY_BINARY or Format = XYY_BINARY), then the size of a binary floating point value must be specified (e. g., Size = 8). For example, the following header identifies an XY_BINARY formatted file consisting of two curves. Curve 0 contains 200 points and curve 1 contains 300 points. The size of each binary floating point value is 4 bytes.

CURVES=2 POINTS=200,300 FORMAT=xy_binary SIZE=4

The values for **Title=**, **Xaxis=**, **Yaxis=**, or **Label=** are specified as a sequence of characters terminated by a semicolon (;). In the case of **Label=**, multiple sequences are specified (i. e., one for each curve). For example, the following header specifies XY_ASCII format, three curves containing 200 points each, a title of Filter Response, an x axis labeled Frequency (Mhz), a y axis labeled Magnitude (dB), the first curve labeled Filter 1, the second curve labeled Filter 2, and the third curve labeled Filter 3.

> **format=XY_ASCII curves=3 points=200,200,200 title=Filter **
> **Response; xaxis=Frequency (Mhz); yaxis=Magnitude (dB); label=**
> **Filter 1;Filter 2;Filter 3;**

The organization of the header information is quite flexible. Keywords may appear in any order within the header. The case of the keywords is not significant. Except within semicolon-terminated strings, blank spaces are not significant. However, since the '=' character is part of the keyword, there must be no spaces between the keyword and the '=' character. For ASCII formats only, the header may consist of one or more lines. The '\\' character at the end of a line indicates that the header information continues on the next line. The *xywrite* function encloses each line of the header within quotes so that spreadsheets will treat the header information as strings. The *xyread* function does not require the surrounding quotes, and simply discards them if present.

The header must be terminated by the end-of-line character ('\n') if the format is ASCII (i. e., XY_ASCII or XYY_ASCII), or by the Ctrl Z character (0x1A) if the format is binary (i. e., XY_BINARY or XYY_BINARY). If the format is ASCII, the header must <u>not</u> contain a Ctrl Z character. If the format is binary, the header must <u>not</u> contain an end-of-line character.

For ASCII files (i. e., Format=XY_ASCII or Format=XYY_ASCII), a comment line must appear above the actual x-y data values, but may contain any text, or even be totally blank. The *xywrite* function uses the comment lines to identify the x and y vectors (i. e., the vectors are identified as Xn or Yn, where n is the curve number). The *xyread* function simply ignores the contents of these comment lines. For example, either of the following two files would be acceptable.

```
Format= XY_ASCII  Curves= 2  Points= 2,3
This comment line must be present but may be blank
 1.000000e+001   1.000000e+001
 2.000000e+001   2.000000e+001
This comment line must be present but may be blank
 1.000000e+001   1.000000e+001
 2.000000e+001   2.000000e+001
 3.000000e+001   3.000000e+001
```

```
"format=xyy_ascii curves=3 points=3 title=Example;"
This comment line must be present but may be blank
 1.000000e+001   1.000000e+001   1.100000e+001   1.200000e+001
 2.000000e+001   2.000000e+001   2.100000e+001   2.200000e+001
 3.000000e+001   3.000000e+001   3.100000e+001   3.200000e+001
```

RETURNS

0 if successful or −1 if an error occurs. If an error occurs, **math_errno** is set to
E_OPEN if *xyread* is unable to open the file, or E_READ if an error occurs while
reading from the file, or E_HEADER if the file header is invalid, or E_CURVES
if the number of curves in the file is greater than **CURVES_MAX**, or E_RSIZE if
the file format is binary (i. e., XY_BINARY or XYY_BINARY) and the **Size=n**
specified by the header does not equal the sizeof(Real) in the program.

SEE ALSO

xywrite(), xyinfo(), vread(), vwrite().

```c
#include "mathlib.h"
#include <string.h>
#include <stdio.h>
char prompt[] = "Enter name of data file to view --> ";

main()
{
    Real_Vector x[CURVES_MAX], y[CURVES_MAX];
    int format, curves;
    unsigned points[CURVES_MAX];
    Labels labels;
    char type[11], buffer[80];
    unsigned i, j;
    math_errmsg = OFF;   /* turn off error messages */
    math_abort = OFF;    /* turn off program abort */
    printf("%s", prompt);
    /* Keep getting file names until only the ENTER key is pressed */
    while (strlen(gets(buffer)) != 0) {
        /* read the file */
        if (xyread(buffer, &format, &curves, points, &labels, x, y) == -1) {
            printf("Attempted to process data file  --> %s\n", buffer);
            math_err();
            printf("\n\n%s", prompt);
            continue;
        }
        /* determine the file format */
        switch (format) {
            case XY_ASCII  : strcpy(type, "XY_ASCII");   break;
            case XY_BINARY : strcpy(type, "XY_BINARY");  break;
            case XYY_ASCII : strcpy(type, "XYY_ASCII");  break;
            case XYY_BINARY: strcpy(type, "XYY_BINARY"); break;
        }
        /* Output information about the file */
        for (i=0; i < 79; i++) putchar('*'); putchar('\n');
        printf("%s contains %d curve(s), %s format\n",
                    buffer, curves, type);
        if (labels.title[0] != '\0') printf("Title = %s\n", labels.title);
        else printf("No Title text\n");
```

```
    if (labels.xaxis[0] != '\0') printf("Xaxis = %s\n", labels.xaxis);
    else printf("No Xaxis text\n");
    if (labels.yaxis[0] != '\0') printf("Yaxis = %s\n", labels.yaxis);
    else printf("No Yaxis text\n");
    for (i=0; i < 79; i++) putchar('*'); putchar('\n');
    /* Output the x-y data points for each curve */
    for (i=0; i < curves; i++) {
        printf("\n***> Press ENTER key to view curve %d", i);
        gets(buffer);
        printf("\n\n        <<< Curve %d x-y data points >>> ", i);
        if (labels.curve[i][0] == '\0') printf("No Label text\n\n");
        else printf(" Label = %s\n\n", labels.curve[i]);
        for (j=0; j < points[i]; j++)
            printf("%4d) %15.6e  %15.6e\n", j, x[i][j], y[i][j]);
    }
    /* Free the memory allocated for each of the x and y vectors */
    for (i=0; i < curves; i++) {
        vfree(x[i]);
        vfree(y[i]);
    }
    printf("\n\n%s", prompt);
}
return 0;
}
```

xywrite

#include "mathlib.h"
int xywrite(char **filename,* **int** *format,* **int** *curves,* **unsigned** *points*[ ],
                    Labels **labels,* **Real__Vector** *x*[ ], **Real__Vector** *y*[ ]);

DESCRIPTION

The *xywrite* function writes one or more curves to a data file, where each curve is represented by an x and y vector (i. e., a set of x and y data points). The files created by the *xywrite* function may subsequently be read by the *xyread* function or by the GRAFIX program. The output format may be ASCII or binary. If the output format is ASCII, then the file can also be read by most editors and spreadsheets. For example, the file import command in Lotus 123 may be used to read the file.

The argument *filename* is the address of a character string that specifies the name of the data file to create.

The argument *format* is an integer value that specifies the format of the data file. The *format* must be one of the following values defined in **mathlib.h**: XY__ASCII, XY_BINARY, XYY_ASCII, or XYY_BINARY.

If the format is ASCII (i. e., XY__ASCII or XYY__ASCII), then floating point values are converted to ASCII characters before being written to the file. This format should be used when portability of the data is important. Most programs, including editors and spreadsheets, have no problem dealing with ASCII formatted files.

If the format is BINARY (i. e., XY_BINARY or XYY_BINARY), then floating point values are written in binary (i. e., nonreadable) format. This format should be used when speed or file size is important. Since no data conversion is necessary with binary format, the data is written much faster and requires much less disk space.

If the format is XY (i. e., XY__ASCII or XY__BINARY), then an independent x and y vector is written for each curve. When multiple curves are stored, this format should be used if any curve has an x vector that differs from the x vector of any other curve.

If the format is XYY (i. e., XYY__ASCII or XYY__BINARY), then an independent y vector is written for each curve, but only a single x vector is written. When multiple curves are stored, this format should be used if all of the curves share an identical x vector.

The argument *curves* is an integer value that specifies the number of curves to write. The value of *curves* must be greater than or equal to 1 and less than or equal to **CURVES__MAX**, an integer constant defined in **mathlib.h**. Since array indexing in C starts at 0, it is convenient to think of the curves as being numbered from $0 \ldots curves -1$. That is, the first curve is 0, the second 1, and so on with the last curve being

curves -1.

The argument *points*[ ] is an array of integers that specifies the number of points for each curve. The number of points for curve 0 is specified by *points*[0], the number of points for curve 1 by *points*[1], and so on. If the format is XYY (i. e., XYY__ASCII or XYY_BINARY), then all of the curves contain the same number of points (i. e., only *points*[0] is required).

The argument *labels* is the address of a structure of type **Labels** that provides a textual description of the data. This argument specifies labels for the <u>title</u>, <u>xaxis</u>, and <u>yaxis</u>, as well as labels for each <u>curve</u>. The NULL constant may optionally be passed as the *labels* argument. If NULL is passed, then no labels are output. The **Labels** structure is defined in **mathlib.h** as follows:

```
typedef struct {
          char title[TITLE_MAX+1];
          char xaxis[AXIS_MAX+1];
          char yaxis[AXIS_MAX+1];
          char curve[CURVES_MAX][LABEL_MAX+1];
      } Labels;
```

The title is specified by the string *labels*.title, the xaxis by the string *labels*.xaxis, and the yaxis by the string *labels*.yaxis. The label for curve 0 is specified by the string *labels*.curve[0], the label for curve 1 by the string *labels*.curve[1], and so on. These strings may contain any characters except the semicolon (;), the end-of-line character ('\n'), or Ctrl Z (0x1A). The title string is limited to TITLE_MAX characters. The xaxis and yaxis strings are limited to AXIS_MAX characters. The number of curves is limited to CURVES_MAX, and the label for each curve is limited to LABEL_MAX characters. All these constants are defined in **mathlib.h**. Strings that are longer than these constants are truncated before being output to the file.

The arguments $x[\,]$ and $y[\,]$ are **Real_Vector** arrays that specify the x and y vectors for each curve in the file. The x and y vectors for curve 0 are specified by $x[0]$ and $y[0]$ respectively, the x and y vectors for curve 1 by $x[1]$ and $y[1]$ respectively, and so on. If the format is XYY (i. e., XYY_ASCII or XYY_BINARY), then all of the curves use the same x vector (i. e., only $x[0]$ is required).

For the ASCII formats (i. e., XY_ASCII or XYY_ASCII), the floating point values are written in exponential format (i. e., $\pm d.ddddd e\pm ddd$). The number of significant digits that are output for each floating point value is controlled by the external integer variable named **math_digits**. By default, **math_digits** is initialized to 6. Therefore, each number output will contain six significant digits. Before calling the *xywrite* function, you may change the number of digits by changing the value of **math_digits**.

For the binary formats (i. e., XY_BINARY or XYY_BINARY), the floating point values are written as a binary sequence of bytes. The size of each binary floating point value is **sizeof(Real)** bytes.

DATA FILE FORMAT

The first line written to the file is a header that defines information about the data. The header defines values for several keywords. **Format=** specifies the format (i. e., XY_ASCII, XY_BINARY, XYY_ASCII, or XYY_BINARY), **Curves=** specifies the number of curves, **Points=** specifies the number of points in each curve, and for binary formats **Size=** specifies the sizeof(Real) (i. e., the size of each floating point value). Optionally, the header may define values for the following additional keywords. **Title=** describes the data as a whole, **Xaxis=** describes the x axis, **Yaxis=** describes the y axis, and **Label=** describes each curve.

For example, the following header specifies data stored in the XYY_ASCII format, consisting of five curves, with each curve containing 25 x and y data points, and no labels.

"Format=XYY_ASCII Curves=5 Points=25"

For XYY format (i. e., XYY_ASCII or XYY_BINARY), each curve must contain the same number of points. Therefore, **Points** is assigned a single value.

For XY format (i. e., XY_ASCII or XY_BINARY), the number of points must be defined for each curve. This is handled by listing a sequence of integer values separated by commas, where the first value defines the number of points in curve 0, the second the number of points in curve 1, and so on. For example, the following header specifies that the file contains data stored in the XY_BINARY format, the size of each binary floating point value being 8 bytes, consisting of three curves, with curve 0 containing 100 points, curve 1 containing 150 points, and curve 2 containing 200 points.

"Format=XY_BINARY Size=8 Curves=3 Points=100,150,200"

A value for **Title=, Xaxis=, Yaxis=,** or **Label=** is specified as a sequence of characters terminated by a semicolon (;). In the case of **Label=**, multiple sequences are specified (i. e., one for each curve). For example, the following header specifies XYY_ASCII format, two curves containing 12 points each, a title of My Plot, an x axis labeled Month, a y axis labeled Monthly Sales, the first curve labeled Product 1, and the second curve labeled Product 2.

"Format=XYY_ASCII Curves=2 Points=12 Title=My Plot; Xaxis=Month;"
"Yaxis=Monthly Sales; Label=Product 1;Product 2;"

By default, the *xywrite* function outputs the entire header as a single line. The line can become quite long if the header contains labels. This can potentially be a problem if you attempt to edit the file, since some editors have a limit on line length. Therefore, for ASCII formats only, the *xywrite* function has the ability to break the header line into multiple lines. The maximum number of characters per line in the header is controlled by an external integer variable named **math_linesize**. By default, **math_linesize** is initialized to HEADER_MAX, which causes the entire header to be written as a single line. Before calling the *xywrite* function, you may change the value of **math_linesize** to change the maximum number of characters per line. The '\' character at the end of a line is used to indicate that the header information continues on the next line.

The header is terminated by the end-of-line character ('\n') if the format is ASCII (i. e., XY_ASCII or XYY_ASCII), or by the Ctrl Z character (0x1A) if the format is binary (i. e., XY_BINARY or XYY_BINARY).

If the format is binary, the data itself immediately follows the header. For a single curve, (i. e., **Curves=1**), the XY_BINARY and XYY_BINARY formats are identical. The x vector is first, followed by the y vector. For multiple curves, the two formats are different. For XY_BINARY format, the x vector for curve 0 is first, followed by the y vector for curve 0, followed by the x vector for curve 1, followed by the y vector for curve 1, and so on. For XYY_BINARY format, a single x vector for all the curves is first, followed by the y vector for curve 0, followed by the y vector for curve 1, and so on.

If the format is ASCII, then immediately following the header is a comment line that identifies each of the x and y vectors. For a single curve (i. e., **Curves=1**), the XY_ASCII and XYY_ASCII formats are identical. The comment line identifies the x vector as "X0" and the y vector as "Y0." Following the comment line, each subsequent line contains a pair of x and y values. There is one pair of x and y values for each point in the curve. For example, the XY_ASCII format for one curve containing five points might look as follows:

```
"Format=XY_ASCII Curves=1 Points=5 Title=Simple Plot; Xaxis=X Data Points;"\
"Yaxis=Y Data Points; Label=Straight Line;"
      "X0"            "Y0"
  1.000000e-001   1.000000e-001
  2.000000e-001   2.000000e-001
  3.000000e-001   3.000000e-001
  4.000000e-001   4.000000e-001
  5.000000e-001   5.000000e-001
```

For multiple curves, the XY_ASCII and XYY_ASCII formats are different. For XY_ASCII format, each subsequent curve is formatted similar to curve 0. That is, each curve begins with a comment line that identifies the x and y vectors, and is followed by a pair of x and y values (one pair per line) for each point in the curve. The comment lines identify curve 1 as "X1" and "Y1," curve 2 as "X2" and "Y2," and so on. For example, the XY_ASCII format for three curves, with each curve containing three points, might look as follows:

```
"Format=XY_ASCII Curves=3 Points=3,3,3"
      "X0"            "Y0"
  1.000000e+001   1.000000e+001
  2.000000e+001   2.000000e+001
  3.000000e+001   3.000000e+001
      "X1"            "Y1"
  1.000000e+001   1.100000e+001
  2.000000e+001   2.100000e+001
  3.000000e+001   3.100000e+001
      "X2"            "Y2"
  1.000000e+001   1.200000e+001
  2.000000e+001   2.200000e+001
  3.000000e+001   3.200000e+001
```

For multiple curves in XYY_ASCII format, only the x vector for curve 0 appears, since all curves have identical x vectors. The y vectors for each subsequent curve appear in columns to the right of the y vector for curve 0. For example, the previous data in XYY_ASCII format would appear as follows:

```
"Format=XYY_ASCII Curves=3 Points=3"
     "X0"              "Y0"              "Y1"              "Y2"
 1.000000e+001  1.000000e+001  1.100000e+001  1.200000e+001
 2.000000e+001  2.000000e+001  2.100000e+001  2.200000e+001
 3.000000e+001  3.000000e+001  3.100000e+001  3.200000e+001
```

RETURNS

0 if successful or -1 if an error occurs. If an error occurs, **math_errno** is set to E_OPEN if *xywrite* is unable to create the file, or E_WRITE if an error occurs while writing to the file, or E_CURVES if the number of *curves* is greater than **CURVES_MAX**.

SEE ALSO

xyread(), xyinfo(), vread(), vwrite().

```c
#include <stdio.h>
#include <stdlib.h>
#include "mathlib.h"

#define CURVES 3
#define POINTS 9

Labels labels={"Text for Title",
               "Text for Xaxis",
               "Text for Yaxis",
               "Text for Curve 0",
               "Text for Curve 1",
               "Text for Curve 2"};
main()
{
    int i, j;
    unsigned points[CURVES];
    Real_Vector x[CURVES], y[CURVES];
```

(continued on next page)

(continued)

```c
    math_errmsg = OFF;    /* turn off error messages */
    math_abort = OFF;     /* turn off program abort */
    math_digits = 4;      /* set digits right of decimal point */
    math_linesize = 80;   /* set maximum characters per line for header */
    for (i = 0; i < CURVES; i++) {
        points[i] = POINTS;
        x[i] = valloc(NULL, POINTS);
        y[i] = valloc(NULL, POINTS);
        if (x[i] != NULL && y[i] != NULL)
            for (j = 0; j < POINTS; j++) {
                x[i][j] = j + 1;
                y[i][j] = x[i][j] + i/10.0;
            }
        else {
            math_err();
            exit(1);
        }
    }
    if (xywrite("xyasc.dat", XY_ASCII, CURVES, points, &labels, x, y) == -1)
        math_err();
    else
    if (xywrite("xyyasc.dat", XYY_ASCII, CURVES, points, NULL, x, y) == -1)
        math_err();
    else
    if (xywrite("xybin.dat", XY_BINARY, CURVES, points, NULL, x, y) == -1)
        math_err();
    else
    if (xywrite("xyybin.dat", XYY_BINARY, CURVES, points, &labels, x, y) == -1)
        math_err();
    else {
        printf("xyasc.dat, xyyasc.dat, ");
        printf("xybin.dat, and xyybin.dat files created");
    }
    return 0;
}
```

Appendix A

Function Categories

Complex Arithmetic Functions	
Cadd	Add two complex numbers
Cdiv	Divide two complex numbers
Cmag	Compute the magnitude of a complex number
Cmul	Multiply two complex numbers
Complx	Form a complex number from its imaginary and real parts
Conjg	Conjugate a complex number
CscalR	Multiply a complex number by a real number
Csqrt	Compute the principal square root of a complex number
Csub	Subtract two complex numbers

Real Matrix and Vector Functions

levinson	Levinson's recursion to solve a Toeplitz system of equations
lineq	Solve system of simultaneous linear equations
mxadd	Add two matrices
mxcentro	Centroid a unimodal 2-D array of data
mxcopy	Copy one matrix to another
mxdeterm	Compute the determinant of a matrix
mxdup	Duplicate and allocate a matrix
mxeigen	Compute the eigenvalues and eigenvectors of a real symmetric matrix
mxhisto	Compute the histogram of 2-D data to any resolution
mxident	Create the identity matrix of specified size
mxinit	Initialize a matrix to a floating point value
mxinv	Invert a matrix of any size with LU decomposition
mxinv22	Invert a two by two matrix
mxinv33	Invert a three by three matrix
mxmaxval	Find the max value of a matrix and the indices where it occurs
mxminval	Find the min value of a matrix and the indices where it occurs
mxmul	Multiply two matrices
mxmul1	Multiply the transpose of a matrix by another matrix
mxmul2	Multiply a matrix by the transpose of another matrix
mxscale	Multiply a matrix by a floating point scalar
mxsub	Subtract two matrices
mxtrace	Compute the trace of a matrix
mxtransp	Transpose a general matrix (in place, if square, and if desired)
pseudinv	Solve an overconstrained system of simultaneous linear equations
vadd	Add two vectors
v_centro	Centroid a unimodal vector of data
vcopy	Copy one vector to another
vcross	Form the cross product between two vectors
vdot	Compute the dot product between two vectors
vdup	Duplicate and allocate a vector
vectomat	Multiply two vectors into a matrix
vinit	Initialize a vector to a floating point value
vmag	Compute the magnitude of a real vector
vmaxval	Find the max value of a vector and the index where it occurs
vminval	Find the min value of a vector and the index where it occurs
vmxmul	Multiply a matrix by a vector to form another vector
vmxmul1	Multiply the transpose of a matrix by a vector into another vector
vscale	Multiply a vector by a floating point scalar
vsub	Subtract two vectors

Complex Matrix and Vector Functions

Clineqn	Solve system of simultaneous linear equations
Cmxadd	Add two matrices
Cmxconjg	Conjugate a complex matrix
Cmxcopy	Copy one matrix to another
Cmxeigen	Compute the eigenvalues and eigenvectors of a Hermitian matrix
Cmxdeter	Compute the determinant of a matrix
Cmxdup	Duplicate and allocate a matrix
Cmxinit	Initialize a matrix to a complex value
Cmxinv	Invert a matrix of any size with LU decomposition
Cmxinv22	Invert a two by two matrix
Cmxinv33	Invert a three by three matrix
Cmxmag	Compute the magnitude of a complex matrix
Cmxmaxvl	Find the max value of a matrix and the indices where it occurs
Cmxminvl	Find the min value of a matrix and the indices where it occurs
Cmxmul	Multiply two matrices
Cmxmul1	Multiply the transpose of a matrix by another matrix
Cmxmul2	Multiply a matrix by the transpose of another matrix
CmxscalC	Multiply a matrix by a complex scalar
CmxscalR	Multiply a matrix by a real scalar
Cmxsub	Subtract two matrices
Cmxtrace	Compute the trace of a matrix
Cmxtrans	Transpose a general matrix (in place, if square, and if desired)
Cvadd	Add two vectors
Cvconjg	Conjugate a complex vector
Cvcopy	Copy one vector to another
Cvdot	Compute the dot product between two vectors
Cvdup	Duplicate and allocate a vector
Cvectomx	Multiply two vectors into a matrix
Cvinit	Initialize a vector to a complex value
Cvmag	Compute the magnitude of a complex vector
Cvmaxval	Find the max value of a vector and the index where it occurs
Cvminval	Find the min value of a vector and the index where it occurs
Cvmxmul	Multiply a matrix by a vector to form another vector
Cvmxmul1	Multiply the transpose of a matrix by a vector into another vector
CvscalC	Multiply a vector by a complex scalar
CvscalR	Multiply a vector by a real scalar
Cvsub	Subtract two vectors

Probability and Statistics Functions

binomdst	Cumulative Binomial distribution function; given k0, find P(k < k0)
curvreg	Polynomial regression routine, y(x) = a + b*x + c*x^2 + d*x^3+ ...
hyperdst	Cumulative Hypergeometric distrib. function; given k0, find P(k < k0)
invprob	Inverse cumulative normal distribution funct; given P(x), find x
least_sq	Generalized least-squares regression, y(x) = a*f0(x) + b*f1(x) + ...
linreg	Linear regression routine, y(x) = m*x + b
normal	Normal (Gaussian) random number of specified mean and variance
nprob	Cumulative normal distribution function; given x0, find P(x < x0)
poissdst	Cumulative Poisson distribution function; given k0, find P(k < k0)
poisson	Poisson random number of specified mean and variance
stats	Compute the mean and variance of an array of data
urand	Uniform random number generator (0...1)

Numerical Analysis Functions

conjgrad	Minimize n-dimensional differentiable function f(x1, x2,..., xn)
deriv	Derivative of equidistant data array with differentiating filter
deriv1	Differentiate a user-defined (analytic) function f(x) at x = x0.
integrat	Integrate equidistant array of data with Simpson's rule
interp	Interpolate equidistant array with 5th degree Lagrange polynomial
p_roots	Compute real and imaginary roots of polynomials with real coefficients
interp1	Interpolate data with nth degree Lagrange polynomial, 0 < n < 10
newton	Find the zeros of the 1-dimensional function; i.e. x, where f(x) = 0
romberg	Romberg integration of a user-defined function f(x) over [x0, x1]
spline	Interpolate data with a natural cubic spline

Signal Processing Functions

auto2dft	2D-autocorrelation of an array via 2D-FFT
autocor	Autocorrelation of a real-time series
autofft	Fast autocorrelation of a real-time series via FFT
bandpass	Design a (FIR) bandpass filter with a Kaiser–Bessel window
bilinear	Bi-linear transformation of general s-plane function to z-plane
conv2dft	2-D convolution of two real arrays via 2-D FFT
convofft	Fast convolution of two time real series via FFT
convolve	Implement a finite impulse response (FIR) filter with convolution
Cpowspec	Frequency analysis via Welch modified periodogram power spectrum
cros2dft	Fast 2D cross correlation of two real arrays via 2D-FFT

Signal Processing Functions (Continued)

crosscor	Cross correlation of two real-time series
crossfft	Fast cross correlation of two time real series via FFT
dct	1-dimensional discrete cosine transform
dct2d	2D-discrete cosine transform via fast block matrix approach
downsamp	Lower the sampling rate of (i.e., decimate) an equidistant array
fft2d	2-D fft generalized for complex data
fft2d_r	Forward 2-D fft optimized for real data
fft42	Cooley-Tukey radix-"4 + 2" Fast Fourier Transform
fftrad2	Cooley-Tukey radix-2 Fast Fourier Transform
fftreal	Forward fft optimized for real data
fftr_inv	Inverse fft optimized for real data
highpass	Design a (FIR) highpass filter with a Kaiser-Bessel window
iirfiltr	Filter real data with an IIR type filter
lmsadapt	Least-mean-square algorithm (for line-enhancement and noise cancelling)
lowpass	Design a (FIR) lowpass filter with a Kaiser-Bessel window
median	Remove shot-noise or noise from digital drop-out with a median filter
powspec	Frequency analysis via Welch modified periodogram power spectrum
resample	Resample a digital signal to any new integral rate
smooth	Smooth equidistant array of data with a lowpass filter
tdwindow	9 spectral windows: Hanning, Hamming, Blackman, etc.
whitnois	White noise generator

Input/Output Functions

Cvread	Read a complex vector from a binary disk file
Cvwrite	Write a complex vector to a binary disk file
hwdclose	Close a Hypersignal Waveform Data file
hwdcreate	Create a new Hypersignal Waveform Data file
hwdopen	Open an existing Hypersignal Waveform Data file
hwdread	Read data from a Hypersignal Waveform Data file
hwdwrite	Write data to a Hypersignal Waveform Data file
iir_read	Read coefficients from a Hypersignal IIR or FIR data file
vread	Read a real vector from a binary disk file
vwrite	Write a real vector to a binary disk file
xyinfo	Read the header information of a binary or ASCII (LOTUS) disk file
xyread	Read data from a binary or ASCII (LOTUS) disk file
xywrite	Write data to a binary or ASCII (LOTUS) disk file

Miscellaneous Functions

acosh	Inverse hyperbolic cosine function
asinh	Inverse hyperbolic sine function
atanh	Inverse hyperbolic tangent function
besi0	Modified Bessel function of zeroth order, $I0(x)$
besi1	Modified Bessel function of first order, $I1(x)$
besin	Modified Bessel function of nth order, $In(x)$
besk0	Modified Bessel function of zeroth order, $K0(x)$
besk1	Modified Bessel function of first order, $K1(x)$
beskn	Modified Bessel function of nth order, $Kn(x)$
besj0	Bessel function of first kind and zeroth order, $J0(x)$
besj1	Bessel function of first kind and first order, $J1x)$
besjn	Bessel function of first kind and nth order, $Jn(x)$
besy0	Bessel function of second kind and zeroth order, $Y0(x)$
besy1	Bessel function of second kind and first order, $Y1(x)$
besyn	Bessel function of second kind and nth order, $Yn(x)$
chebser	Evaluate a Chebyshev series with finite coefficients
combin	Combinations function, $N!/(r!*(N - r)!)$
fact	Factorial function, $N!$
logn	Logarithm function of general base, n
permut	Permutations function, $N!/(N - r)!$
pseries	Evaluate a power series with finite coefficients

Appendix B

References

ABRAMOWITZ, M., AND I. A. STEGUN, *Handbook of Mathematical Functions*, Applied Mathematics Series, vol. 55, Washington: National Bureau Standards, 1964. Reprinted by Dover Publications, N.Y., 1970.

AHMED, N., AND T. NATARAJAN, *Discrete-Time Signals and Systems*. Reston, Va.: Reston Publishing Company, Inc., A Prentice-Hall Company, 1983.

AYRES, F., *Theory and Problems of Matrices*. New York: Schaum's Outline Series, McGraw-Hill Inc., 1962.

BRIGHAM, E. O., *The Fast Fourier Transform*. Englewood Cliffs, N.J.: Prentice-Hall, 1974.

CHILDERS, D. G., ed., *Modern Spectrum Analysis*. IEEE Press, New York, 1978.

CLENSHAW, C. W., "Chebyshev Series for Mathematical Functions," *National Physical Laboratory Mathematical Tables*, vol. 5. London, England: Her Majesty's Stationery Office, 1963.

COOLEY, J. W., AND J. W. TUKEY, "An Algorithm for the Machine Calculation of Complex Fourier Series," *Math. of Comput.*, vol. 19, April 1965, pp. 297–301.

DUDGEON, D. E., AND R. M. MERSEREAU, *Multidimensional Signal Processing*. Englewood Cliffs, N.J.: Prentice Hall, Inc., 1984.

ELLIOTT, D. F., AND K. R. RAO, *Fast Transforms; Algorithms, Analysis, Applications*. Orlando, Fla.: Academic Press, Inc., 1982.

EMBREE, P. M., AND B. KIMBLE, *C Language Algorithms for Digital Signal Processing*. Englewood Cliffs, N.J.: Prentice Hall, Inc., 1991.

FELLER, W., *An Introduction to Probability Theory and its Applications*. New York: John Wiley and Sons, Inc., 1968.

FILLIPI, S., AND H. ENGELS, "Altes und Neues zur Numerischen Differentiation," *Elektronische Datenverarbeitung*, iss. 2, 1966, pp. 57–65.

FLETCHER, R., AND C. M. REEVES, "Function Minimization by Conjugate Gradients," *Computer Journal*, vol. 7, iss. 2, 1964, pp. 149–154.

GOLD, B., AND C. RADER, *Digital Processing of Signals*. New York: Lincoln Laboratory Publications, McGraw-Hill, Inc., 1969.

GOLUB, G. H., AND C. F. VAN LOAN, *Matrix Computations*. Baltimore, Md.: The Johns Hopkins University Press, 1983.

GONZALEZ, R. C., AND P. WINTZ, *Digital Image Processing*. Reading, Mass.: Addison-Wesley Publishing Co., 1977.

GRAHAM, R. L., D. E. KNUTH, AND O. PATASHNIK, *Concrete Mathematics*. Menlo Park, Calif.: Addison-Wesley Publishing, 1989.

HAMMING, R. W., *Numerical Methods for Engineers and Scientists*. New York: McGraw-Hill, 1973. Reprinted by Dover Publications, 1986.

HARDY, G. H., *A Mathematician's Apology*. Cambridge, England: Cambridge University Press, 1989.

HART, J. F., ET AL., *Computer Approximations*. New York: John Wiley and Sons, Inc., 1968.

HARRIS, F. J., "On the Use of Windows for Harmonic Analysis with the Discrete Fourier Transform," *Proc. IEEE*, vol. 66, Jan. 1978, pp. 51–83.

HASTINGS, C., *Approximations for Digital Computers*. Princeton, N.J.: Princeton University Press, 1955.

HILDEBRAND, F. B., *Introduction to Numerical Analysis*. New York: McGraw-Hill, Inc., 1974. Reprinted by Dover Publications, 1987.

HORNBECK, R. W., *Numerical Methods*. Englewood Cliffs, N.J.: Prentice-Hall, Quantum Publishers, 1975.

HOUSEHOLDER, A. S., *The Theory of Matrices in Numerical Analysis*. New York: Dover Publications, 1974.

"IBM 360 Scientific Subroutine Package—Programmer's Manual," IBM Technical Publications, 1968.

JAYANT, N. S., AND P. NOLL, *Digital Coding of Waveforms, Principles and Applications to Speech and Video*. Englewood Cliffs, N.J.: Prentice Hall, Inc., 1984.

JURY, E. I., *Theory and Application of the z-Transform Method*. Malabar, Fla.: Robert E. Krieger Publishing Co., 1964.

KAHANER, D., C. MOLER, AND S. NASH, *Numerical Methods and Software*. Englewood Cliffs, N.J.: Prentice-Hall, Inc., 1989.

KALMAN, R. E., AND N. DeCLARIS, EDS., *Aspects of Network and System Theory*. New York: Holt, Rinehart, and Winston Publishing, Inc., 1971.

KERNIGHAN, B. W., AND D. M. RITCHIE, *The C Programming Language*. Englewood Cliffs, N.J.: Prentice-Hall, Inc., 1978.

KNUTH, D. E., *Seminumerical Algorithms*, vol. 2 of *The Art of Computer Programming*. Reading, Ma.: Addison-Wesley, 1981.

KUESTER, J. L., AND J. H. MIZE, *Optimization Techniques with FORTRAN*. New York: McGraw-Hill, Inc., 1973.

LANCZOS, C., *Applied Analysis*. Englewood Cliffs, N.J.: Prentice-Hall, 1956. Reprinted by Dover Publications, 1988.

LAPIN, L. L., *Probability and Statistics for Modern Engineering*. Boston: PWS-KENT Publishers, 1983.

LEVINSON, N., "The Weiner RMS (Root Mean Square) Error Criterion in Filter Design and Prediction," *Journal of Math and Physics*, vol. 25, 1946, pp. 261–278.

MEYER, P. L., *Introductory Probability and Statistical Applications*. Reading, Mass.: Addison-Wesley, Inc., 1970.

NELSON, M., *The Data Compression Book*. Redwood City, Calif.: M&T Publishing, Inc., 1991.

OPPENHEIM, A. V., AND R. W. SCHAFER, *Digital Signal Processing*. Englewood Cliffs, N.J.: Prentice Hall, Inc., 1975.

OPPENHEIM, A. V., AND R. W. SCHAFER, *Discrete Time Processing*. Englewood Cliffs, N.J.: Prentice Hall, Inc., 1989.

OPPENHEIM, A. V., ET AL., *Signals and Systems*. Englewood Cliffs, N.J.: Prentice Hall, Inc., 1983.

PAPOULIS, A., *Probability, Random Variables, and Stocastic Processes*. New York: McGraw-Hill 1965.

PARK, S. K., AND K. W. MILLER, "Random Number Generators: Good Ones are Hard to Find," *Communications of the ACM*, vol. 31, no. 10, October 1988, pp. 1192–1201.

PRATT, W. K., *Digital Image Processing*. New York: John Wiley and Sons, Inc., 1991.

PRESS, W. H., ET AL., *Numerical Recipes—The Art of Scientific Programming*. Cambridge, England: Cambridge University Press, 1986.

"Programs for Digital Signal Processing," ed. by DSP-ASSP-Committee, IEEE Press. New York: John Wiley and Sons, Inc., 1979.

RABINER, L. R., AND B. GOLD, *Theory and Application of Digital Signal Processing*, Englewood Cliffs, N.J.: Prentice Hall, Inc., 1975.

RABINER, L. R., AND C. M. RADER, EDS., *Digital Signal Processing* (selected papers). Piscataway, N.J.: IEEE Press, 1972.

ROBINSON, E. A., *Multichannel Time Series Analysis with Digital Computer Programs*. Goose Pond Press, 1983.

SCHAFER, R. W., AND L. R. RABINER, "A Digital Signal Processing Approach to Interpolation," *Proc. IEEE*, vol. 61, No. 6, June 1973, pp. 692–702.

"Selected Papers in Digital Signal Processing, II," ed. by Digital Signal Processing Committee IEEE-ASSP. New York: IEEE Press, 1975.

STEARNS, S. D., AND R. A. DAVID, *Signal Processing Algorithms*. Englewood Cliffs, N.J.: Prentice Hall, Inc., 1988.

TUKEY, J. W., *Exploratory Data Analysis*. Reading, Mass.: Addison-Wesley, 1971.

WIDROW, B., "Adaptive Filters" in *Aspects of Network and System Theory*, ed. by R. E. Kalman and N. DeClaris. New York: Holt, Rinehart, and Winston Publishing, Inc., 1971, pp. 563–587.

WIDROW, B., AND S. D. STEARNS, *Signal Processing Algorithms*. Englewood Cliffs, N.J.: Prentice Hall, Inc., 1985.

WILKINSON, J. H., *The Algebraic Eigenvalue Problem*. New York: Oxford University Press, 1965.

ZIOLKOWSKI, A., *Deconvolution*. Boston: International Human Resources Development Corp. Publishing, 1984.

Index

A

acosh function, 130
adaptive filtering, 122, 332
adaptive interference canceling, 122, 126, 332
adaptive line enhancement, 126
adaptive signal processing, 122
allocation, 161, 232, 350, 426
asinh function, 130
atanh function, 130
auto2dft function, 121, 132
autocor function, 121, 134
autocorrelation, 120, 132, 134, 136
autocovariance, 120
autofft function, 121, 136
averaging, n-point, 210
AXIS_MAX constant, 454, 456, 462

B

BANDPASS.DAT, 29
bandpass filtering, 112
bandpass function, 114, 138
Bernoulli trial, 57

BESSEL.C, 26
BESSEL.DAT, 26
bessel functions:
    besi0 function, 141
    besi1 function, 141
    besin function, 141
    besj0 function, 144
    besj1 function, 144
    besjn function, 144
    besk0 function, 141
    besk1 function, 141
    beskn function, 141
    besy0 function, 144
    besy1 function, 144
    besyn function, 144
bilinear function, 116, 147
bilinear transformation, 115, 147, 305
binomdst function, 58, 149
binomial coefficients, 93
binomial distribution, 57
binomial expansion, 56
binomial probability, 149
biquad filter, 115
Blackman–Harris window, 420
Blackman window, 420

C

Cadd function, 54, 151
Cdeterm_ variable, 23
Cdiv function, 54, 152
central differences, 265, 267
Central limit theorem, 62
centroiding, 90, 352, 427
chebser function, 153
Chebyshev approximation, 153, 403
Chebyshev's theorem, 70
Clineqn function, 83, 155
Cmag function, 54, 157
Cmul function, 54, 158
Cmxadd function, 78, 159
Cmxalloc function, 161
Cmxconjg function, 163
Cmxcopy function, 165
Cmxdeter function, 81, 166
Cmxdup function, 168
Cmxeigen function, 87, 170
Cmxfree function, 172
Cmxinit function, 173
Cmxinv function, 82, 174
Cmxinv22 function, 82, 176
Cmxinv33 function, 82, 178
Cmxmag function, 180
Cmxmaxvl function, 181
Cmxminvl function, 182
Cmxmul function, 78, 183
Cmxmul1 function, 78, 185
Cmxmul2 function, 78, 187
CmxscalC function, 78, 189
CmxscalR function, 78, 191
Cmxsub function, 78, 193
Cmxtrace function, 81, 195
Cmxtrans function, 79, 196
coefficient of determination, 73
combinations function, 198
combinations, 56
combinatorial analysis, 55
combin function, 56, 198
Complex, 10
complex add, 53
complex arithmetic, 53, 151–52, 157–59,
        183, 193, 199–200, 224–25, 227,
        230, 255
complex conjugate, 53, 163, 200, 234
complex divide, 53
complex linear equations, 155
complex matrix determinant, 166
complex matrix functions:
        Cmxadd function, 159
        Cmxalloc function, 161
        Cmxconjg function, 163

Cmxcopy function, 165
Cmxdup function, 168
Cmxfree function, 172
Cmxinit function, 173
Cmxinv function, 174
Cmxinv22 function, 176
Cmxinv33 function, 178
Cmxmag function, 180
Cmxmaxvl function, 181
Cmxminvl function, 182
Cmxmul function, 183
Cmxmul1 function, 185
Cmxmul2 function, 187
CmxscalC function, 189
CmxscalR function, 191
Cmxsub function, 193
Cmxtrace function, 195
Cmxtrans function, 196
Cvectomx function, 238
Cvmxmul function, 245
Cvmxmul1 function, 246
Complex_Matrix type, 12
complex multiply, 53
Complex_Ptr type, 11, 13
complex scalar functions:
        Cadd function, 151
        Cdiv function, 152
        Cmag function, 157
        Cmul function, 158
        Complx function, 199
        Conjg function, 200
        CscalR function, 224
        Csqrt function, 225
        Csub function, 227
complex subtract, 53
complex vector functions:
        Cvadd function, 230
        Cvalloc function, 232
        Cvconjg function, 234
        Cvcopy function, 235
        Cvdup function, 237
        Cvectomx function, 238
        Cvfree function, 240
        Cvinit function, 241
        Cvmag function, 242
        Cvmaxval function, 243
        Cvminval function, 244
        Cvmxmul function, 245
        Cvmxmul1 function, 246
        CvscalC function, 251
        CvscalR function, 253
        Cvsub function, 255
Complex_Vector type, 10
Complx function, 54, 199
compression, 102, 103, 262
Conjg function, 54, 200

conjgrad function, 99, 201
conjugate gradient approach, 99
conjugate gradient technique, 202
constants:
    AXIS__MAX, 454, 456, 462
    CURVES__MAX, 453, 457, 461
    LABEL__MAX, 454, 456, 462
    TITLE__MAX, 454, 456, 462
continuous random variable, 60
conv2dft function, 105, 205
convofft function, 104, 207
convolution theorem, 104
convolution, 94, 104, 205, 207
    conv2dft function, 205, 207
    defined, 104
    via 2-D FFT, 104, 205
    via FFT, 207
convolve function, 104, 210
Cooley and Tukey, 119
coordinates of centroid, 90, 352, 427
correlation:
    2-D, 121
    auto2dft function, 132
    autocor function, 134
    autofft function, 136
    cros2dft function, 216
    crosscor function, 219
    crossfft function, 221
    defined, 120
    via 2-D FFT, 121, 132, 216
    via FFT, 136, 221
correlation coefficient, 73
cosine transform, 102, 260, 262
covariance, 120, 132, 134, 136, 216, 219, 221
Cpowspec function, 120, 213
Cramer's rule, 82
cros2dft function, 121, 216
cross product, 84, 430
crosscor function, 121, 219
crosscorrelation, 120, 216, 219, 221
crossfft function, 121, 221
CscalR function, 54, 224
Csqrt function, 54, 225
Csub function, 54, 227
cubic spline interpolation, 89, 415
cumulative distributions:
    binomial, 149
    hypergeometric, 303
    Poisson, 394
CURVES__MAX constant, 453, 457, 461
curvreg function, 72, 228
Cvadd function, 78, 230
Cvalloc function, 232
Cvconjg function, 234
Cvcopy function, 235

Cvdot function, 236
Cvdup function, 237
Cvectomx function, 238
Cvfree function, 240
Cvinit function, 241
Cvmag function, 85, 242
Cvmaxval function, 243
Cvminval function, 244
Cvmxmul function, 245
Cvmxmul1 function, 246
Cvread function, 248
CvscalC function, 78, 251
CvscalR function, 78, 253
Cvsub function, 78, 255
Cvwrite function, 257

D

data files, 24
data types:
    Complex, 10
    Complex__Matrix, 12
    Complex__Ptr, 11, 13
    Complex__Vector, 10
    Real, 9
    Real__Matrix, 12
    Real__Ptr, 11, 13
    Real__Vector, 10
DAT extension, 24
dB, 109, 113
dct function, 103, 260
dct2d function, 103, 262
decibels, 109, 113
decimation, 117, 269, 407
definite integral, 311, 410
derivative, 93, 94
deriv function, 96, 265
deriv1 function, 96, 267
determinant, 81, 166, 355
determ__ variable, 23
difference filter, 265
difference formulas, 93, 94
differentiation, 93, 265, 267
digital filtering, 106, 114–15, 138, 147, 210,
    300, 305, 338
discrete cosine transform, 102, 260
discrete cosine transform 2-D, 262
discrete Fourier transform (DFT)
    defined, 101
discrete Fourier transform 2-D, 101–2,
    274, 276, 278, 281, 284, 286
discrete random variable, 57
dot product, 84, 236, 432
downsamp function, 117, 269
dynamic memory mangagement, 15

E

Eigenanalysis, 85
eigenvalues, 170, 358
eigenvectors, 170, 358
error handling, 17
error notification, 341
euler__ variable, 22
e__ variable, 22
expectation, 68
expectation, mean, and variance, 69
expectation of a function, 69
exponential and uniform relationship, 66
exponential distribution, 66
exponential regression, 74

F

fact function, 56, 273
factorial, 56
factorial function, 273
far keyword, 16
FAR__OBJECT constant, 16
fast Fourier transform (FFT), 74, 118, 274,
        276, 278, 281, 284, 286
        radix-2, 119
        radix-4, 119
        radix-4 + 2, 119
        real data, 276, 284, 286
        two-dimensional, 102, 274, 276
fft2d function, 102, 274
fft2d__r function, 102, 276
fft42 function, 119, 278
fftrad2 function, 119, 281
fftreal function, 119, 284
fftr__inv function, 119, 286
file formats:
        Hypersignal FIR or IIR
          Data, 308
        Hypersignal Waveform Data,
        288, 291, 293, 297
        XY__ASCII, 453, 456, 461
        XY__BINARY, 453, 456, 461
        XYY__ASCII, 453, 456, 461
        XYY__BINARY, 453, 456, 461
FILTER.C, 29
filtering, 114, 138, 210, 300, 305, 338, 412
        adaptive, 122, 332
        median, 116, 346
        prediction, 123
finite impulse response (FIR), 107, 138,
        210, 300, 338
Fletcher and Reeves, 99
Fourier analysis, 213, 274, 276, 278, 281, 284,
        286, 398

Fourier regression, 73
freeing dynamic memory, 172, 240, 344, 360,
        437
frequency analysis, 213, 274, 276, 278, 281,
        284, 286, 398
frequency response, 114

G

Gauss, 71
Gaussian distribution, 61, 321, 391
Gaussian elimination, 82, 83
Gaussian random number, 389
geometric power curve regression, 75
Gibbs phenomenon, 108
GPRINT program, 39
gradient method, 98
gradient minimization, 201
GRAFIX.LPT, 38
GRAFIX program, 24

H

Hamming window, 420
Harris, 106
header, 462
Hermitian matrix, 80, 85, 170
HIGHPASS.DAT, 29
highpass filtering, 111
highpass function, 111, 300
histogram, 361
Horner's formula, 403
Householder, 80
huge keyword, 16
HUGE__OBJECT constant, 16
hwdclose function, 288
hwdcreate function, 288
hwdopen function, 291
hwdread function, 293
hwdwrite function, 297
hyperbolic functions, 130
hyperbolic regression, 75
HYPER.C, 28
HYPER.DAT, 28
hyperdst function, 60, 303
hypergeometric distribution, 58
hypergeometric probability, 303

I

i0(x), 141
i1(x), 141
identity matrix, 79, 363
iirfiltr function, 115, 305
iir__read function, 308

image compression, 102–3, 262
impulse response, defined, 107
infinite impulse response (IIR), 114–15, 147, 305
inner product, 84, 236, 432
input, hwdread function, 293
input/output:
    hwdread function, 293
    hwdwrite function, 297
    xyinfo function, 453
    xyread function, 456
    xywrite function, 461
integer resampling, 407
integrat function, 93, 311
integration, 91, 410
interp function, 89, 118, 314
interp1 function, 89, 118, 318
interpolation, 89, 118, 314, 318, 407, 415
inverse hyperbolic functions:
    acosh function, 130
    asinh function, 130
    atanh function, 130
inverse interpolation, 90, 352, 427
inverse of a matrix, 82
invprob function, 64, 321
in(x), 141

J

j0(x), 144
j1(x), 144
Jacobi method, 86
Jacobi transformation method, 85, 170, 358
jn(x), 144
JPEG, 102–3, 262

K

k0(x), 141
k1(x), 141
Kaiser window filter, 138, 300, 338, 407
Kaiser window filter design, 108
kn(x), 141

L

LABEL_MAX constant, 454, 456, 462
labels, curve, 453, 456, 462
labels, title, 453, 456, 462
labels, xaxis, 453, 456, 462
labels, yaxis, 453, 456, 462
Labels structure, 453, 456, 462
Lagrange interpolation, 89, 314, 318
least mean square (LMS) algorithm, 122, 125, 332

least_squares, 71
least_squares approximation:
    curvreg function, 228
    linreg function, 330
    pseudinv function, 405
least_squares generalized, 71, 323
least_squares approximation
    least_sq function, 72, 323
Levinson function, 123–24
Levinson, 81
Levinson's recursion, 326
linear algebra, 77
linear equations, 82, 405
linear independent basis, 74, 82
linear interpolation, 89
linear regression, 72
lineqn function, 73, 83, 328, 330
lmsadapt function, 332
logarithmic function, 336
logarithmic regression, 73
logn function, 336
lower diagonal matrix, 79
LOWPASS.DAT, 29
lowpass filtering, 109
lowpass function, 111, 338
LPT n, 38
LU decomposition, 82–83, 155, 166, 328, 355

M

magnitude, 180, 439
magnitude of a complex number, 53
magnitude of a matrix, 180
magnitude of a vector, 85, 242, 439
matched filtering, 120
math_abort variable, 18, 22
math_digits variable, 23, 462
math_err function, 19, 341
matherr_ function, 19
math_errmsg variable, 18, 22
math_errno variable, 19, 22
math_errs variable, 20
math_errs[] variable, 22
mathfree function, 344
mathlib.h file, 3, 4
math_linesize variable, 23, 463
matrices, 77
matrix addition, 77, 159, 348
matrix allocation, 161, 168, 350, 357
matrix copy, 165, 354
matrix determinant, 81, 355
matrix duplication, 168, 357
matrix functions:
    Cmxeigen function, 170
    mxadd function, 348
    mxalloc function, 350

matrix functions (*cont.*)
    mxcentro function, 352
    mxcopy function, 354
    mxdup function, 357
    mxeigen function, 358
    mxfree function, 360
    mxhisto function, 361
    mxident function, 363
    mxinit function, 364
    mxinv function, 365
    mxinv22 function, 367
    mxinv33 function, 369
    mxmaxval function, 371
    mxminval function, 372
    mxmul function, 373
    mxmul1 function, 375
    mxmul2 function, 377
    mxscale function, 379
    mxsub function, 381
    mxtrace function, 383
    mxtransp function, 384
    vectomat function, 435
    vmxmul function, 442
    vmxmul1 function, 443
matrix inverse, 174, 176, 178, 365, 367, 369
matrix multiplication, 78, 183, 185, 187, 373, 375, 377
matrix scaling, 78
    CmxscalC function, 189
    CmxscalR function, 191
    mxscale function, 379
matrix subtraction, 77, 193, 381
matrix transpose, 78, 185, 187, 375, 377
maximum value, 181, 243, 371, 440
mean, 55, 69, 418
mean of a random variable, 69
median function, 117, 346
memory management, 10, 15
method of steepest descent, 98, 201
minimization in 1-D, 386
minimization in n-dimensions, 98, 201
minimization of functions, 96
minimization of nonlinear equations, 201, 386
minimum value, 182, 244, 372, 441
modified periodograms, 213, 398
mxadd function, 78, 348
mxalloc function, 350
mxcentro function, 91, 352
mxcopy function, 354
mxdeterm function, 81, 355
mxdup function, 357
mxeigen function, 86, 358
mxfree function, 360
mxhisto function, 361

mxident function, 81, 363
mxinit function, 364
mxinv function, 82, 365
mxinv22 function, 82, 367
mxinv33 function, 82, 369
mxmaxval function, 371
mxminval function, 372
mxmul function, 78, 373
mxmul1 function, 78, 375
mxmul2 function, 78, 377
mxscale function, 78, 379
mxsub function, 78, 381
mxtrace function, 81, 383
mxtransp function, 79, 384

N

newton function, 98, 386
Newton–Raphson technique, 97–98, 386, 401
noise generator, 451
noise removal, 116, 122, 332, 346, 412
nonlinear regression, 74
normal and uniform relationship, 65
normal approximation, 63
normal distribution, 61, 63, 321, 391
normal distribution, inverted, 64
normal equations, 72
normal function, 64, 66, 389
normal probabiltity, 321, 391
normal random number, 389
nprob function, 64, 391
Nyquist frequency, rate, 101

O

OBJECT constant, 16
output, xywrite function, 297, 461

P

p_roots function, 98, 401
Pascal's Triangle, 56, 93
periodogram method, 213, 398
permutations, 56
permutations function, 393
permut function, 56, 393
piecewise interpolation, 89
pi_ variable, 22
poissdst function, 60, 394
Poisson, 60
Poisson and uniform relationship, 66
Poisson distribution, 60
Poisson function, 60, 67, 396
Poisson probability, 394

Poisson random number, 396
polynomial approximations:
        chebser function, 153
        curvreg function, 228
        linreg function, 330
        pseries function, 403
polynomial regression, 72
polynomial roots, 98, 401
powers of a complex number, 54
power spectral analysis, 118
power spectrum, 120
power spectrum estimation, 120, 213, 398
powspec function, 120, 398
prediction filtering, 123
principal root, 53
printing graphics screen, 37
probability, 55
p_roots function, 98, 401
properties of expectation, 69
properties of variance, 70
pseries function, 403
pseudinv function, 84, 405
pseudo-inverse, 83, 405

R

radix-2 FFT, 119, 281
radix-4 FFT, 119
radix-4+2 FFT, 119, 278
random number generators:
        Poisson, 396
        uniform, 422
        normal (Gaussian), 389
random seed, 422
random white noise, 451
rank of a matrix, 82
REAL constant, 9
Real_Matrix type, 12
Real_Ptr type, 11
Real_Ptr type, 13
Real type, 9
Real_Vector type, 10
rectangular rule, 91
regression, 71
resample function, 117–18, 407
Richardson's improvement, 92, 95
Romberg extrapolation, 267
romberg function, 92, 410
Romberg integration, 92, 96, 410
root-finding, 96–98, 386, 401
root of a complex number, 54
roots of polynomials, 98, 401
Runge, 89

S

sample rate conversion, 117
sample space, 60

sample standard deviation, 70
sample variance, 70
sampling rate conversion, 269, 314, 318, 407, 415
sampling theorem, 101
sampling with replacement, 57
sampling without replacement, 58
scalar multiplication, 78, 189, 191, 251, 253, 379, 446
seed value, 422
Shannon, 101
shot noise, 116, 346
signal function, 20
similarity transform, 85
Simpson's rule, 92, 311
simultaneous linear equations, 82, 155, 326, 328, 405
singular matrix, 82
smooth function, 116, 412
smoothing, 116, 412
spectral analysis, 118
spectral estimation, 213, 274, 276, 278, 281, 284, 286, 398, 420
spline function, 90, 118, 415
sqrt2_ variable, 22
sqrt3_ variable, 22
square matrices, 79
standard deviation, 55
stat function, 70
statistical sampling, 418
statistics, 68
stats function, 418
STDC constant, 16
symmetric matrix, 80, 84–85, 170, 326, 358

T

Taylor series, 403
tdwindow function, 106, 420
TITLE_MAX constant, 454, 456, 462
Toeplitz matrix, 80, 123, 326
trace of a matrix, 81, 195, 383
transpose of a matrix, 78, 196, 246, 384, 443
transpose of a vector, 238, 435
Trapezoidal Rule, 92, 410
tridiagonal matrix, 79, 415
two-dimensional convolution, 104
twopi_ variable, 22

U

unbiased estimate, 70
uniform and exponential relationship, 66
uniform and normal relationship, 65
uniform and Poisson relationship, 66
uniform distribution, 64
uniform random number, 422

uniform white noise, 65, 451
upper diagonal matrix, 79
urand function, 65-66, 67, 422
uraninit function, 422
useinput__ variable, 23

V

vadd function, 78, 424
valloc function, 426
variables:
    Cdeterm__, 23
    determ__, 23
    e__, 22
    euler__, 22
    math__abort, 22
    math__digits, 23, 462
    math__errmsg, 22
    math__errno, 22
    math__errs[], 22
    math__linesize, 23, 463
    pi__, 22
    sqrt2__, 22
    sqrt3__, 22
    twopi__, 22
    useinput__, 23
variance, 55, 70, 418
variance of a random variable, 70
v__centro function, 91, 427
vcopy function, 429
vcross function, 85, 430
vdot function, 84, 432, 434
vectomat function, 435
vector addition, 230, 424
vector allocation, 232, 237, 426, 434
vector copy, 235, 429
vector cross product, 430
vector dot product, 236, 432
vector duplication, 237, 434
    Cvdot function, 236
    Cvread function, 248
    Cvwrite function, 257
    v__centro function, 427
    vadd function, 424
    valloc function, 426
    vcopy function, 429
    vcross function, 430
    vdot function, 432
    vdup function, 434
    vectomat function, 435
    vfree function, 437
    vinit function, 438
    vmag function, 439
    vmaxval function, 440
    vminval function, 441

vmxmul function, 442
vmxmul1 function, 443
vread function, 444
vscale function, 446
vsub function, 447
vwrite function, 449
vector magnitude, 85
vector read, 248, 444
vectors, 84
vector scaling:
    CvscalC function, 251, 253
    vscale function, 446
vector subtraction, 255, 447
vector write, 257, 449
vfree function, 437
vinit function, 438
vmag function, 85, 439
vmaxval function, 440
vminval function, 441
vmxmul function, 442
vmxmul1 function, 443
void type, 19
vread function, 444
vscale function, 78, 446
vsub function, 78, 447
vwrite function, 449

W

Welch, 120
Welch periodogram method, 213, 398
white noise, 451
whitnois function, 451
Wiener filtering, 122
Wiener–Hopf equation, 123
Wiener prediction filter, 123
windowing, 105

X

XY format, 25
xyinfo function, 453
xyread function, 456
xywrite function, 461
XYY format, 25

Y

y0(x), 144
y1(x), 144
yn(x), 144